Quantum Computing (Second Edition)

A pathway to quantum logic design

Online at: https://doi.org/10.1088/978-0-7503-5683-1

Quantum Computing (Second Edition)

A pathway to quantum logic design

Hafiz Md Hasan Babu
Department of Computer Science and Engineering, University of Dhaka, Dhaka, Bangladesh

IOP Publishing, Bristol, UK

ISBN 978-0-7503-5683-1 (ebook)
ISBN 978-0-7503-5681-7 (print)
ISBN 978-0-7503-5684-8 (myPrint)
ISBN 978-0-7503-5682-4 (mobi)

DOI 10.1088/978-0-7503-5683-1

Version: 20230601

IOP ebooks

British Library Cataloguing-in-Publication Data: A catalogue record for this book is available from the British Library.

Published by IOP Publishing, wholly owned by The Institute of Physics, London

IOP Publishing, No.2 The Distillery, Glassfields, Avon Street, Bristol, BS2 0GR, UK

US Office: IOP Publishing, Inc., 190 North Independence Mall West, Suite 601, Philadelphia, PA 19106, USA

To my respected, wonderful parents and also to my beloved wife, daughter, and son who made it possible for me to write this book

Contents

Part IV QCA fault tolerance

Preface to the Second Edition

The rise of nanotechnology has led to an increasing role for quantum computing in the development of smaller and more energy-efficient computers. By leveraging the principles of quantum mechanics, certain computations can be performed at significantly higher speeds. Quantum computing represents one of the most promising nanotechnologies for creating modern and compact computer systems. Unlike traditional computers that rely on binary logic, quantum computers operate on a fundamentally different concept. They exploit quantum entanglement to simultaneously evaluate multiple probabilities. The processing and memory unit of a quantum computer system is becoming small like an atom and the switching techniques have reached the peak of their development. Consequently, a completely fresh mindset is needed for constructing computing machines. Recently, advancements in mathematics, materials science, and computer engineering have brought quantum computing from the realm of theory into practical implementation.

This quantum computing book covers basic themes of quantum computing in the first part, fault-tolerant quantum computing in the second part, quantum-dot cellular automata (QCA) in the third part and QCA fault-tolerant circuits in the fourth part. Each part has a critical thinking question section for the readers.

The book covers the design aspects of quantum computing, specifically focusing on the design of various quantum circuits. These circuits include important components such as adder, multiplier, divider, encoder, decoder, barrel shifter, comparator, and arithmetic logic unit (ALU) as well as the processor itself. These quantum circuits serve as the essential building blocks of a functioning quantum computer. By studying the content of this book, readers will gain a comprehensive understanding of quantum computing, ranging from basic concepts to advanced topics. This knowledge will empower them to create new and innovative quantum circuits. Additionally, the book explores the significance of quantum error correction, which has greatly enhanced the potential of quantum computing technology in the long run. With the implementation of quantum error correction, the encoded quantum information can be safeguarded against errors resulting from uncontrolled interactions with the environment or imperfect execution of quantum logical operations.

This book also covers the area of fault-tolerant quantum computing, focusing on the designs of various fault-tolerant quantum circuits, including adder, multiplier, and divider. Achieving fault tolerance in quantum circuits is a highly intricate task. It involves the development of circuits that can effectively protect qubits from quantum errors resulting from insufficient control over environmental interactions. By studying this section, readers will get a greater level of knowledge of fault-tolerant quantum circuits and acquire the knowledge necessary to design their own fault-tolerant circuits. QCA (Quantum-dot Cellular Automata) represents a novel computational paradigm that encodes binary information through charge configurations within individual cells, departing from the conventional approach of using current switches. In QCA, computation is achieved solely through Coulombic interactions, eliminating the need for current flow. This revolutionary paradigm offers a potential solution for transistor-less computation at the nanoscale. By studying this book, the readers will achieve a deep knowledge of QCA and the design processes involved in creating various quantum circuits using QCA design software. Additionally, they will become familiar with the costs associated with QCA circuits, including cell area, delay, kink energy, and power consumption.

In the last part, the QCA fault-tolerant circuits are discussed. The necessity of QCA fault-tolerant circuits, fault-tolerant QCA majority gate, fault-tolerant QCA 1-to-2 demultiplexer, fault-tolerant QCA full-adder, fault-tolerant QCA SRAM cell, fault-tolerant QCA subtractor and fault-tolerant QCA multiplier are presented in this part.

This comprehensive book serves as an invaluable resource for both quantum computing researchers, professionals and students. Given the scarcity of books in this particular field, it fills an important gap and provides a wealth of knowledge. Whether you're a beginner or an advanced reader, this book caters to a wide range of expertise levels, making it suitable for individuals at any stage of their quantum computing journey.

Acknowledgments

I would like to express my sincerest gratitude and particular appreciation to the various researchers in the field of quantum computing. The contents in this quantum computing book have been compiled from a wide variety of research works, where the researchers are pioneers in their respective fields. All the research articles related to the contents are listed at the end of each chapter.

I am grateful to my parents and family members for their endless support. Most of all, I want to thank my lovely wife Mrs Sitara Roshan, my sweet daughter Ms Fariha Tasnim, and my sweet son Md Tahsin Hasan for their invaluable cooperation in completing this book.

Finally, I am also grateful to all of those who have provided immense support and their valuable time to finish this book, in particular my beloved students Nitish Biswas, Md Tareq Hasan, and Rownak Borhan Himel.

Author biography

Dr Hafiz Md Hasan Babu

Dr Hafiz Md Hasan Babu is currently working as a Professor in the Department of Computer Science and Engineering as well as the Dean in the Faculty of Engineering and Technology of the University of Dhaka, Bangladesh. In addition, at present, he is a member (part-time) of the Bangladesh Accreditation Council, Ministry of Education of the Government of the People's Republic of Bangladesh. He is also the Director of the Board of Directors of the Bangladesh Submarine Cable Company Limited. Dr Hasan Babu was the Chairman of the Department of Computer Science and Engineering of the University of Dhaka from 2003 to 2006 and Pro-Vice-Chancellor of The National University of Bangladesh from 2016 to 2020. He was also a Professor and the founding Chairman of the Department of Robotics and Mechatronics Engineering, University of Dhaka, Bangladesh. Dr Hasan Babu obtained his PhD in Electronics and Computer Science in Japan under a Japanese Government Scholarship and received his MSc degree in Computer Science and Engineering in the Czech Republic under a Czech Government Scholarship. He also received a DAAD Research Fellowship from Germany.

Dr Hafiz Md Hasan Babu was awarded the Dr M O Ghani Memorial Gold Medal by the Bangladesh Academy of Sciences in 2017 for his excellent research work in the progress of physical sciences in Bangladesh. In addition, he was awarded the UGC Gold Medal Award-2017 in the Mathematics, Statistics and Computer Science category for his research work on quantum multiplier–accumulator devices. He is currently an Associate Editor of the well-known research journal *IET Computers and Digital Techniques*, published by the Institution of Engineering and Technology of the United Kingdom. He was a member of the Prime Minister's ICT Task Force in Bangladesh. Dr Hasan Babu was also the President of the Bangladesh Computer Society for the session 2017–20. At present, he is the President of the International Internet Society, Bangladesh chapter.

Professor Dr Hafiz Md Hasan Babu has published more than a hundred research papers. Three of his research papers have received top research awards at international conferences.

In addition, he has published the following four textbooks for graduate and post-graduate students with three well-known publishers in the UK and USA :

- Hasan Babu H M 2020 *Quantum Computing: A Pathway to Quantum Logic Design* (Bristol: IOP Publishing)
- Hasan Babu H M 2021 *Reversible and DNA Computing* (Chichester: Wiley)
- Hasan Babu H M 2022 *VLSI Circuits and Embedded Systems* (Boca Raton, FL: CRC Press)
- Alam M J, Hu G, Hasan Babu H M and Xu H 2022 *Control Engineering Theory and Applications* (Boca Raton, FL: CRC Press)

Acronyms

ALU	arithmetic logic unit
CPLD	complex programmable logic device
CPU	central processing unit
CSM	carry shift multiplication
MSD	most significant digit
ODQ	overflow detection qubit
PLA	programmable logic array
PLD	programmable logic device
PNS	partial numerator subtractor
PPA	partial product addition
PPG	partial product generation
QCA	quantum-dot cellular automata
QCSA	quantum carry skip adder
QFS	quantum full-subtractor
QIP	quantum information processing
QPAL	quantum programmable array logic
RAM	random access memory

Part I

Quantum logic

An overview of quantum circuits

Quantum computation and quantum information are the study of the information processing tasks that can be accomplished using quantum mechanical systems. Quantum computation is an application of quantum mechanics, where the evolution of a quantum system is described by a quantum algorithm. This observation suggests that quantum computing can be used as an introduction to quantum mechanics, because quantum computation and quantum information offer an excellent conceptual laboratory for understanding the basic concepts and unique aspects of quantum mechanics.

Advancements in higher-level integration and fabrication processes have resulted in better logic circuits and energy loss has also been reduced dramatically over the last few decades. This trend of the reduction of heat in computation also has its physical limit. It is well understood that in logic computation every bit of information loss generates $kT \ln 2$ joules of heat energy, where k is Boltzmann's constant of 1.38×10^{-23} J K^{-1} and T is the absolute temperature of the environment. At room temperature, the dissipating heat is around 2.9×10^{-21} J. The energy loss limit is also important as it is likely that the growth of heat generation as a cause of information loss will be noticeable in the future.

Quantum computing is an emerging technology that has the potential to change the perspectives and applications of computing in general. A wide range of applications can be enabled, from faster algorithmic solutions of still-difficult classical problems to theoretically more secure communication protocols. A quantum computer uses the quantum mechanical effects of particles or particle-like systems, and a major similarity between quantum and classical computers is that both are abstracted as information processing machines. Whereas a classical computer operates on classical digital information, a quantum computer processes quantum information, which shares similarities to analog signals. One of the central differences between the two types of information is that classical information is more fault-tolerant compared to its quantum counterpart.

It is possible to visualize a quantum circuit as a series of gates acting on the qubits which are abstracted as wires. The wires do not have a direct physical representation, but are associated with a temporal axis. The inputs are on the left of the diagram, the outputs on the right, and in each time step a quantum gate is applied to a qubit's state abstracted by the wire. It is standard to assume that the inputs are being initialized in the computational basis $|0\rangle$, and it will be indicated if the case is otherwise.

Reliable quantum circuits are the result of designing circuits that operate directly on encoded quantum information, but the circuit's reliability is also increased by supplemental redundancies, such as sub-circuit repetitions. Reliable quantum circuits have not been widely used, and one of the major obstacles is their vast associated resource overhead, however, recent quantum computing architectures show promising scalabilities. Consequently, the number of particles used for computing can be increased more easily and the classical control hardware (inherent for quantum computation) is also more reliable.

The number of output bits is relatively small compared to the number of input bits in most computing tasks. For example, in a decision problem, the output is only one bit (yes or no) and the input can be as large as desired. However, computational tasks in quantum computing require that all of the information encoded in the input should be preserved in the output. One might expect to obtain further speed-ups by adding instructions to allow the computation of a quantum circuit.

Quantum communication and computation study information transmission and processing as physical phenomena that follow the laws of quantum mechanics. Considering quantum mechanics introduces new possibilities, such as private communication with quantum cryptography or efficient factoring algorithms. Most quantum information protocols and algorithms can be explained as a sequence of transformations applied to a known initial state and a final measurement stage. The intermediate evolution is usually the key to the procedure. This state evolution can be studied from different perspectives.

This part of the book starts with some background and preliminary studies on quantum logic and some characteristics, as well as the evolution of quantum computers, which are discussed in chapter 1. Many of the quantum gates are included in multiple chapters. Basic definitions of the different efficiency parameters of quantum logic which are needed to design quantum circuits are described in chapter 2. Chapter 3 presents the characteristics and design procedure of a quantum bit string comparator. The designs for a quantum full-adder and quantum subtractor circuits are shown in chapter 4. In this chapter, the design techniques for a quantum full-adder as well as a quantum half-subtractor and full-subtractor are described. The design for a quantum multiplexer and quantum demultiplexer are discussed in chapter 5. In this chapter, a quantum 2-to-1 multiplexer, a quantum 4-to-1 multiplexer, and a quantum 2^n-to-1 multiplexer are shown. In addition, a quantum 1-to-2 demultiplexer, a quantum 1-to-4 demultiplexer, and a quantum 1-to-2^n demultiplexer are also provided in this chapter. Chapter 6 presents the design of a quantum adder circuit which is optimized in terms of quantum cost. The four-bit quantum carry skip adder and eight-bit quantum carry skip adder circuits are described in this chapter. The design for a quantum multiplier accumulator is described in chapter 7. The reduction of garbage outputs and ancillary inputs of quantum circuits is also described in this chapter. Chapter 8 presents the quantum divider circuit. The tree-based quantum division technique is also described in this chapter.

The design procedure and analysis of the properties of a quantum binary coded decimal (BCD) priority encoder circuit are introduced in chapter 9. Chapter 10 describes the design procedure and analysis of the properties of a quantum decoder circuit. The quantum square root circuit is described in chapter 11. In this chapter, the design of a quantum adder/subtractor circuit is also shown. The designs for a quantum SR latch, quantum D latch, quantum T latch, and quantum J–K latch are discussed in chapter 12. The quantum asynchronous and quantum synchronous counters are also introduced in this chapter. Chapter 13 presents the design of a quantum controlled ternary barrel shifter circuit. In this chapter, quantum ternary Peres gates and quantum ternary modified Fredkin gates are used. The construction procedure of quantum random access memory is shown in chapter 14. The construction procedure for a quantum memory unit is also given in this chapter. Three approaches for designing a quantum arithmetic logic unit are described in chapter 15. Chapter 16 presents the quantum programmable logic devices such as quantum PAL, quantum PLA, quantum FPGA, quantum CPLD and chapter 17 presents the details of quantum processor circuit. Finally, several real-life applications of quantum computing technology are given in chapter 18.

IOP Publishing

Quantum Computing (Second Edition)
A pathway to quantum logic design
Hafiz Md Hasan Babu

Chapter 1

Quantum logic

Learning objectives

- Learn about quantum computers and quantum logic.
- Describe the operation of a quantum computer.
- Study the superposition of the states that a qubit can represent.
- Understand how quantum and reversible logic interact with one another.
- Discuss the significance of quantum computers.
- Know the benefits and drawbacks of quantum computers.
- Obtain information about the background of a quantum computer.
- Understand the differences between conventional and quantum computers.

In the age of developing nanotechnology, quantum computing can play an incredibly important role in developing more compact and lower power consumption computers. The main appeal of quantum logic originates in its reflection of the physical law of energy conservation, in which the creation or destruction of energy is impossible, and only transformation from one form to another is possible. Hence, the fundamental law of energy conservation is incorporated into the logic design of circuits and systems by the quantum logic. The motivation to implement circuits and systems using quantum computing is the fact that, theoretically, the internal computations in quantum logic systems consume no power.

1.1 Overview

Quantum computing is a technology which consumes less power and for which the design is compact. In irreversible logic, energy dissipation is a common phenomenon since every bit loss causes energy loss in the irreversible operation. Quantum logic circuits are necessarily reversible and hence there is no dissipation of energy while processing a bit in quantum computation.

Quantum technology is one of the most promising nanotechnologies which are useful for designing modern circuits. Logic design with quantum logic is of great interest in recent technologies which allow scaling to atomistic dimensions. In this particular logic design approach, quantum cells are arranged in a particular fashion

doi:10.1088/978-0-7503-5683-1ch1

to define the logic. A classical gate cannot handle the superposition of states represented by a qubit (discussed in the section 2.1 of chapter 2). Thus, this forms the special case of the quantum device.

Quantum registers, which are necessary for the implementation of a quantum electronics device, combine n qubits to form larger Hilbert spaces H_n using the tensor product ($\otimes$) operator to form

$$|\Psi\rangle = |\Psi_1\rangle \otimes |\Psi_2\rangle \otimes \cdots \otimes |\Psi_n\rangle = \sum_{i=0}^{n} |\alpha_i|i\rangle;$$

where $\alpha_i \in C$, $|\Psi_i\rangle$ represents a qubit and $\sum_{i=0}^{n} |\alpha_i|^2 = 1$. High-speed multiplication has always been a fundamental requirement of high-performance processors and systems. In quantum signal processing (QSP) applications, multiplication is one of the most utilized arithmetic operations. Improving multiplier design directly benefits the high-performance embedded processors and QSP applications used in consumer and industrial electronic products. Moreover, quantum information processing (QIP) is a high-impact research area in quantum information science to construct a quantum computer. The main goal of QIP is to harness the fundamental laws of quantum mechanics to improve all aspects (e.g. acquisition, transmission, and processing) of information processing dramatically as well as to enhance the performance of quantum computers.

There are several tasks for which a quantum computer will be useful. The first, which is mentioned most frequently, is that quantum computers will be able to read secret messages communicated over the Internet using current technologies such as Rivest–Shamir–Adleman (RSA), Diffie–Hellman, and other cryptographic protocols; these protocols are based on the difficulty of number theoretic problems such as factoring and discrete logarithms. In addition, quantum computers are also useful for scientists conducting virtual experiments and searching huge amounts of data.

1.2 Motivations towards quantum computing

Quantum logic is a great achievement for very-large-scale integrated (VLSI) circuit design, and can work faster than classical logic circuits. Quantum circuits are used to build quantum super computers and can solve complex problems in polynomial time. Quantum algorithms are used to implement quantum circuits.

Quantum bits have the important property of superposition, which means that the values of quantum bits can stay in more than one position at the same time, which is impossible in the case of the classical logic designs. Reversible computation supports binary (0, 1) values, whereas quantum logic can hold multiple values (binary, ternary, quaternary, etc). The multi-value support makes quantum circuits more compact and efficient with optimal delay. Unlike the two fixed values 0 or 1 of classical logic, quantum bits can take the values of a linear combination of 0 and 1. Quantum circuits are inherently reversible and there is no dissipation of energy in quantum circuits. Thus these circuits prevent the loss of information, since energy dissipation causes bit or information loss. The above-mentioned properties of quantum logic circuits have motivated researchers to design circuit components using quantum logic.

1.3 The relationship between reversible and quantum logic

Reversible logic circuits have one-to-one mapping between inputs and outputs. In other words, if the number of outputs in a logic circuit is equal to the number of inputs, and any input pattern may map to a unique output pattern, it is called a reversible logic circuits.

All reversible circuits can be represented by quantum logic gates. Quantum circuits maintain the rules of reversible logic. Quantum logic circuits must have a one-to-one relationship between the input and output vectors like reversible logic. Quantum logic gates have unique unitary matrices which are also present in reversible logic. However, there are also differences between these two types of logic. Unlike in quantum logic, the superposition of bits and multiple values is not possible in reversible logic. Thus both similarities and differences exist between reversible and quantum logic.

Reversible circuits have broad applications in nanocomputing. Nanoelectronics engineering that would enable device scaling down to a molecular levels will almost surely imply a cellular architecture with near-neighbor connectivity. The scheme which has been developed to physically realize such a concept is called the quantum-dot cellular automaton (QCA). QCA have drawn a lot of attention for their very small feature size and ultra-low power consumption, which make them one candidate suitable for replacing complementary metal-oxide semiconductor (CMOS) technology.

1.4 Quantum computers

In 1959, Richard Feynman noted that 'as electronic components begin to reach microscopic scales, effects projected by quantum mechanics occur'. He advised that it might be used in the design of more powerful computers. In particular, quantum researchers hope to bind the superposition. In the quantum mechanical world, objects do not necessarily have visibly defined states. A traditional digital computer works with binary digits that can be in one of two states, denoted as 0 and 1; thus, for example, a four-bit computer register can hold any one of 16 (2^4) possible numbers. However, a quantum bit (qubit) exists in a wavelike superposition of values from 0 to 1; thus, for example, a four-qubit computer register can hold 16 different numbers concurrently. In theory, a quantum computer can therefore control many values in parallel, so that a 30-qubit quantum computer would be equivalent to a digital computer proficient at performing 10 trillion floating-point operations per second.

During the 1980s and 1990s, the theory of quantum computers advanced considerably beyond Feynman's early assumptions. In 1985, David Deutsch designated the structure of quantum logic gates for a universal quantum computer. In 1994, Peter Shor established an algorithm to factor numbers with a quantum computer that would involve six qubits (although more qubits would be essential for factoring large numbers in a rational time).

In 1998, Isaac Chuang, Neil Gershenfeld, and Mark Kubinec produced the first quantum computer (two-qubit) that could be burdened with data and output a result. Although their system was rational for only a few nanoseconds and slight from the perspective of solving meaningful problems, it verified the principles of quantum computation. This type of quantum computer can be prolonged by using molecules with more individually addressable elements. In March 2000, Emanuel Knill, Raymond Laflamme, and Rudy Martinez reported that they had generated a seven-qubit quantum computer. However, many researchers are skeptical about extending magnetic techniques much beyond 10 to 15 qubits because of diminishing consistency among the elements.

Quantum computers established on semiconductor technology are yet another opportunity. In a collective approach a discrete number of free electrons (qubits) reside within incredibly small sections known as quantum dots. Although relying on decoherence, such quantum computers are constructed on well-established, solid-state procedures and offer the prospect of readily applying integrated circuit 'scaling' technology. In addition, large groups of identical quantum dots could theoretically be contrived on a single silicon chip. The chip controls an external magnetic field that controls the electron spin states, while neighboring electrons are weakly coupled through quantum mechanical effects. An array of covered wire electrodes allows individual quantum dots, which have been discussed.

1.5 The working principles of quantum computers

The huge amount of processing power created by computer designers has not yet been able to satisfy the desire for speed and computing ability. In 1947, Howard Aiken claimed that only six electronic digital computers would satisfy the computing needs of the United States. Others have made similar inaccurate predictions about the amount of computing power that would satisfy the increasing needs of technological. Of course, Aiken did not take into account the large amounts of data produced by scientific research, the explosion of personal computers, or the appearance of the Internet, which further driven the need for computing power.

Scientists have already constructed basic quantum computers that can carry out certain calculations. Quantum computers will harness the power of atoms and molecules to achieve memory and processing activities. Quantum computers have the potential to achieve certain controls significantly faster than any silicon-based computer. In this book, the reader will learn what the design mechanisms of a quantum computer and just how they will be adapted for the next era of computing.

It is not necessary to look back too far to find the origins of quantum computing, as it was first conceived of less than 30 years ago. On the other hand, classical computers have been used, with many difficulties, for a comparatively long time. Paul Benioff is credited with being the first to apply quantum theory to computers, in 1981, when he posited creating a quantum Turing machine.

1.6 The evolution of quantum computers

Quantum computing is still in its emerging stages. There is a long way to go before a functionally running quantum computer can be built, let alone brought to the market. However, progress in this new technology is occurring regularly and no chronological record can ever be complete. What follows is a brief timeline clarifying key areas of progress in quantum computing. Much of the technical development has been achieved in this century, while most of the primary theoretical perspectives were laid down in the late twentieth century.

In 1980, Paul Benioff was the first to design a computer which operated under quantum mechanical principles. His idea of a quantum computer was based on Alan Turing's famous paper tape computer, described in his article published in 1936. In 1981, Feynman proved that it was impossible to simulate quantum systems on a classical computer. His argument hinged on Bell's theorem, written in 1964. In 1985, David Deutsch published a report describing the world's first universal quantum computer. He showed how such a quantum machine could reproduce any realizable physical system. Enthusiasm for creating the first quantum computer was stimulated by Paul Shor's algorithm in 1994. Shor described a method for factorizing large integers. This had serious implications for cryptography, which relies on this operation being difficult in order to keep codes secure. Shor's algorithm searched for periodicities in long integer-sequences of repeated digits. It used the quantum principles of superposition to scour for periodicities in the astonishingly fast time of a few minutes. To perform this same computation on a classical computer would take longer than the age of the Universe. In 1996, Lov Grover used quantum mechanics to solve an old unstructured search problem. For example, if someone wants to match a large database of names with a long list of telephone numbers, a classical computer could only solve this problem by querying each name with a telephone number until it found the right one, which is not fewer than $O(N)$. Grover's quantum algorithm, however, produces the output value using only $O(\sqrt{N})$ evaluations of the function.

In 2000, the first working five-qubit nuclear magnetic resonance (NMR) computer was put through its paces at the Technical University of Munich. Shortly after, the Los Alamos National Laboratory surpassed this feat with a working seven-qubit NMR quantum computer.

The year 2001 is known for the demonstration of Shor's algorithm. A team at the IBM Almaden Research Center in California succeeded in factorizing the integer 15 into 5 and 3. They used a thimbleful of a bespoke liquid containing billions of molecules. The molecules were constructed from five fluoride and two carbon atoms, each with their own nuclear spin state. The molecules worked as a seven-qubit quantum computer when pulsed with electromagnetic waves and monitored using NMR.

In 2006, researchers offered a new functioning standard by monitoring a 12-qubit quantum system with only minimal decoherence. NMR quantum information processors were used to decrypt the computation. These quantum controls led to the hope that higher-qubit quantum computers might be available one day. The same year, scientists took a step closer to the building of a quantum gate, the quantum representation of a mathematical rule. Also in 2006, scholars created molecules of quantum-dot pairs. These have great potential for quantum computers, in particular if more complex elements can be created.

This book focuses on how quantum gates and quantum representations of different circuits are to be designed to make a complete quantum computer. Readers will be motivated to design such circuits and to make quantum representations of different sequential and logical circuits of their own with different examples and designs of the circuit.

1.7 Why pursue quantum computing?

Despite the most remarkable wave of technological inventions, there are definite computational problems that the digital revolution still cannot seem to solve, even though some of these computational problems could be solved by scientific advances. Although conventional computers have doubled in processing speed and power nearly every two years for decades, they still do not appear to be fast enough to solve these enduring problems. In the long run, to solve the world's most tenacious computing problems competently, we will need to turn to an utterly new and more capable machine: the quantum computer.

Finally, the dissimilarity between a classical computer and a quantum computer is not like the difference between an old car and a new one. Instead, it is like the difference between a horse and a hawk: while one can run, the other can fly. Classical computers and quantum computers certainly have that difference. In this chapter we have taken a careful look at where the key differences lie and have take a profound plunge into what makes quantum computers unique.

1.8 Summary

This chapter mainly introduces researchers to quantum logic and simply delineates why future generations will choose quantum computing instead of conventional computers. In this chapter, the relation between reversible and quantum logic, the salient principles of quantum computers, their evolution, and necessity are described briefly.

Critical thinking questions

1. Why do quantum computers perform better than conventional computers?
2. Give an explanation of the distinction between bits and qubits in terms of quantum computing and conventional computing.
3. How do reversible and quantum logic relate to each other?
4. What makes quantum computing superior to conventional computers for future generations?

References

[1] Bonsor K and Strickland J 2000 How quantum computers work *How Stuff Works* https://computer.howstuffworks.com/quantum-computer.htm (Accessed: 4 December 2018)

[2] Wikipedia Quantum cellular automaton https://en.wikipedia.org/wiki/Quantum_cellular_automaton (Accessed: 4 December 2018)

[3] Holton W C Quantum computer *Encyclopædia Britannica* https://www.britannica.com/technology/quantum-computer (Accessed: 4 December 2018)

[4] Timeline of quantum computers *Quantum Computing 101* http://quantumly.com/timeline-of-quantum-computing-history-of-quantum-computers//-dates.html (Accessed: 4 December 2018)

[5] Akbar E P A, Haghparast M and Navi K 2011 Novel design of a fast reversible Wallace sign multiplier circuit in nanotechnology *Microelectron. J.* **42** 973–81

[6] Barenco A, Bennett C H, Cleve R, DiVincenzo D P, Margolus N, Shor P, Sleator T, Smolin J A and Weinfurter H 1995 Elementary gates for quantum computation *Phys. Rev.* A **52** 3457

[7] Cai X-D, Wu D, Su Z-E, Chen M-C, Wang X-L, Li L, Liu N-L, Lu C-Y and Pan J-W 2015 Entanglement-based machine learning on a quantum computer *Phys. Rev. Lett.* **114** 110504

[8] Cohen D W 2012 *An Introduction to Hilbert Space and Quantum Logic* (Berlin: Springer)

[9] Dibbo S V, Babu H M H and Jamal L 2016 An efficient design technique of a quantum divider circuit *IEEE Int. Symp. on Circuits and Systems (ISCAS)* (Piscataway, NJ: IEEE) pp 2102–5

[10] Kitaev A Y 1997 Quantum computations: algorithms and error correction *Russ. Math. Surv.* **52** 1191–249

[11] Landauer R 1961 Irreversibility and heat generation in the computational process *IBM J. Res. Dev.* **5** 183–91

[12] Li D-F, Wang R-J, Zhang F-L, Deng F-H and Baagyere E 2015 Quantum information splitting of arbitrary two-qubit state by using four-qubit cluster state and bell-state *Quantum Inf. Process.* **14** 1103–16

[13] Nielsen M A and Chuang I 2002 *Quantum Computation and Quantum Information* 10th edn (Leiden: Cambridge University Press)

[14] Schubert M and Rana F 2006 Analysis of terahertz surface emitting quantum-cascade lasers *IEEE J. Quantum Electron.* **42** 257–65

[15] Tóth G, Timler J and Lent C S 1998 Quantum computing with quantum-dot cellular automata using coherence vector formalism *Proc IEEE Int. Workshop on Computational Electronics (IWCE-6)*

IOP Publishing

Quantum Computing (Second Edition)
A pathway to quantum logic design
Hafiz Md Hasan Babu

Chapter 2

Basic definitions of quantum logic

Learning objectives

- Discuss quantum qubits.
- Understand how the gate's area affects the circuit's size.
- Define several quantum gates.
- Acquire knowledge of the strength, depth, and delay of any quantum circuit.
- Obtain information about the garbage outputs and constants inputs.
- Determine the cost of a quantum circuit and the complexity of quantum gates.

Quantum computing was advanced as a technological attempt to build a propositional structure that would permit relating the events of interest in quantum mechanics. Quantum computing seeks to replace the Boolean structure which, although appropriate for addressing classical physics, is insufficient for representing atomic elements. The scientific structure of the propositional linguistics of classical systems is a power set, partly ordered by set enclosure, with a pair of operations that denote conjunction and disjunction.

During the progress of quantum computing, numerous lines of study have tried to address quantum mechanics from a logical viewpoint. This book offers a map of these multiple methods in order to familiarize the reader with the very different approaches and problems deliberated in the quantum computing literature. When possible, redundant theories, algorithms, and examples are avoided in order to provide an instinctive understanding of the ideas before designing or presenting the associated mathematics.

The procedure of a two-qubit 'controlled-NOT' quantum logic gate has been designed which, in concurrence with simple single-qubit operations, forms a common quantum logic gate for quantum computation. The two quantum qubits placed in the internal and external degrees of freedom of a single trapped element which is first laser cooled to the zero-point energy. Decoherence properties are acknowledged for the operation, and the prospect of expanding the system to more qubits seems favorable.

doi:10.1088/978-0-7503-5683-1ch2

2.1 The quantum qubit

A quantum bit (qubit) is typically derived from the state of a two-level quantum system, such as the ground and excited states of an atom or the vertical and horizontal polarizations of a single photon. A qubit, represented by $|A\rangle$, is the basic unit of information in a quantum computer, which can hold two states, 0 or 1, simultaneously or at different times. A qubit can also be a superposition (both states at the same time) of these two states, i.e. a linear combination of the binary values $|1\rangle$ and $|0\rangle$ ($\alpha|0\rangle_i + \beta|1\rangle_i$, where α and β are the probabilities of being in the $|1\rangle$ and $|0\rangle$ states, respectively), whereas classical bits or binary bits are in one of two possible states, labeled 1 and 0.

2.2 The quantum gate

A quantum gate is a basic quantum circuit operating on a small number of qubits. Previously, various quantum gates with different functionalities have been designed. Among them, the NOT, CNOT, controlled-V, and controlled-V$^+$ gates represent an important class of quantum gates. These gates are shown in figure 2.1. In this figure the control, target, and contact qubits are represented by the •, ⊕, and | symbols, respectively. In a quantum gate the number of outputs must be equal to the number of inputs.

2.2.1 The quantum Feynman gate

The quantum Feynman gate is a 2 × 2 quantum gate which implements the logical functions of $P = A$ and $Q = A \oplus B$, and is illustrated in figure 2.2. The quantum Feynman gate can be used for copying a bit. When B is set to zero, then the output vector will be $P = A$ and $Q = A$, which ensures copying of the input A.

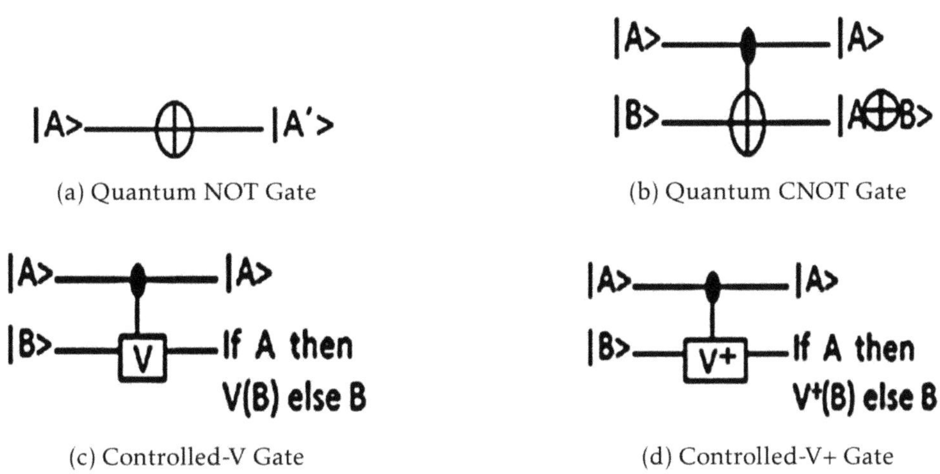

Figure 2.1. Basic quantum gates. (a) Quantum NOT gate. (b) Quantum CNOT gate. (c) Controlled-V gate. (d) Controlled-V+ gate.

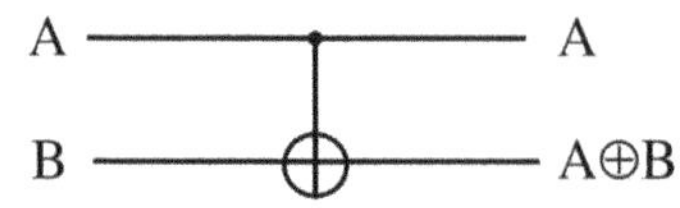

Figure 2.2. The quantum Feynman gate.

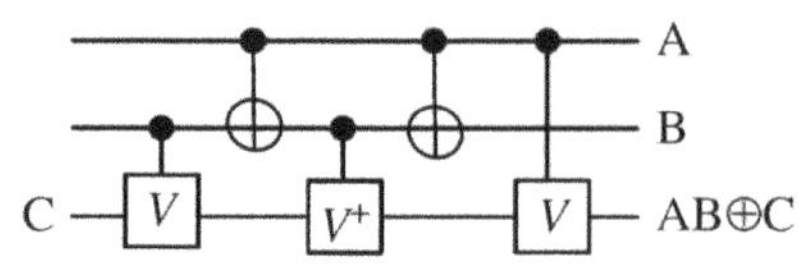

Figure 2.3. The quantum Tofolli gate.

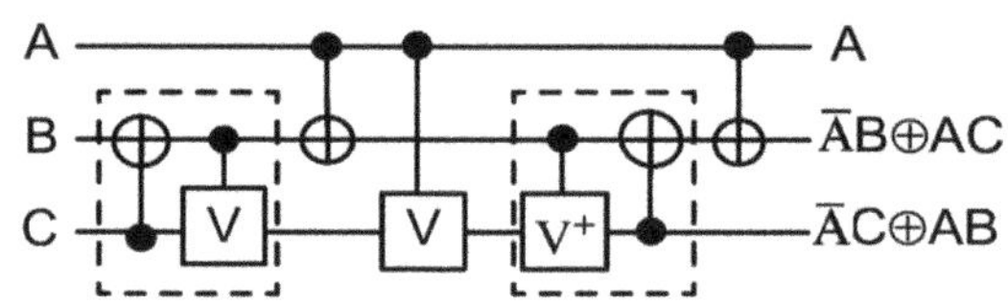

Figure 2.4. The quantum Fredkin gate.

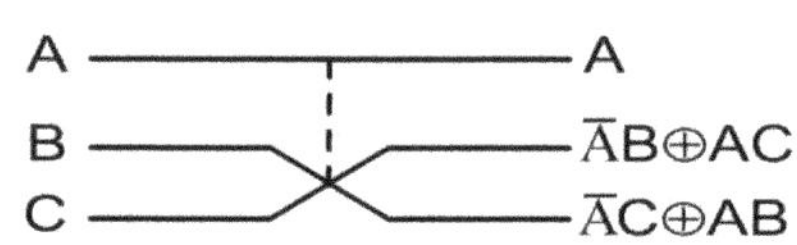

Figure 2.5. Swapping of the quantum Fredkin gate.

2.2.2 The quantum Tofolli gate

The quantum Tofolli gate is a 3×3 gate in which A, B, and C are input vectors and the output vectors are $P = A$, $Q = B$, and $R = AB \oplus C$ logical functions. The quantum Tofolli gate can implement the 'AND' operation when C is set to zero. The quantum Tofolli gate is presented in figure 2.3. The Tofolli gate is an important quantum gate and is widely applied in the construction of quantum circuits.

2.2.3 The quantum Fredkin gate

The quantum Fredkin gate is illustrated in figure 2.4, which is a 3×3 quantum gate in which A, B, and C are the input vectors and the output vectors are $P = A$, $Q = \bar{A}B \oplus AC$, and $R = \bar{A}B \oplus AB$ logical functions. One of the important applications of the quantum Fredkin gate is swapping. The input bit A is used as a control bit which decides whether the other two bits will swap or not, which is illustrated in figure 2.5. It shows that the quantum Fredkin gate can swap the values when A is set to 1.

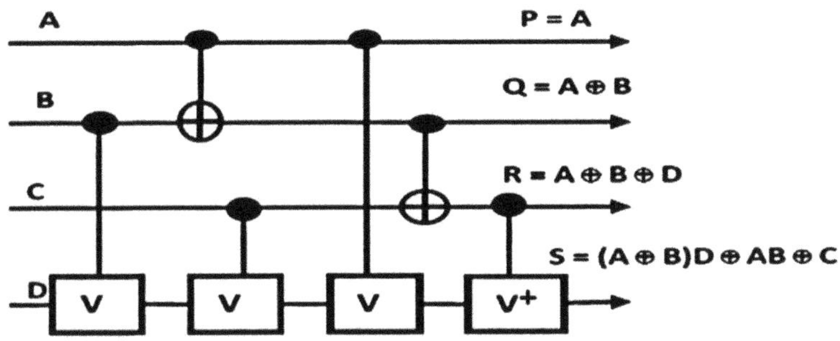

Figure 2.6. Quantum realization of the DPG circuit.

2.3 Garbage outputs

In a quantum gate the number of outputs must be equal to the number of inputs. As a consequence, there are usually some outputs which are not required further. They are called garbage outputs. Each garbage output incurs a heavy price. In figure 2.1(b), the output $|A\rangle$ is the garbage output for the CNOT quantum gate.

2.4 Constant inputs

Constant inputs are the inputs which are added to a function to make the one-to-one mapping between inputs and outputs. For example, to perform the adder operation using a double Peres gate (DPG) quantum circuit, the C input bit in figure 2.6 has to remain 0 and this input is called the constant input.

2.5 Area

The area of a gate is defined by the circuit size. This size varies according to the number of quantum gates of the circuit. As the basic quantum gates are fabricated with quantum dots with size ranging from several to tens of nanometers (10^{-9} m) in diameter, the size of basic quantum gates ranges from 50 Å–300 Å. The angstrom (Å) is a unit equal to 10^{-10} m (one ten-billionth of a meter) or 0.1 nm. Its symbol is the Swedish letter Å. Quantum circuits can be implemented with the basic quantum gates and the quantum cost of a gate depends on the number of basic quantum gates needed to implement it. Thus the area of a gate can be defined as follows: $A = N_q \times S_q$, where A= area, N_q = number of quantum gates, and S_q = size of basic quantum gates. According to the circuit size, the area of the quantum Toffoli gate is ((50 × 5) Å–(300 × 5) Å) = (250 Å–1500 Å), where the number of quantum gates of the quantum Toffoli gate is five.

2.6 Power

The power of a gate is defined by the energy consumed. The energy of basic quantum gates is 142.3 meV (microelectronvolts). Quantum circuits can be implemented with basic quantum gates and the quantum cost of a gate depends on the number of basic quantum gates needed to implement it. Thus the power of a gate can be defined as follows, for example: the energy of the Toffoli gate is (5 × 142.3) meV = 711.5 meV, where the number of quantum gates of the quantum Toffoli gate is five.

2.7 Delay

The delay represents the critical delay of the circuit. In delay calculations, the logical depth is used as the measure of the delay. The delays of all 1 × 1 and 2 × 2 quantum gates are taken as the unit delay, designated by Δ. Any quantum gate can be designed from 1 × 1 and 2 × 2 quantum gates, such as the CNOT quantum, controlled-V, or Controlled-V^+ gates. Thus the delay of a quantum gate can be computed by calculating its logical depth when it is designed using smaller 1 × 1 and 2 × 2 quantum gates. Each 2 × 2 quantum gate in the logic depth contributes to 1Δ delay. For example, the quantum Toffoli gate requires 5Δ delay, as shown in figure 2.3.

2.8 Depth

The depth of a quantum circuit is the maximum number of stages or slices where each stage or slice represents a quantum gate or a number of quantum gates along the same vertical line.

A quantum circuit is constructed using different quantum gates. These quantum gates are placed in different input lines. To find the depth of any quantum circuit, it is necessary to divide it into some slices. There may be more than one quantum gate in any slice. The maximum number of slices is considered as the depth of that quantum circuit. For example, consider the Thapliyal Ranganathan (TR) gate-based half subtractor, shown in figure 2.7. This quantum circuit needs four quantum gates to be implemented. Now, to find the depth of this circuit, it will be divided into stages according to the quantum gates. The vertical lines are used to split the circuit into slices. In the first slice there is a controlled-V^+ gate, in the second slice there is a CNOT gate, and in the third and fourth slices there are controlled-V gates. This circuit has a maximum of four slices, as numbered in the figure. Thus it can be said that the depth of the circuit is 4.

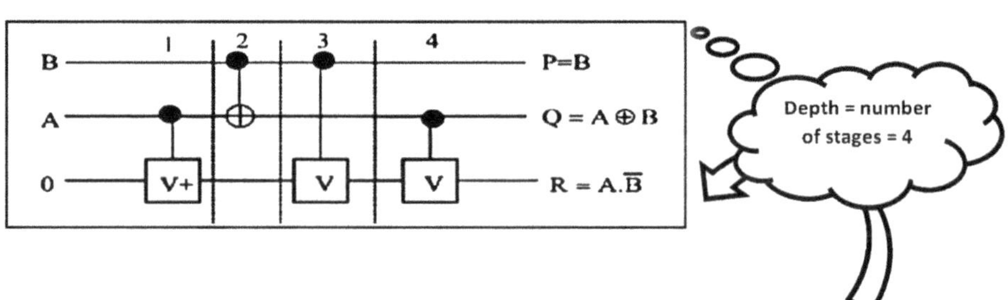

Figure 2.7. Illustration of the depth of a quantum circuit.

2.9 Quantum cost

The quantum cost of a quantum circuit is the number of basic quantum gates in the circuit. Quantum cost is an important measure of the performance of a quantum circuit. The quantum cost of the basic quantum gates, such as the NOT, CNOT, controlled-V, and controlled-V^+ gates, is considered to be 1. The quantum circuit in figure 2.7 consists of four basic quantum gates and thus the quantum cost of this circuit is 4.

2.10 Quantum gate calculation complexity

The quantum gate calculation complexity refers to the number of quantum gates (NOT, CNOT, controlled-V, and controlled-V^+) used to synthesize the given circuit, with ρ being the NOT quantum gate calculation complexity, σ the CNOT quantum gate calculation complexity, and Ω the controlled-V (controlled-V^+) gate calculation complexity. For example, the DPG quantum circuit has two CNOT quantum gates and four controlled-V (controlled-V^+) gates. Therefore, the quantum gate calculation complexity of the DPG quantum circuit is $2\sigma + 4\Omega$, which is depicted in figure 2.6.

2.11 Summary

Quantum gates and quantum networks offer a very useful language for constructing any quantum computer or quantum multi-particle circuits (which are basically the same). Now the question is, whether it possible to build quantum logic gates or not.

Single-qubit quantum gates are viewed as comparatively easy to implement. For example, a classic quantum optical realization uses elements as qubits and switches their states with laser light pulses of appropriately selected frequency, strength, and duration; any recommended superposition of two selected logical states can be prepared in this way.

Research into quantum computation and all of its all possible variations has become vigorously active and any comprehensive review of the field cannot help but be obsolete as soon as it is written. Here, only some of the very basic knowledge has been provided, hoping that this will serve as a good starting point to enter into the field.

Critical thinking questions

1. What are the characteristics of qubits?
2. How can qubits be controlled?
3. How much more powerful is a qubit than a bit?
4. How many qubits are there in a quantum computer?
5. Design circuits for implementing the quantum AND and quantum OR operations using the basic quantum gates. Write descriptions of the circuits implementing quantum AND and quantum OR operations.

References

[1] Oxford quantum http://oxfordquantum.org/ (Accessed: 5 December 2018)

[2] Balandin A A and Wang K L 1999 Implementation of quantum controlled-not gates using asymmetric semiconductor quantum dots *Quantum Computing and Quantum Communications* (Berlin: Springer) pp 460–7

[3] Landaurer R 1961 Irreversibility and heat generation in the computational process *IBM J. Res. Dev.* **5** 183–91

[4] Li X, Steel D, Gammon D and Sham L J 2004 Quantum information processing based on optically driven semiconductor quantum dots *Opt. Photonics News* **15** 38–43

[5] Mohammadi M and Eshghi M 2009 On figures of merit in reversible and quantum logic designs *Quantum Inf. Process.* **8** 297–318

[6] Monroe C, Meekhof D M, King B E, Itano W M and Wineland D J 1995 Demonstration of a fundamental quantum logic gate *Phys. Rev. Lett.* **75** 4714

[7] Nielsen M A and Chuang I L 2000 *Quantum Computation and Quantum Information* 10th edn (Leiden: Cambridge University Press)

[8] Shende V V, Prasad A K, Markov I L and Hayes J P 2003 Synthesis of reversible logic circuits *IEEE Trans. Comput.-Aided Des. Integr. Circuits Syst.* **22** 710–22

[9] Toffoli T 1980 Reversible computing *International Colloquium on Automata, Languages, and Programming* (Berlin: Springer) pp 632–44

[10] Zhou R-G, Li Y-C and Zhang M-Q 2014 Novel designs for fault tolerant reversible binary coded decimal adders *Int. J. Electron.* **101** 1336–56

IOP Publishing

Quantum Computing (Second Edition)
A pathway to quantum logic design
Hafiz Md Hasan Babu

Chapter 3

The quantum qubit string comparator

Learning objectives

- Understand the characteristics of a quantum comparator.
- Understand how a circuit for a two-qubit quantum comparator is built.
- Learn how to use a quantum magnitude comparator.
- Study the comparison between different algorithms for an n-qubit string comparator.
- Learn how to construct an n-qubit quantum comparator.
- Mention applications of a quantum comparator.

The quantum qubit string comparator enables the implementation of a quantum algorithm using conditional statements, a fundamental structure for designing algorithms. This widens the number of applications where quantum algorithms can be used and at the same time it brings quantum programmers close to some of the successful techniques used in classical computation based on comparisons. For example, constructing a database composed of only prime numbers is possible using the quantum qubit string comparator.

3.1 Characteristics of a quantum comparator

A comparator, as the name suggests, compares a signal on one input of an op-amp, as shown in figure 3.1, with a known voltage called the reference voltage on the other input. The comparator is nothing other than an open loop op-amp with two analog inputs (differential input) and one digital output (signal ended output). The op-amp has a very large gain used in an open loop. Hence the output may be in the positive or negative saturation voltage depending upon which input is larger. An op-amp is perfectly suited for comparator applications because of its high input impedance and large open loop gain.

The important characteristics of a quantum comparator are:

1. Speed of operation.
2. Accuracy.
3. Compatibility of output.

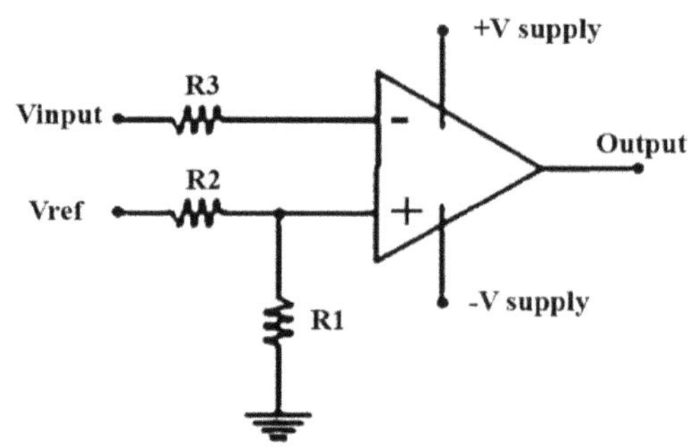

Figure 3.1. An open loop op-amp.

3.2 The quantum magnitude comparator

A quantum magnitude comparator is a logic circuit that first compares the sizes of A and B and then determines the result among $A > B$, $A < B$, and $A = B$. When the two numbers in the comparator circuit are two one-qubit numbers, the result will be only one bit from 0 and 1. Thus the circuit is called a one-qubit magnitude comparator, which is the basis of comparison of the two numbers of the n qubits. A quantum qubit string comparator is designed for the quantum qubit string comparator circuit. In this chapter two quantum states are identified by providing a comparison status, such as equality or larger or smaller, after performing a comparison between these states. In addition, this chapter shows that the quantum qubit string comparator enables the implementation of conditional statements in quantum computation, which is a basic structure for designing a comparison algorithm. However, the design requires a huge number of quantum gates, garbage outputs, and constant inputs. Moreover, the design does not show the area and power requirements of the circuit. A quantum comparator circuit is designed by using quantum dot cellular automata. The design requires a huge number of quantum gates and garbage outputs.

An ASIC implementation of a low power area efficient folded binary comparator circuit was invented in 2014. This comparator consists of a pre-computation unit and an encoder block. The basic principle is to group the binary inputs into digit sets. The digit sets are sent to the pre-computation unit starting from the most significant digit (MSD) to check for equality, and the computations in the pre-computation unit are stopped at the first digit set which produces a 1 as the output. The corresponding digit set is then sent to the carry look-ahead (CLA) encoder block to find the greater of the two inputs. However, as this is an irreversible comparator circuit, this circuit has huge dynamic power dissipation.

3.3 The design of a quantum comparator

There are two important properties to define a quantum gate, which are as follows:

Property 3.1. Each quantum gate has an unique unitary matrix. A complex square matrix U is *unitary* only if

$$U^{*}U = UU^{*} = I,$$

where I is the identity matrix and U^{*} is the conjugate transpose of U.

Property 3.2. A quantum state is represented by a state vector in a Hilbert space over complex numbers.

The matrices for the controlled-E and controlled-E^{+} gates are

$$M_{\text{controlled-E}} = \begin{pmatrix} -i/2 & -i/2 \\ -i/2 & i/2 \end{pmatrix}; \quad M_{\text{controlled-E}^{+}} = \begin{pmatrix} i/2 & i/2 \\ i/2 & -i/2 \end{pmatrix}.$$

The conjugate transpose of $M_{\text{controlled-E}}$ is $M_{\text{controlled-E}^{+}}$ and vice versa. After multiplying $M_{\text{controlled-E}}$ and $M_{\text{controlled-E}^{+}}$, an identity matrix is obtained which is

$$M_{\text{controlled-E}} \times M_{\text{controlled-E}^{+}} = \begin{pmatrix} -i/2 & -i/2 \\ -i/2 & i/2 \end{pmatrix} \times \begin{pmatrix} i/2 & i/2 \\ i/2 & -i/2 \end{pmatrix} = \begin{pmatrix} 1 & 0 \\ 0 & 1 \end{pmatrix} = I.$$

The controlled-E and controlled-E^{+} gates have a unique matrix which is unitary. Figures 3.2(a) and (b) show diagrams of the controlled-E and controlled-E^{+} gates.

The matrices for the XN (M_{XN}) gate and its conjugate transpose ($M_{XN}{}^{*}$) are given below:

$$M_{XN} = \begin{pmatrix} 0&1&0&0 \\ 1&0&0&0 \\ 0&0&1&0 \\ 0&0&0&1 \end{pmatrix} \quad M_{XN^{*}} = \begin{pmatrix} 0&1&0&0 \\ 1&0&0&0 \\ 0&0&1&0 \\ 0&0&0&1 \end{pmatrix}.$$

After multiplying M_{XN} and $M_{XN}{}^{*}$, an identity matrix is obtained which is given below:

$$M_{XN} \times M_{XN^{*}} = \begin{pmatrix} 0&1&0&0 \\ 1&0&0&0 \\ 0&0&1&0 \\ 0&0&0&1 \end{pmatrix} \times \begin{pmatrix} 0&1&0&0 \\ 1&0&0&0 \\ 0&0&1&0 \\ 0&0&0&1 \end{pmatrix} = \begin{pmatrix} 1&0&0&0 \\ 0&1&0&0 \\ 0&0&1&0 \\ 0&0&0&1 \end{pmatrix} = I.$$

(a) Controlled-E^{+} Gate　　(b) Controlled-E Gate

Figure 3.2. Controlled gates. (a) Controlled-E+ gate. (b) Controlled-E gate.

Thus the *XN* gate has a unique matrix which is unitary. Figure 3.3 shows a diagram of the *XN* gate.

Algorithm 3.1 describes the computational process for the designed method for the proof of the time complexity. The technique of comparison of the comparator has four basic steps. First, sort the two numbers and also calculate their middle bit

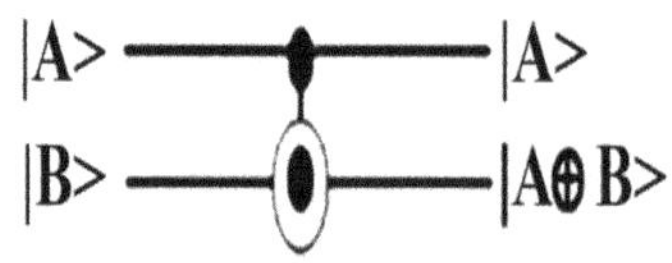

Figure 3.3. *XN* gate.

INPUT: INPUT: Two n-qubits strings $|A> = |A_n>|A_{n-1}> \dots |A_0>$ and $|B> = |B_n>|B_{n-1}> \dots |B_0>$

OUTPUT: $|A>> |B>$ or $|A>< |B>$ or $|A>= |B>$

1: Begin
2: $|A[i]>$=Array containing $n$qubits of first string
3: $|B[i]>$=Array containing $n$qubits of second string
4: $|SA[i]>$=Sorted $|A[i]>$
5: $|SB[i]>$=Sorted $|B[i]>$
6: n=Size of array
7: mid= $|\frac{n}{2}|$
8: $Ex - OR[mid] = |SA[mid]> \oplus |SB[mid]>$
9:     If .Ex-OR[mid]=0 then
10:         Find Ex-OR values of left half of $|SA[i]>$ and $|SB[i]>$s.t.i<mid
11:             If Ex-OR of left side $|SA[i]>$ and $|SB[i]>$ is 0 then
12:                 Find Ex-OR values of left half of $|A[i]>$ and $|B[i]>$s.t.i<mid
13:                     If Ex-OR of any position is 1 and $|A[i]>> |B[i]>$, then
14:                         $|A>> |B>$
15:                     Else If Ex-OR of any position is 1 and $|A[i]><|B[i]>$, then
16:                         $|B>> |A>$
17:                     Else goto Step 19
18:                     End If
19:                 Repeat Steps12 to 16 for right side, i.e., mid<i<=n
20:                     Else goto Step26
21:             End If
22:     End If
23: Else
24:     Repeat Steps 10-20 for right half of the numbers $|SA[i]>$ and $|SB[i]>$s.t.i>mid
25: End Else
26: If X-NOR of both halves of $|A[i]>$ and $|B[i]>$ is 1, where $|SA[mid]> \oplus |SB[mid]>=0$, then
27:     $|A> = |B>$
28: End If
29: End

Algorithm 3.1. Algorithm of the comparison technique.

position. Second, find Ex-OR of the middle bit position and decide based on that value whether to go to the left half or to the right half of the numbers. Third, repeat the second step after entering into the left or right half and also calculate Ex-OR of the other bits at the same time. Finally, take a decision if any value of Ex-OR is $|1\rangle$ as to which number is greater, otherwise carry out Ex-NOR to find whether they are equal or not.

Property 3.3. Algorithm for the comparison technique for an n-quantum bit string comparator which requires a time complexity of $O(2(2 + \log n))$.

Proof. Let n be the number of qubits of two strings. To sort two strings of n qubits in parallel using merge sorting, the total time complexity in the best case is $O(\log n)$. Now each string is divided into half ($n/2$) of the original string when the comparison technique works in each time. Thus it reduces the working space by half. In the worst case, the comparison technique operations need to be executed on both the left and right halves of the two n-qubit strings, where the time required for one half is the run time ($T(|n/2|)$) for half ($n/2$) of the n-qubit string using the comparison technique, and the time required for the other half is the same, which is treated as the time for copying the same operations in the other half of the string. In contrast, in the best case only the left/right half operations need to be executed. The positions of the qubit strings are stored in two different arrays. When comparing two qubits, the two arrays containing the qubit positions need to be accessed in parallel and the complexity of an array access is $O(1)$. According to the algorithm, Ex-OR of the midpoint of the sorted strings is performed and depending on the result, the left/right half of unsorted strings is used for comparison. Hence we need to access the two unsorted arrays of strings again and this also has time complexities. Thus the recurrence of the comparison technique in the best case can be specified as

$$T(n) = T(|n/2|) + 1,$$

where $T(n)$ denotes the run time of the comparison technique, $T(|n/2|)$ is the time required to perform operations on the left/right half of the n-qubit strings, n is the length of the qubit strings, and a constant 1 (one) is the time for midpoint qubit operation of the qubit string.

Guess. It is assumed that the solution of the recurrence is $T(n) = O(\log n)$, i.e. it is true iff $T(n) \leqslant c(\log(n))$, where $c > 0$ is a constant.

Proof by substitution. Assume that this bound holds for all positive $m < n$ in the particular $m = [n/2]$, where n is the number of qubits and m is a constant term. It yields that

$$T(n/2) \leqslant c(\log(|n/2|)).$$

By substituting into the recurrence,

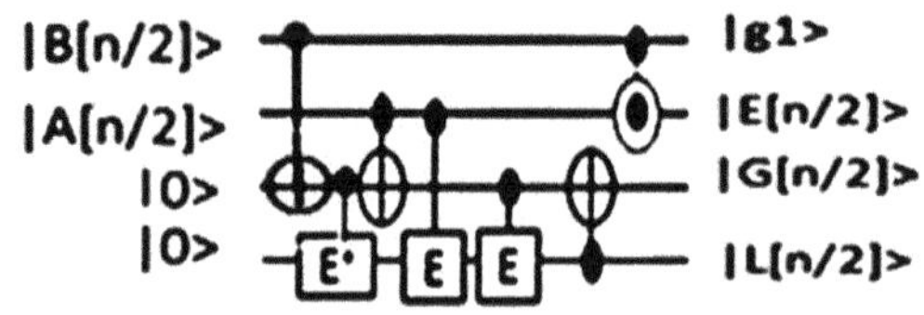

Figure 3.4. The MQC circuit. Reproduced with permission from [2]. Copyright 2017 IEEE.

$$\begin{aligned} T(n) &\leqslant c(\log(|n/2|)) + 1 \\ &= c(\log(n)) - c(\log(2)) + 1 \\ &= c(\log(n)) - c + 1 \\ &\leqslant c(\log(n)) \text{ as long as } c \geqslant 1. \end{aligned}$$

Thus, $T(n) \leqslant c(\log(n))$, that is, $T(n) = O(\log n)$. Therefore, the total time complexity of the comparison algorithm in the best case is sorting time + comparison time + array of position access time + array of unsorted qubits access time:

$$\begin{aligned} &= O(\log n) + O(\log n) + O(2) + O(2) \\ &= O(2(2 + \log n)). \end{aligned}$$

Thus this completes the proof and the property 3.3 is true.

The design of a quantum comparator consists of two circuits: the midpoint qubit comparison (MQC) circuit and rest qubit comparison (RQC) circuit. First, the most significant quantum bits are compared using the MQC circuit. Second, the rest of the qubits are compared using the RQC circuit, where each qubit comparison is performed by one RQC circuit. In figure 3.4 the MQC circuit consists of two controlled-E gates, one controlled-E^+ gate, three CNOT gates, and one XN gate. This circuit generates the following three outputs and produces one garbage output which is $|g_1\rangle$:

$$\begin{aligned} &|E_{n/2}\rangle = |A_{n/2}\rangle \odot |B_{n/2}\rangle \\ &|G_{n/2}\rangle = |A_{n/2}\overline{B_{n/2}} \\ &|L_{n/2}\rangle = |\overline{A_{n/2}}B_{n/2}\rangle. \end{aligned} \tag{3.3.1}$$

In figure 3.5 the RQC circuit consists of four controlled-E gates, three controlled-E^+ gates, one XN gate, and eight CNOT gates. This circuit generates three outputs, and it produces two garbage outputs which are $|g_1\rangle$ and $|g_2\rangle$:

$$\begin{aligned} &|AGB_{n/2-1}\rangle = |G_{n/2}\rangle + |E_{n/2}\rangle.\ |A_{n/2-1}\overline{B_{n/2-1}}\rangle \\ &|ALB_{n/2-1}\rangle = |L_{n/2}|E_{n/2}\rangle.\ |\overline{\ A_{n/2}}_{-1}B_{n/2-1}\rangle \\ &|AEB_{n/2-1}\rangle = |AGB_{n/2-1}\rangle \odot |ALB_{n/2-1}\rangle. \end{aligned} \tag{3.3.2}$$

3.3.1 Example

To construct a two-qubit quantum comparator circuit, one MQC circuit and one RQC circuit are needed to perform the greater than, less than, and equality

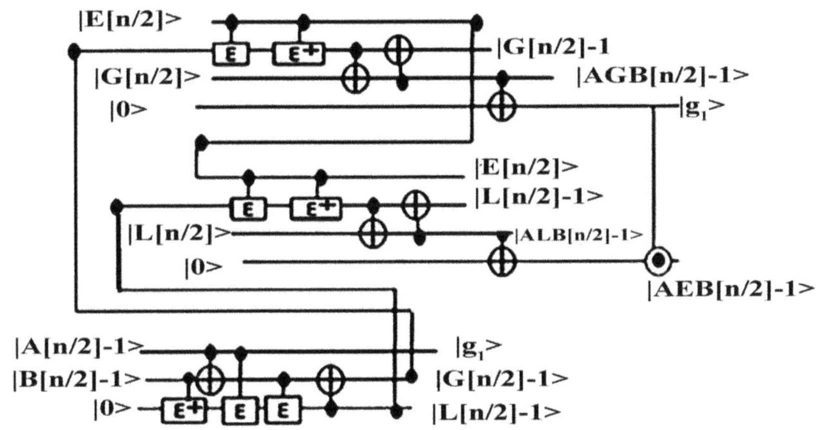

Figure 3.5. The RQC circuit. Reproduced with permission from [2]. Copyright 2017 IEEE.

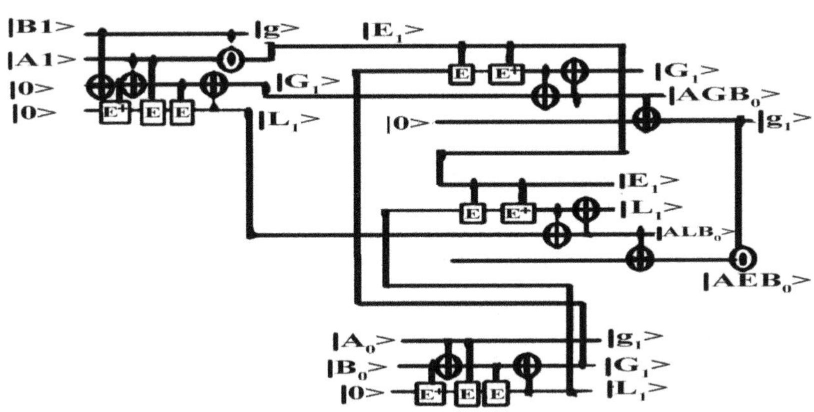

Figure 3.6. The two-qubit quantum comparator circuit. Reproduced with permission from [2]. Copyright 2017 IEEE.

operations. In figure 3.6 the detailed quantum circuit for a two-qubit comparator is shown. The midpoint is the qubits in position 1. At first, the midpoint qubits $|A_1\rangle$ and $|B_1\rangle$ are applied in an MQC circuit to generate the following outputs:

$$
\begin{aligned}
|E_1\rangle &= |A_1\rangle \odot |B_1\rangle \\
|G_1\rangle &= |A_1\overline{B_1}\rangle 3 \\
|L_1\rangle &= |\overline{A_1}B_1\rangle.
\end{aligned}
\tag{3.3.1.1}
$$

Then the outputs of the MQC circuit ($|E_1\rangle$, $|G_1\rangle$, $|L_1\rangle$) and the rest of the qubits ($|A_0\rangle$, $|B_0\rangle$) are used as inputs to the next circuit which is the RQC circuit. This circuit produces the final outputs

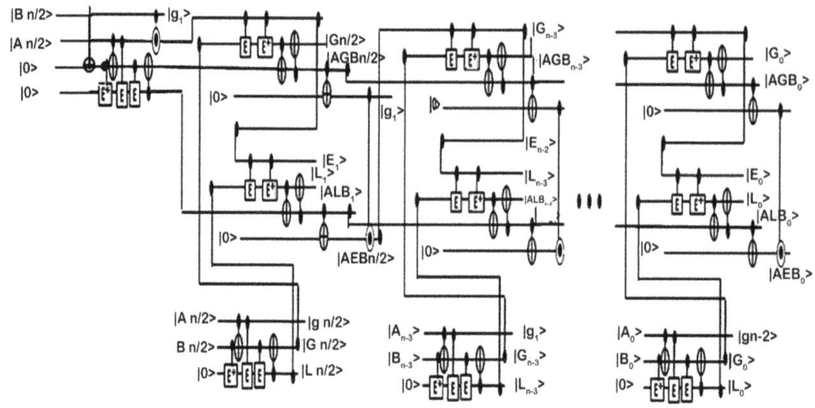

Figure 3.7. The n-qubit quantum comparator circuit. Reproduced with permission from [2]. Copyright 2017 IEEE.

$$
\begin{aligned}
|AGB_0\rangle &= |G_1\rangle + |E_1\rangle.\ |A_0(B_0)\rangle \\
|ALB_0\rangle &= |L_1\rangle + |E_1\rangle.\ |(A_0)B_0\rangle \\
|AEB_0\rangle &= |AGB_0\rangle \odot |ALB_0\rangle.
\end{aligned} \tag{3.3.1.2}
$$

Similarly, an n-qubit quantum comparator circuit can be constructed using one MQC circuit and maximum (n – 1) RQC circuits or minimum ((n/2) – 1) RQC to perform the greater than, less than, and equality operations. In figure 3.7 the detailed quantum circuit for an n-qubit quantum comparator is shown.

3.4 Summary

This chapter presents the design methodology for a noble quantum n-qubit comparator using a fast comparison technique. The time complexity of the comparison technique is $O(\log n)$ where n is the number of qubits. The comparator was constructed in two steps. First, bit comparisons for greater than and less than operations were performed in parallel. Then a bit comparison for the equality operation was performed. The quantum comparator circuit can be used in designing different quantum circuits, such as quantum processing units, complex arithmetic circuits, and communication systems. Thus the design which has been shown in this chapter should help the reader understand the next three chapters better. In addition, the concept of designing a quantum bit string comparator will help in designing the quantum subtraction, multiplier, divider, and adder circuits.

Critical thinking questions

1. What are the characteristics of a quantum comparator?
2. Describe the quantum magnitude comparator.
3. How many steps are required to build a quantum comparator? Explain in detail.

4. Is it possible to draw a four-qubit comparator? If possible then write the pseudocode for a four-qubit comparator.
5. Design a four-qubit comparator circuit and describe the design procedure.

References

[1] Barwad R 2015 The characteristics of comparator *Polytechnic Hub* http://www.polytechnichub.com/characteristics-comparator/ (Accessed: 5 December 2018)

[2] Babu H M H, Jamal L, Dibbo S V and Biswas A K 2017 Area and delay efficient design of a quantum bit string comparator *IEEE Computer Society Annual Symp. on VLSI* (Piscataway, NJ: IEEE) pp 51–6

[3] Das J C and De D 2016 Reversible comparator design using quantum dot-cellular automata *IETE J. Res.* **62** 323–30

[4] Dibbo S V, Babu H M H and Jamal L 2016 An efficient design technique of a quantum divider circuit *IEEE Int. Symp. on Circuits and Systems* (Piscataway, NJ: IEEE) pp 2102–5

[5] Nielsen M A and Chuang I L 2000 *Quantum Computation and Quantum Information* (Leiden: Cambridge University Press)

[6] Phaneendra P S, Vudadha C, Sreehari V and Srinivas M B 2014 An optimized design of reversible quantum comparator *27th Int. Conf. VLSI Design and 13th Int. Conf. Embedded Systems* (Piscataway, NJ: IEEE) pp 557–62

[7] Saravanakumar N, Nirmalkumar A, Nandhakumar A and Kanyakumari G E 2013 ASIC implementation of low power area efficient folded binary comparator *Int. J. Eng. Technol.* **5** 4582

[8] Velagaleti S 2009 A novel high speed dynamic comparator with low power dissipation and low offset *Doctoral dissertation* National Institute of Technology, Rourkela.

Chapter 4

The quantum full-adder and subtractor

Learning objectives

- Become familiar with the quantum adder.
- Investigate the quantum subtractor.
- Acquire knowledge of a quantum full-adder with a suitable circuit.
- Learn how to construct quantum full- and half-subtractors.

In circuit design the quantum adder and quantum subtractor are the circuits that are capable of adding or subtracting numbers. In this chapter quantum adder, namely quantum half-adder and quantum full-adder, circuits and quantum subtractor, namely quantum half-subtractor and quantum full-subtractor, circuits are presented with their quantum representations.

4.1 The quantum adder

A quantum adder is a circuit in electronics that implements the addition of numbers. In many computers and other types of quantum processors, quantum adders are used to calculate addition and similar operations in the quantum arithmetic logic unit (ALU) and also in other parts of quantum processors. These can be constructed for many numerical representations, such as excess-3 or binary coded decimal. Quantum adders are classified into two types: quantum half-adder and quantum full-adder. The quantum half-adder circuit has two inputs, A and B, which add two input digits and generate a carry and a sum. The quantum full-adder circuit has three inputs, A, B, and C, which add three input numbers and generate a carry and a sum. Table 4.1 is the truth table of a quantum half-adder. From this table we can obtain the quantum half-adder circuit's outputs, which are

$$\text{Sum} = A \oplus B$$
$$\text{Carry} = AB.$$

doi:10.1088/978-0-7503-5683-1ch4

Table 4.1. The truth table of the quantum half-adder.

A	B	Carry	Sum
0	0	0	0
0	1	0	1
1	0	0	1
1	1	1	0

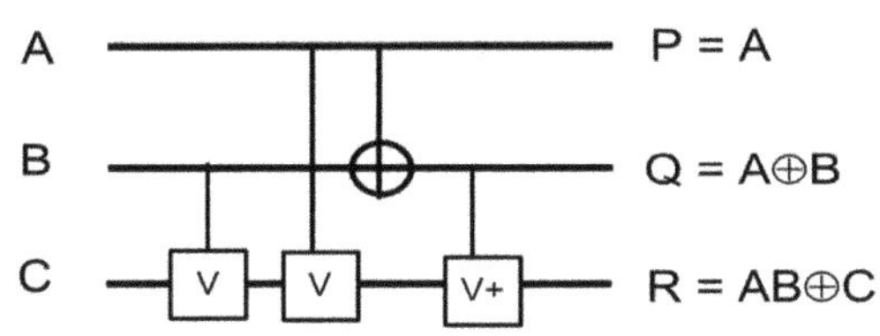

Figure 4.1. The quantum Peres gate.

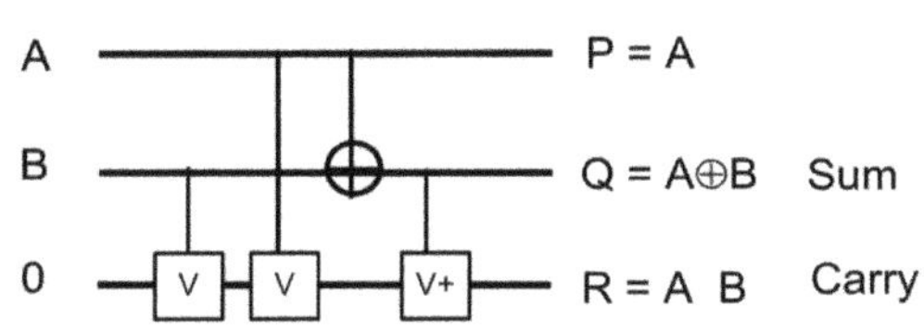

Figure 4.2. The quantum Peres gate as a quantum half-adder.

The output equations of the quantum half-adder can be mapped with a quantum Peres gate, as shown in figure 4.1, and the quantum half-adder can be obtained by putting $C = 0$, which is illustrated in figure 4.2. The quantum cost of the quantum half-adder is 4 and the delay is 4Δ.

4.1.1 The quantum full-adder

Table 4.2 is the truth table of a quantum full-adder. From this table we can obtain the output of the quantum full-adder circuit:

$$\text{Sum} = A \oplus B \oplus C_{\text{in}}$$
$$\text{Carry} = (A \oplus B)C_{\text{in}} \oplus AB.$$

The output equations of the quantum full-adder can be mapped with the quantum modified Thapliyal Srinivas gate (MTSG), as shown in figure 4.3, and the quantum full-adder can be obtained by putting $D = 0$, which is shown in figure 4.4. The cost of the quantum full-adder is 6 and the delay is 5Δ.

Table 4.2. The truth table of the quantum full-adder.

A	B	C_{in}	Carry	Sum
0	0	0	0	0
0	0	1	0	1
0	1	0	0	1
0	1	1	1	0
1	0	0	0	1
1	0	1	1	0
1	1	0	1	0
1	1	1	1	1

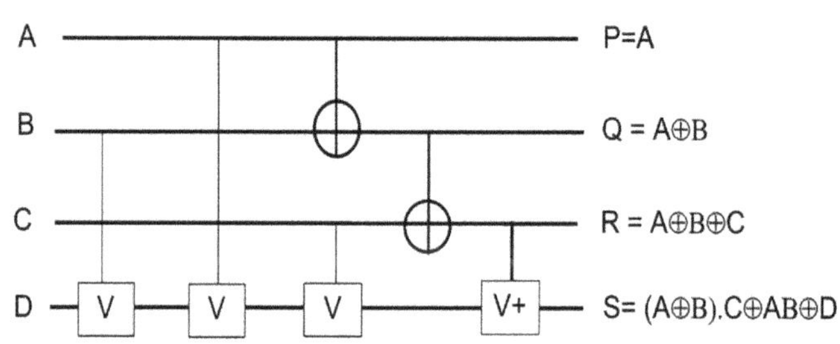

Figure 4.3. The quantum MTSG.

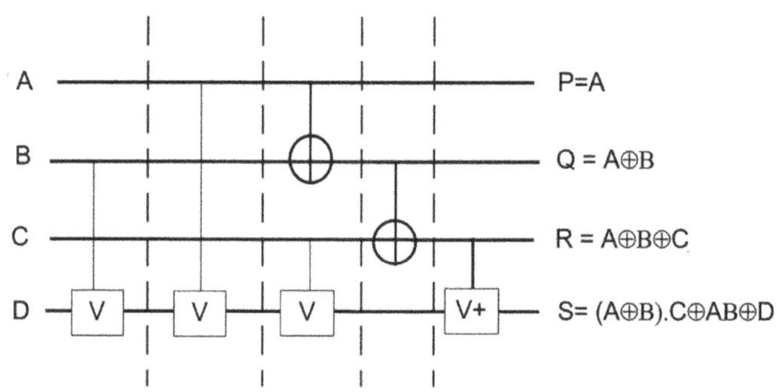

Figure 4.4. The quantum MTSG as a quantum full-adder.

4.2 The quantum subtractor

A quantum subtractor is also an important logic component in circuit design. Quantum subtractors are classified into two types: quantum half-subtractors and quantum full-subtractors. The quantum half-subtractor circuit has two inputs, A and B, where the half-subtractor performs the $A - B$ operation.

4.2.1 The quantum half-subtractor

Table 4.3 is the truth table of a quantum half-subtractor. From this table we can obtain the quantum half-subtractor circuit:

$$\text{Difference} = A \oplus B$$
$$\text{Borrow} = \overline{A}B$$

The output equations of the quantum half-subtractor can be mapped with a quantum Thapliyal Ranganathan (TR) gate, as shown in figure 4.5, and the quantum half-subtractor can be obtained by putting $C = 0$, which is shown in figure 4.6. The cost of the quantum half-subtractor is 4 and the delay is 4Δ.

4.2.2 The quantum full-subtractor

The quantum full-subtractor circuit has three inputs, A, B, and C_{in}, which realizes the operation $Y = A - B - C$. Table 4.4 is the truth table of a quantum full-

Table 4.3. The truth table of the quantum half-subtractor.

A	B	Borrow	Difference
0	0	0	0
0	1	1	1
1	0	0	1
1	1	0	0

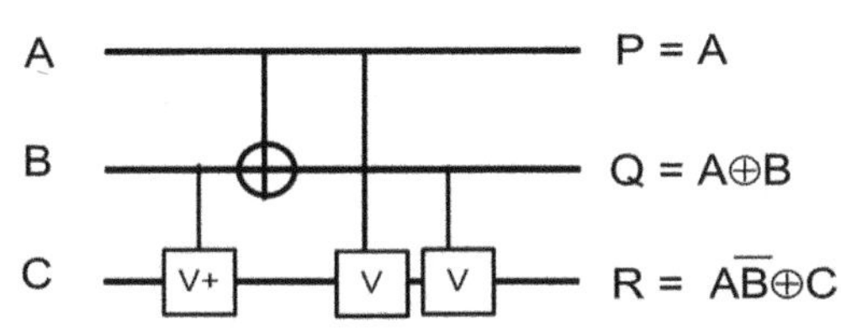

Figure 4.5. The quantum TR gate.

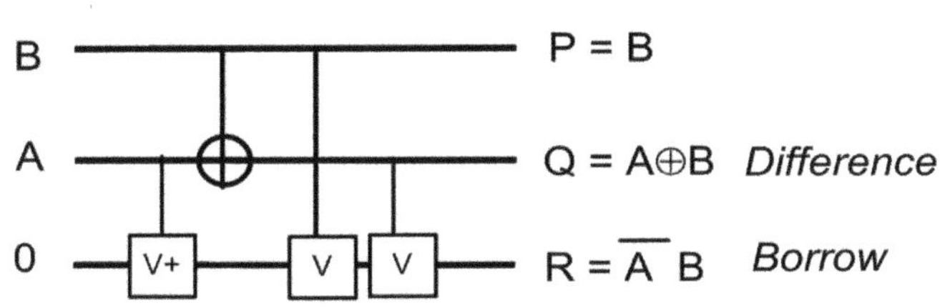

Figure 4.6. The quantum half-subtractor.

Table 4.4. The truth table of the quantum full-subtractor.

A	B	C_{in}	Borrow	Difference
0	0	0	0	0
0	0	1	1	1
0	1	0	1	1
0	1	1	1	0
1	0	0	0	1
1	0	1	0	0
1	1	0	0	0
1	1	1	1	1

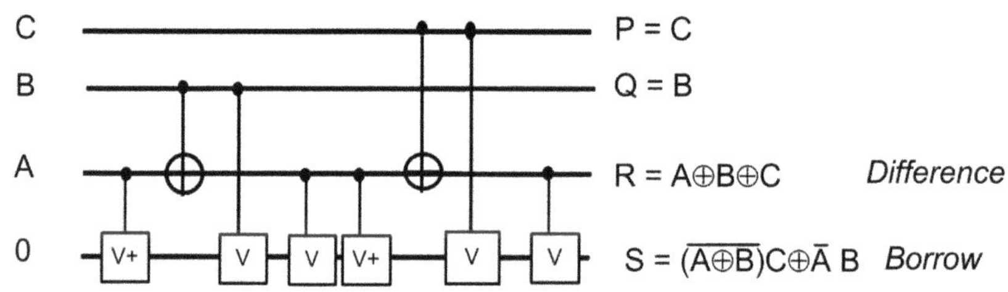

Figure 4.7. The quantum full-subtractor by cascading two quantum TR gates.

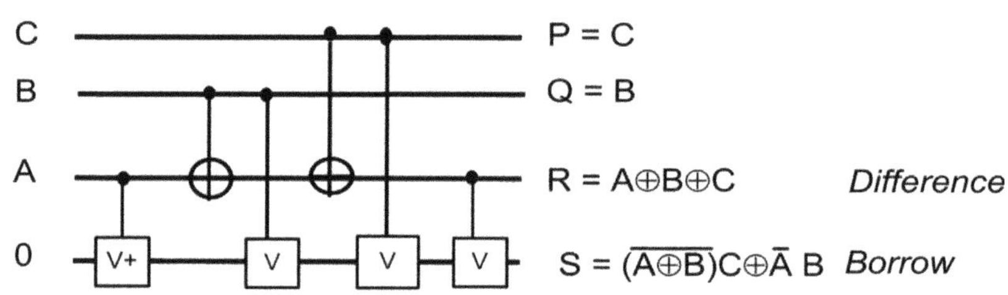

Figure 4.8. The optimized quantum full-subtractor.

subtractor. From this table we can obtain the output of the quantum full-subtractor circuit:

$$\text{Difference} = A \oplus B \oplus C_{\text{in}}$$

$$\text{Borrow} = (\overline{A \oplus B})C_{\text{in}} \oplus AB.$$

A quantum full-subtractor can be designed using two quantum TR gates, shown in figure 4.7. Figure 4.8 is the optimized version of the quantum full-subtractor. The cost of the quantum full-subtractor is 6 and the delay is 4Δ.

4.3 Summary

This chapter presents the quantum full-adder and quantum full-subtractor from the quantum half-adder and quantum half-subtractor circuits. The costs of the quantum adder and quantum subtractor circuits are discussed for better understanding of the circuits.

Critical thinking questions

1. What are the applications of quantum adders and quantum subtractors?
2. Distinguish between quantum subtractors and quantum adders.
3. In terms of the semantics of its inputs and outputs, what distinguishes a quantum half-adder from a quantum full-adder?
4. Using the quantum half-adder, design circuit diagrams for the quantum full-adder.
5. Construct a four-qubit adder–subtractor using quantum full adders and quantum basic gates.
6. What is the purpose of using quantum adders and quantum subtractors in a quantum processor?

References

[1] Babu H M H, Islam Md R, Chowdhury S M A and Chowdhury A R 2004 Synthesis of full-adder circuit using reversible logic *Proc. 17th Int. Conf. on VLSI Design (Mumbai)* pp 757–60

[2] Babu H M H, Islam R, Chowdhury A R and Chowdhury S M A 2003 On the realization of reversible full-adder circuit *Int. Conf. on Computer and Information Technology* pp 880–3

[3] Babu H M H, Islam R, Chowdhury A R and Chowdhury S M A 2003 Reversible logic synthesis for minimization of full-adder circuit *Proc. Euromicro Symp. Digital System Design* pp 50–4

[4] Babu H M H, Jamal L and Saleheen N 2013 An efficient approach for designing a reversible fault tolerant *n*-bit carry look-ahead adder *IEEE Int. SOC Conf.* pp 98–103

[5] Cheng K-W and Tseng C-C 2002 Quantum full adder and subtractor *Electron. Lett.* **38** 1343–4

[6] Cuccaro S A, Draper T G, Kutin S A and Moulton D P 2004 A new quantum ripple-carry addition circuit arXiv: quant-ph/0410184

[7] Khan M H A and Perkowski M A 2007 Quantum ternary parallel adder/subtractor with partially-look-ahead carry *J. Syst. Archit* **53** 453–64

[8] Monfared A T and Haghparast M 2016 Design of new quantum/reversible ternary subtractor circuits *J. Circuits Syst. Comput.* **25** 1650014

[9] Murali K V R M, Sinha N, Mahesh T S, Levitt M H, Ramanathan K V and Kumar A 2002 Quantum-information processing by nuclear magnetic resonance: experimental implementation of half-adder and subtractor operations using an oriented spin-7/2 system *Phys. Rev.* A **66** 22313

[10] Takahashi Y 2009 Quantum arithmetic circuits: a survey *IEICE Trans. Fundam. Electron. Commun. Comput. Sci.* **92** 1276–83

[11] Takahashi Y, Tani S and Kunihiro N 2010 Quantum addition circuits and unbounded fan-out *Quantum Inf. Comput.* **10** 872–90

IOP Publishing

Quantum Computing (Second Edition)
A pathway to quantum logic design
Hafiz Md Hasan Babu

Chapter 5

The quantum multiplexer and demultiplexer

Learning objectives

- Study the quantum multiplexer (MUX).
- Define the quantum demultiplexer (DEMUX).
- Formulate design procedures for a 2-to-1 quantum MUX, 4-to-1 quantum MUX, and 8-to-1 quantum MUX with proper circuits.
- Construct a 1-to-2 quantum DEMUX, 1-to-4 quantum DEMUX, and 1-to-8 quantum DEMUX with suitable circuits.
- Acquire knowledge of designing a 1-to-2n quantum MUX.
- Design a 1-to-2n quantum DEMUX.

In large-scale computing systems, it is necessary for a single line to carry two or more signals. One signal can be placed on one line at a time. To allow us to select and place the signal on a common line we use a circuit called the quantum multiplexer (MUX). The purpose of a quantum MUX is to select the input of any n input lines and feed that to one output line. The function of a quantum demultiplexer (DEMUX) is the inverse of the function of the quantum MUX.

5.1 The quantum multiplexer

This section presents the design of the quantum multiplexer. The quantum multiplexer causes the transmission of a large number of information units over a smaller number of channels. Architecturally, a digital multiplexer is a logic circuit that puts one out of several inputs to a single output. A set of selected inputs controls the inputs. Generally a quantum multiplexer has m inputs and n selected inputs, where $m = 2^n$.

5.1.1 The quantum 2-to-1 multiplexer

A 2-to-1 quantum multiplexer is the smallest unit of architecture of a quantum multiplexer. The characteristic function of a quantum multiplexer is $s_0'I_0 + s_0I_1$. A quantum Fredkin gate, as shown in figures 5.1(a) and (b), can be used as a 2-to-1

doi:10.1088/978-0-7503-5683-1ch5

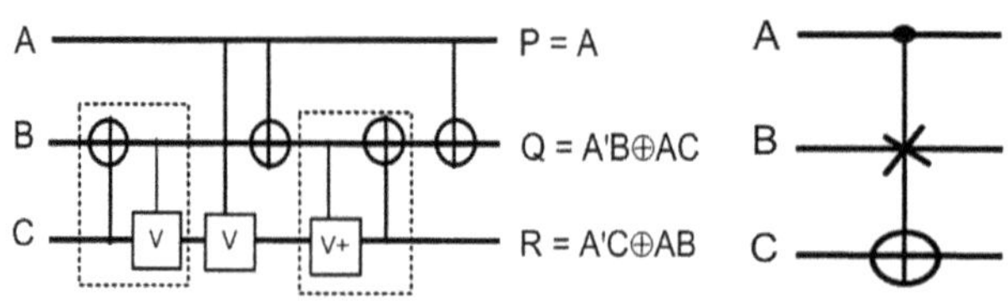

(a) Quantum circuit of quantum Fredkin gate (b) Symbol of Quantum Fredkin gate

Figure 5.1. The quantum Fredkin gate. (a) Quantum circuit of a quantum Fredkin gate. (b) Symbol of a quantum Fredkin gate.

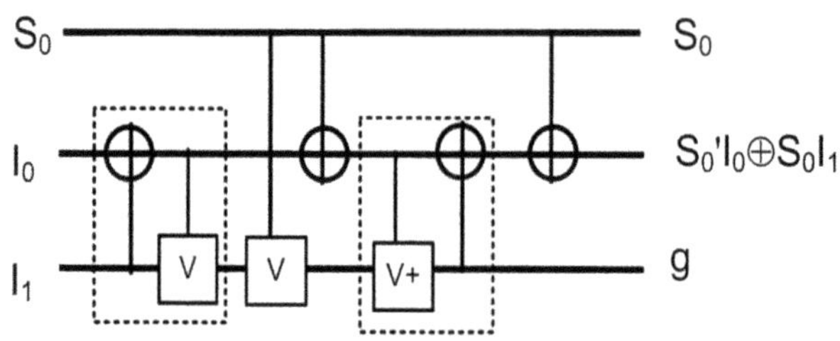

Figure 5.2. The quantum Fredkin gate as a quantum 2-to-1 multiplexer.

quantum multiplexer as it can map the characteristic function of a quantum multiplexer.

Let I_0 and I_1 be the inputs and S_0 the selected input of a 2-to-1 quantum multiplexer. When $S_0 = 0$, then input I_0 transmits to the output Y and when $S_0 = 1$ then input I_1 transmits to the output Y. Figure 5.2 shows the architecture of a quantum 2-to-1 multiplexer using a quantum Fredkin gate. The quantum cost and delay of this quantum 2-to-1 multiplexer are 5 and 5Δ, respectively. Moreover, the number of garbage outputs is 2.

5.1.2 Quantum 4-to-1 multiplexer

The quantum 4-to-1 multiplexer has four inputs, two select lines, and one output. Figure 5.3 illustrates the design of a quantum 4-to-1 multiplexer, where I_0, I_1, I_2, and I_3 are the inputs, and S_0 and S_1 are the select lines. The bit combination of select lines controls the function of the 4-to-1 quantum multiplexer as presented in table 5.1. Three quantum Fredkin gates are used in this design. Thus the quantum cost of the quantum 4-to-1 multiplexer is 15 and the delay is 15Δ in the logic circuits, respectively; the number of garbage outputs is 5.

5.1.3 The quantum 2^n-to-1 multiplexer

Figure 5.4 shows the design of a quantum 8-to-1 multiplexer. As the consequence of the design of quantum multiplexers, a 2^n-to-1 multiplexer can be constructed using

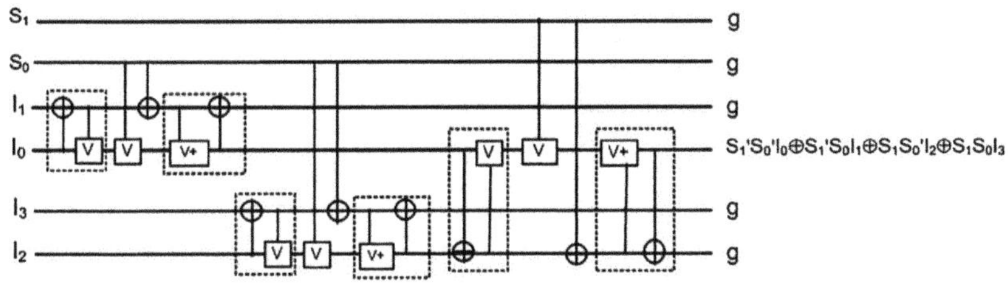

Figure 5.3. The quantum 4-to-1 multiplexer.

Table 5.1. Function of the S_0 and S_1 select lines.

S_0	S_1	Output (O)
0	0	I_0
0	1	I_1
1	0	I_2
1	1	I_3

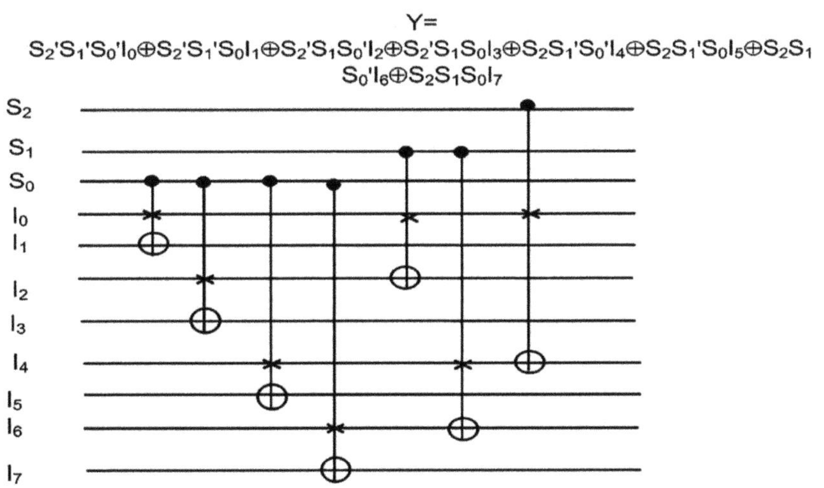

Figure 5.4. The quantum 8-to-1 multiplexer.

two 2^{n-1}-to-1 quantum multiplexers and one 2-to-1 quantum multiplexer. Figure 5.5 presents the 2^n-to-1 multiplexer, and the properties of the 2^n-to-1 quantum multiplexer are given in property 5.1.

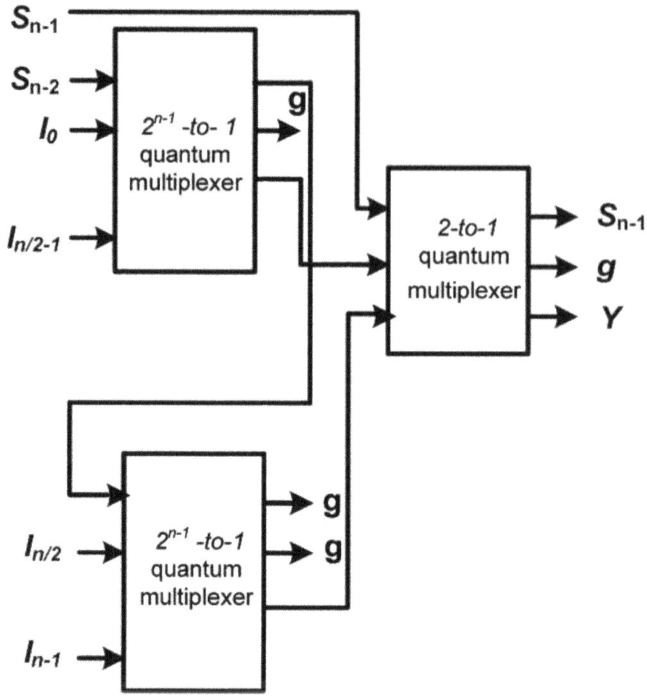

Figure 5.5. The block diagram of 2^n-to-1 multiplexers.

Property 5.1. A quantum 2^n-to-1 multiplexer can be designed with $2^n - 1$ gates, which produces $2^n + n - 1$ garbage outputs. It also requires $5(2^n - 1)$ quantum cost and a delay of $5(2^n - 1)\Delta$, where n denotes the number of selection lines and Δ denotes the unit-delay.

5.2 The quantum demultiplexer

This section presents the design of the quantum demultiplexer. A quantum DEMUX is a device that takes a single input line and routes it to one of several digital output lines. A quantum demultiplexer of 2^n outputs has n select lines which are used to select the output line to which to send the input. A quantum demultiplexer is also called a data distributor. The quantum demultiplexer can be used to implement general purpose logic. By setting the input to true, the DEMUX behaves as a decoder. The reverse of a quantum multiplexer is the quantum demultiplexer.

5.2.1 The quantum 1-to-2 demultiplexer

A 1-to-2 quantum demultiplexer is the smallest unit of the architecture of a quantum demultiplexer. The characteristic function of a 1-to-2 quantum demultiplexer is $s_0'D$ s_0D on the different output lines, as shown in table 5.2. The quantum Fredkin gate

Table 5.2. The truth table of a 1-to-2 quantum demultiplexer.

S	Y_1	Y_0
0	0	D
1	D	0

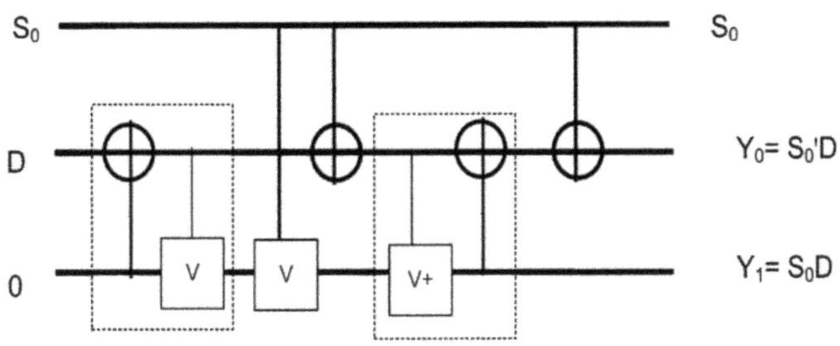

Figure 5.6. The quantum Fredkin gate as a quantum 1-to-2 demultiplexer.

can be used as a 1-to-2 quantum demultiplexer as it can map the characteristic functions of a quantum demultiplexer.

Let D be the inputs and S_0 is the select input of a 1-to-2 demultiplexer. When $S_0 = 0$, then the D input is transmitted to the second output Y_0 and when $S_0 = 1$, then the D input is transmitted to the third output Y_1. Figure 5.6 shows the architecture of a quantum 1-to-2 demultiplexer using a quantum Fredkin gate. The quantum cost and delay of this quantum 1-to-2 demultiplexer are 5 and 5Δ, respectively. Moreover, the number of garbage outputs is 1.

5.2.2 The quantum 1-to-4 demultiplexer

The quantum 1-to-4 demultiplexer has two select lines, one data input, and four outputs. Figure 5.7 shows the design of a quantum 1-to-4 demultiplexer where Y_0, Y_1, Y_2, and Y_3 are the outputs, and S_0 and S_1 are the select lines. The bit combination of select lines controls the function of the 1-to-4 demultiplexer, as shown in table 5.3. Three quantum Fredkin gates are used in this design. Thus the quantum cost of the quantum 1-to-4 demultiplexer is 15 and the delay of the quantum 1-to-4 demultiplexer is 15Δ, respectively; the number of garbage outputs is 2.

5.2.3 Quantum 1-to-2^n demultiplexer

Figure 5.8 shows the design of a quantum 1-to-8 demultiplexer. As the consequence of the design of quantum demultiplexers, a 1-to-2^n quantum demultiplexer can be constructed using 1-to-2^{n-1} quantum demultiplexers and 2^{n-1} quantum Fredkin gates, which is shown in figure 5.9. The properties of the 1-to-2^n quantum demultiplexer are given in property 5.2.

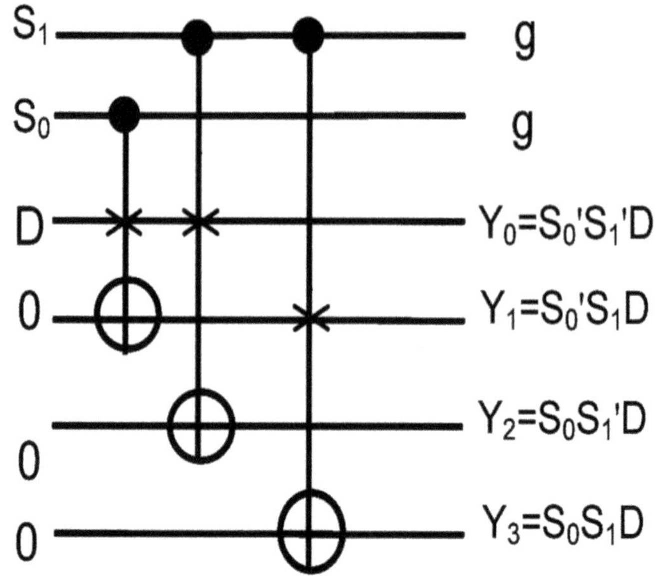

Figure 5.7. The quantum 1-to-4 demultiplexer.

Table 5.3. The truth table of a 1-to-4 demultiplexer.

S_1	S_0	Y_3	Y_2	Y_1	Y_0
0	0	0	0	0	D
0	1	0	0	D	0
1	0	0	D	0	0
1	1	D	0	0	0

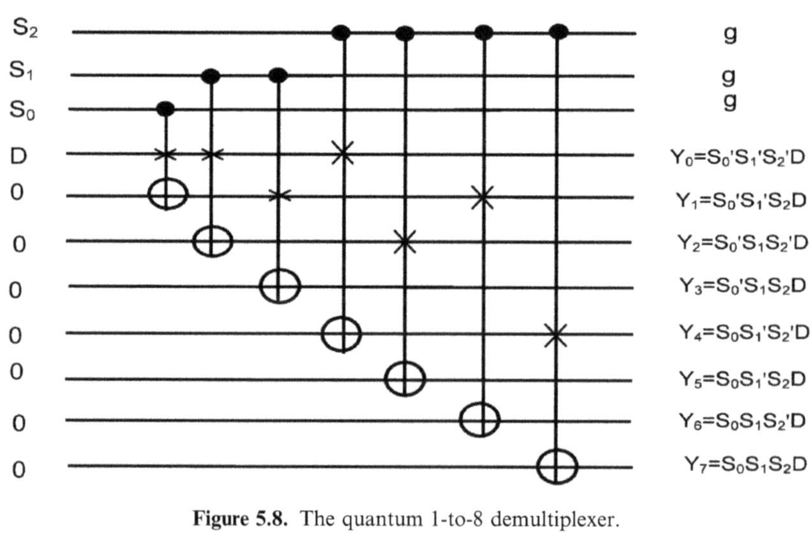

Figure 5.8. The quantum 1-to-8 demultiplexer.

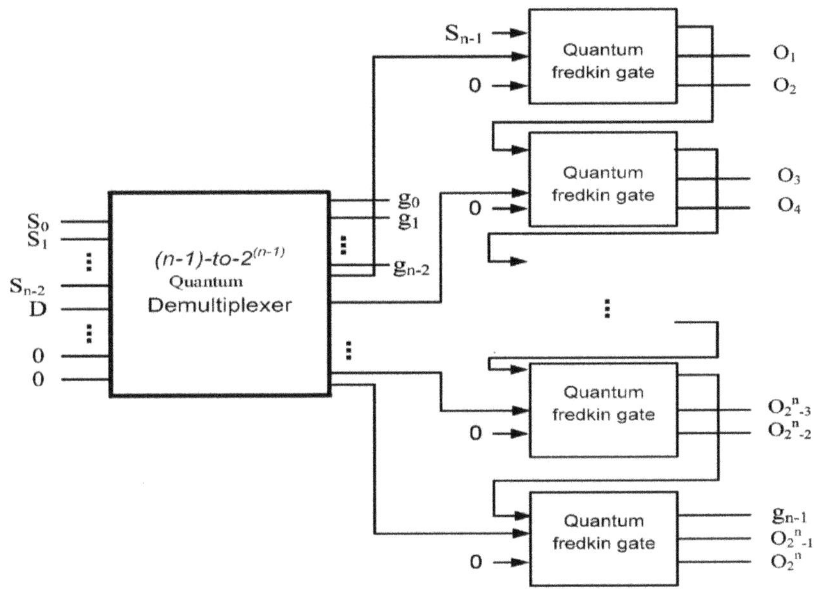

Figure 5.9. The block diagram of a 1-to-2^n demultiplexer.

Property 5.2. A quantum 1-to-2^n demultiplexer can be designed with $2^n - 1$ gates which produce n garbage outputs. It also requires a $5(2^n - 1)$ quantum cost and a delay of $5(2^n - 1)\Delta$, where n denotes the number of selection inputs and Δ denotes the unit-delay.

5.3 Summary

In this chapter the quantum MUX and DEMUX are presented and explained with quantum circuit representations. In the multiplexer section a 2-to-1 MUX, 4-to-1 MUX, and 8-to-1 MUX are described and generalized to the 2^n-to-1 multiplexer. In the demultiplexer section a 1-to-2 DEMUX, 1-to-4 DEMUX, and 1-to-8 DEMUX are described and generalized to the 1-to-2^n demultiplexer. Moreover, the quantum costs and delays of the multiplexers and demultiplexers are provided.

Critical thinking questions

1. What are the major applications of quantum MUXs and quantum DEMUXs?
2. Discuss the distinction between a quantum MUX and quantum DEMUX.
3. What function does the quantum MUX perform when the enable input is high? Explain.
4. Distinguish the difference between the two primary quantum multiplexing methodologies.

5. Which type of concept is used in a quantum DEMUX?
6. For the 1-to-8 quantum DEMUX, how many select lines are required? Explain.
7. For the 4-to-1 quantum MUX and 1-to-4 quantum DEMUX, write down the pseudocode.

References

[1] Haghparast M and Monfared A T 2017 Novel quaternary quantum decoder, multiplexer and demultiplexer circuits *Int. J. Theor. Phys.* **56** 1694–707

[2] Khan M H A 2008 Reversible realization of quaternary decoder, multiplexer, and demultiplexer circuits *38th Int. Symp. on Multiple Valued Logic (ISMVL 2008)* pp 208–13

[3] Mardiris V A and Karafyllidis I G 2010 Design and simulation of modular 2^n to 1 quantum-dot cellular automata (QCA) multiplexers *Int. J. Circuit Theory Appl.* **38** 771–85

Chapter 6

The quantum adder circuits

Learning objectives

- Discuss the quantum carry skip adder (QCSA).
- Learn the procedure to calculate the size and output power of a QCSA circuit.
- Acquire knowledge of the quantum comparison circuit.
- Find out the complexity of the *n*-qubit QCSA.
- Learn how to design a QCSA.
- Describe a quantum binary coded decimal (BCD) adder.
- Construct a quantum 2-to-1 multiplier circuit.
- Design algorithms for describing the four-qubit QCSA and the *n*-qubit QCSA.

Several types of adders are used in computing systems, such as quantum carry skip adders (QCSAs), quantum binary coded decimal (BCD) adders, and so on. Among them the most common is the quantum ripple carry adder in which quantum full-adders are connected in series and the carry is propagated through all the stages and hence requires more carry propagation time to generate the carry output. Carry look-ahead quantum adders are the fastest of all the quantum adders since the carry output is generated in parallel computation, but this requires a large number of gates. The QCSA is the most promising quantum adder which presents a compromise between hardware and performance compared to both quantum adders mentioned above. In the quantum full-adder operation, if either of the inputs is a logical one, then the cell will propagate the carry input to carry output. Hence the *i*th quantum full-adder carry input C_i will propagate to its carry output, C_{i+1}, when $P_i = X_i \oplus Y_i$.

In computers, numbers are stored in straight binary format. Due to the inherent characteristics of floating-point numbers and limitations in storage formats, not all floating-point numbers can be represented with the desired precision. Thus computing in decimal format is gaining popularity as losses due to precision can be avoided in this format. However, the hardware support for binary arithmetic allows it to be performed faster than decimal arithmetic. Faster hardware for decimal floating-point arithmetic is also imminent as it has importance in financial and Internet-based applications. Thus faster circuits for BCD numbers will have a great impact as they are likely to be incorporated in more complex circuits, such as future mathematical processors.

In this chapter the QCSA and quantum BCD adder circuits are presented in sections 6.1 and 6.5, respectively.

6.1 The quantum carry skip adder

Quantum adders are used widely in generic computers because they are very important for adding data in the processor. The simplest quantum adder is the quantum ripple carry adder. It is easy to understand and implement. A more complex quantum adder is the QCSA. It is used widely due to its superior performance compared to the quantum ripple carry adder. In this chapter an n-qubit QCSA is presented which has fewer quantum gates. It also produces fewer garbage outputs and requires less area, power, and delay. With the help of these properties, the efficiency of the quantum logic synthesis of an n-qubit carry skip adder has also been proved in this chapter.

Quantum adders are one of the most common elements in circuit design since addition is a fundamental operation for any kind of digital system. However, traditional ripple carry adders have some defects such as low computing efficiency and long delay. When generalizing and designing large quantum adders, the quantum ripple carry adder can end up with a large number of quantum gates, garbage outputs, and constant inputs, and high hardware complexity. The reason behind this lies in the carry propagation. If the carry propagation is not performed in an effective way a huge summation network is required to design a large scale adder. In order to overcome the disadvantages mentioned above, a new kind of adder called the QCSA is used which has a shorter delay.

6.2 The quantum comparison circuit

The quantum comparison circuit (QCC) is constructed, as depicted in figure 6.1, with 11 inputs comprising three constant inputs. This quantum circuit consists of

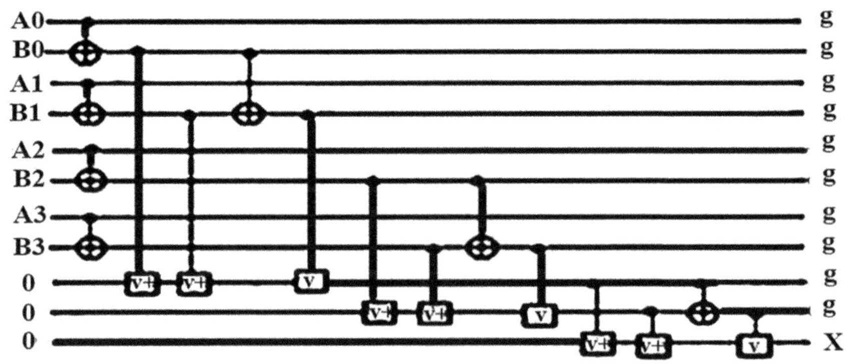

Figure 6.1. The quantum comparison circuit.

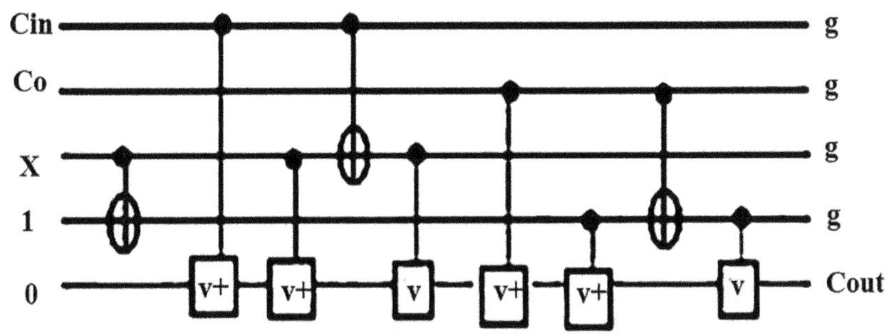

Figure 6.2. The quantum 2-to-1 MUX circuit. Reproduced with permission from [18]. Copyright 2014 IEEE.

four CNOT gates and three quantum Peres gates. The circuit performs the following function:

$$X = (A_3 \oplus B_3) \cdot (A_2 \oplus B_2) \cdot (A_1 \oplus B_1) \cdot (A_0 \oplus B_0).$$

If all $A_i \neq B_i$, X will generate the value 1. The number of quantum gate is calculated as follows: $4 \times 1 + 4 \times 3 = 16$.

6.3 The quantum 2-to-1 multiplier circuit

The quantum multiplier (MUX) block carries out the function as $C_{\text{out}} = C_{\text{in}} \cdot X \cdot C_o \cdot X$ where X comes from the quantum comparison circuit and C_{out} comes from the last quantum full-adder in the four-bit quantum ripple carry adder. Its quantum design is illustrated in figure 6.2 which has a CNOT gate and two quantum Peres gates. The number of quantum gate is $1 \times 1 + 4 \times 2 = 9$.

6.4 The design of a quantum carry skip adder

The QCSA is constructed by cascading a four-qubit carry skip adder. Each four-qubit carry skip adder contains a four-qubit ripple carry adder, a comparison circuit, and a 2-to-1 quantum MUX. If the two inputs are completely reversed, then the carry entering the quantum carry skip adder will simply be propagated to the next four-qubit carry skip adder. Thus there is no need for the next quantum carry skip adder to wait for the carry-in created by this current four-qubit carry skip adder. Therefore, the delay generated by the carry qubit can be shortened. Each comparison block is used for un-equivalence in input qubits. A quantum MUX is responsible for selecting a generated carry or a propagated (previous) carry with its selection line being the output of the comparison circuit just described. If each four-qubit carry skip block $A_i \neq B_i$, then it can be said that a carry can skip over the block. Otherwise, if $A_i = B_i$ it can be said that the carry must be generated in the block.

In the following subsections a four-qubit quantum carry skip adder circuit is constructed and then an n-qubit quantum carry skip adder circuit is shown.

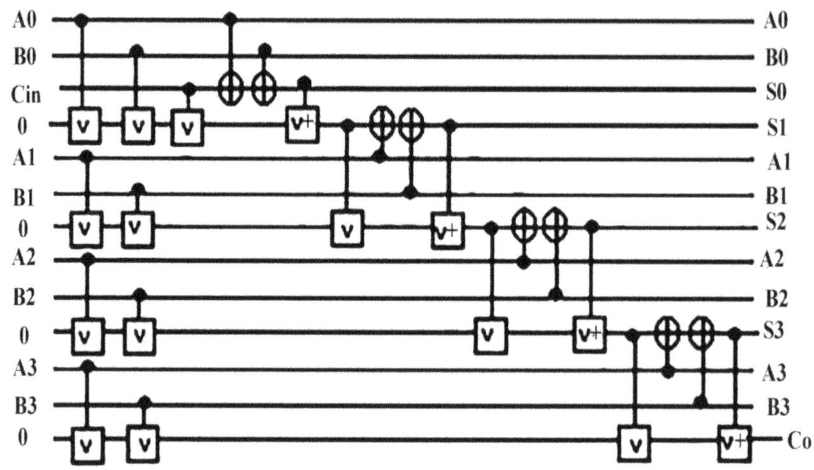

Figure 6.3. A four-bit quantum ripple carry adder. Reproduced with permission from [18]. Copyright 2014 IEEE.

6.4.1 The four-qubit quantum carry skip adder

A four-qubit QCSA circuit is designed using a four-qubit ripple carry adder along with a comparison circuit. The quantum comparison circuit ensures that the comparison of each block is indeed unequal. In this regard, a quantum MUX circuit is used that is responsible for selecting a generated carry or a propagated (previous) carry with its selection line which is considered as the output of the quantum comparison circuit.

The quantum four-qubit ripple carry adder circuit is constructed by cascading a double Peres gate (DPG) quantum circuit, where the DPG works as a quantum full-adder, depicted in figure 6.3, where two four-qubit binary additions are performed, such as A and B together with the input carry C_{in} being first added in the quantum ripple carry adder circuit to produce the binary sum and the carry out. The number of quantum gates of this circuit is $6 \times 4 = 24$ and the delay is 18Δ. Then the quantum comparison gate (QCG) constructs the quantum 2-to-1 MUX module for generating the carry, shown in figures 6.1 and 6.2, respectively. Finally, all circuits are combined and the four-qubit quantum carry skip adder is obtained, shown in figure 6.4. The design procedure of a four-qubit quantum carry skip adder circuit is shown using algorithm 6.1.

6.4.2 The n-qubit quantum carry skip adder

A four-qubit QCSA circuit is constructed using a four-qubit quantum ripple carry adder, a QCG, and a quantum 2-to-1 MUX gate (QMG), as shown in figure 6.4. Therefore, an n-qubit quantum carry skip adder circuit can be constructed by cascading ($N = n/4$) four-qubit QCSA circuits, as shown in figure 6.5. The process is described by algorithm 6.2.

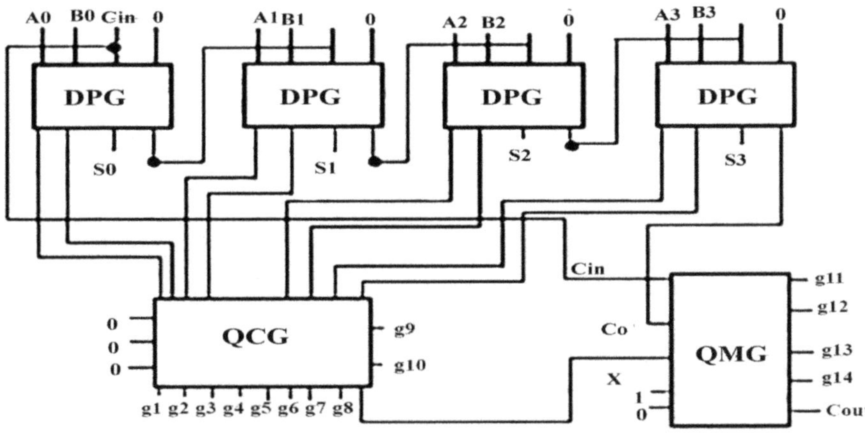

Figure 6.4. The four-qubit QCSA. Reproduced with permission from [18]. Copyright 2014 IEEE.

Input: $A(A_3A_2A_1A_0), B(B_3B_2B_1B_0), C_{in}$
Output: $S(S_3S_2S_1S_0), C_o, X, C_{out}$

START
for i = 0 to 3 **do**
  Apply a DPG quantum circuit with,
  Input:={A_i, B_i, C_{in}, 0}
  Output:={ A_i, B_i, S_i, C_{oi}}
end for
Apply a QCG gate with,
Input:=A_0, B_0, A_1, B_1, A_2, B_2, A_3, B_3, 0, 0, 0
Output:=g_1, g_2, g_3, g_4, g_5, g_6, g_7, g_8, g_9, g_{10}, X
Apply a QMG gate, $(a, b, c, d, e)(p, q, r, s, t)$
Input:=C_{in}, C_{03}, X, 1, 0
Output:=g_{11}, g_{12}, g_{13}, g_{14}, C_{out}
END

Algorithm 6.1. Construction of a quantum four-qubit QCSA.

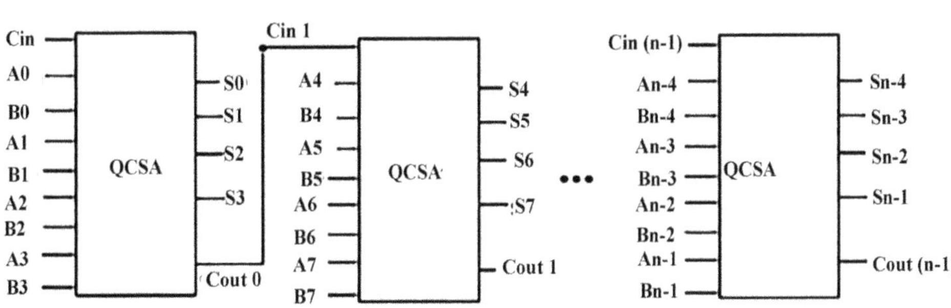

Figure 6.5. The *n*-qubit QCSA circuit. Reproduced with permission from [18]. Copyright 2014 IEEE.

```
START
for i = 1 to n/4 do
    Call Algorithm 6.4.1.1
end for
END
```

Algorithm 6.2. Construction of an n-qubit QCSA.

```
START
Area = 0, Q_QCSA = 0, N = n/4
for i =1 to N do
    Get the Number of Quantum Gates of 4-bit QCSA (Quantum Carry
    Skip Adder) Circuit, QCSA
    Q_QCSA = Q_QCSA + Q.QCSA
end for
Area = 50 × Q_QCSA
return Area
END
```

Algorithm 6.3. Calculation of the area for an n-bit QCSA circuit.

6.4.3 Calculation of the area and power of a quantum carry skip adder circuit

Area is an important issue in designing a circuit. If the area of a circuit is very large, the cost of that circuit will increase. The area of a gate is defined by the feature size. This size varies according to the quantum cost of the gate. As basic quantum gates are fabricated using quantum dots, with sizes ranging from several to tens of nanometers (10^{-9} m) in diameter, the size of the basic quantum gates ranges from 50 Å to 300 Å. The angstrom (Å) is a unit equal to 10^{-10} m (one ten-billionth of a meter) or 0.1 nm. Its symbol is the Swedish letter Å. Quantum circuits can be implemented with the basic quantum gates and the quantum cost of a gate depends on the number of basic quantum gates needed to implement it. Thus the area of a gate can be defined as follows: $A = N_q \times S_q$, where A= area, N_q = number of quantum gates, and S_q= size of basic quantum gates. According to the feature size of the DPG quantum circuit the area is ((50 Å × 6) to (300 Å × 6)) = (300 Å to 1800 Å) where the number of quantum gates of the DPG circuit is 6. Algorithm 6.3 describes the area calculation of the n-bit QCSA circuit.

Power is also an important issue in designing a circuit. One should design a circuit which needs little power. The power of a gate is defined by the energy. The energy of the basic quantum gate is 142.3 meV. Quantum circuits can be implemented with basic quantum gates and the quantum cost of a gate depends on the number of basic quantum gates needed to implement it. Thus the power of a gate can be defined as follows: $P = N_q \times E_q$, where P= power, N_q = number of quantum gates, and

```
START
Power = 0, Q_QCSA = 0, N = n/4
for i =1 to N do
    Get the Number of Quantum Gates of 4-bit QCSA (Quantum Carry
    Skip Adder) Circuit, QCSA
    Q_QCSA = Q_QCSA + Q.QCSA
end for
Power = 142.3 × Q_QCSA
return Power
END
```

Algorithm 6.4. Calculation of power for an *n*-bit QCSA circuit.

E_q = energy of basic quantum gates. For example, the energy of the DPG quantum circuit is (6 × 142.3) meV (microelectronvolts) = 853.8 meV, where the number of quantum gates of DPG circuit is 6. Algorithm 6.4 describes the power calculation of the *n*-bit QCSA circuit.

6.4.4 Complexity of the *n*-qubit quantum carry skip adder circuit

The complexities of a quantum circuit in terms of quantum gates, garbage output, and delay, etc, are very important as they have a great impact during the physical implementation. In this subsection the different complexities of an *n*-qubit QCSA circuit have been presented.

Property 6.1. An *n*-qubit QCSA circuit requires $49N$ quantum gates where $N = n/4$.

Proof The above statement is proved by mathematical induction. As the number of quantum gate of a DPG quantum circuit is six, the number of quantum gates of a four-qubit quantum ripple carry adder is 24. A four-qubit QCSA requires a four-qubit quantum ripple carry adder, a QCC for which the number of quantum gates is 16, and a quantum 2-to-1 MUX for which the number of quantum gates is 9. Thus the total number of quantum gates (NOG) required to construct a four-qubit QCSA circuit is $NOG_4 = NOG_{DPG} \times 4 + NOG_{QCG} + NOG_{QMG} = 6 \times 4 + 16 + 9 = 49$.

Thus the statement holds for base case $n = 4$.

Assume that the statement holds for $n = k$. Thus a k-qubit QCSA can be realized with $49K$ quantum gates, where $K = k/4$.

A $(k + 1)$-qubit QCSA requires $(k + 1)$ number of QCSA blocks. As a result, the total number of quantum gates required to construct a $(k + 1)$-qubit QCSA is $49 \times (K + 1)$, where $K = k/4$. Thus the statement holds for $n = k + 1$.

Therefore an n-qubit QCSA circuit can be realized with $49N$ quantum gates, where $N = n/4$.

Property 6.2. An n-qubit QCSA circuit produces $43N\Delta$ quantum delay, where Δ is the unit delay and $N = n/4$.

Proof A QCSA circuit consists of a quantum ripple carry adder circuit, a quantum comparison circuit, and a quantum 2-to-1 MUX circuit. A one-qubit quantum ripple carry adder circuit requires 6Δ delay. Thus the total delay for an n-qubit quantum ripple carry adder circuit is $6n\Delta$ delay.

A QCC and a quantum 2-to-1 MUX require 16Δ and 9Δ delay, respectively. As there are $2(n - 1)$ controlled-V gates at the same level of an n-qubit quantum ripple carry adder circuit, the $2(n - 1)\Delta$ delay can be deducted from the total delay of the quantum ripple carry adder circuit. Thus the delay of a four-qubit QCSA circuit is

$$6 \times 4\Delta + 16\Delta + 9\Delta - 2(4 - 1)\Delta = 24\Delta + 16\Delta + 9\Delta - 8\Delta + 2\Delta = 49\Delta - 6\Delta.$$

$$= 43\Delta$$

Therefore, the total delay of an n-qubit QCSA circuit is $43N\Delta$ where $N = n/4$.

Example 6.1. When $n = 4$, the total delay of a four-qubit QCSA circuit is $(43 \times 1)\Delta = 43\Delta$, where $N = n/4 = 4/4 = 1$.

Property 6.3. An n-qubit QCSA circuit requires $18N\sigma + 31N\Omega$ quantum gate calculation complexity, where σ is the CNOT gate calculation complexity, Ω is the controlled-V (controlled-V^+) gate calculation complexity, and $N = n/4$.

Proof From property 6.2 it is found that a QCSA circuit consists of a quantum ripple carry adder circuit, a QCC, and a quantum 2-to-1 MUX circuit. A one-qubit quantum ripple carry adder circuit has 2σ CNOT quantum gate calculation complexity and 4Ω controlled-V (controlled-V^+) quantum gate calculation complexity. Thus the quantum gate calculation complexity of the four-qubit quantum ripple carry adder circuit is $(2 \times 4)\sigma = 8\sigma$ for eight CNOT gates, and $(4 \times 4) = 16\Omega$ for 16 controlled-V (controlled-V^+) gates, of the QCC is 7σ for seven CNOT gates and 9Ω for nine controlled-V (controlled-V^+) gates, and of the quantum 2-to-1 MUX is 3σ for three CNOT gates and 6Ω for six controlled-V (controlled-V^+) gates, i.e. $8\sigma + 16\Omega + 7\sigma + 9\Omega + 3\sigma + 6\Omega = 18\sigma + 31\Omega$ quantum gate calculation complexity.

Therefore, the total quantum gate calculation complexity for an n-qubit QCSA circuit is $18N\sigma + 31N\Omega$ where $N = n/4$.

Example 6.2. When $n = 4$, the total quantum gate calculation complexity of a four-qubit QCSA circuit is $(18 \times 1)\sigma + (31 \times 1)\Omega = 18\sigma + 31\Omega$, where $N = n/4 = 4/4 = 1$.

Property 6.4. An n-qubit QCSA circuit requires $2450N$ Å area, where Å is the unit of measuring area and $N = n/4$.

Proof. From property 6.2 it is found that a QCSA circuit consists of a quantum ripple carry adder circuit, a QCC, and a quantum 2-to-1 MUX circuit. A one-qubit quantum ripple carry adder circuit requires 6Δ quantum gates. Thus the total number of quantum gates for the n-qubit quantum ripple carry adder circuit is $6n$.

Moreover, the QCC and the quantum 2-to-1 MUX require 16 and 9Δ quantum gates, respectively. Thus the area for a four-qubit QCSA circuit is $((6 \times 4 + 16 + 9) \times 50)$ Å = 2450 Å. Therefore, the total area for an n-qubit QCSA circuit is $2450N$ Å, where $N = n/4$.

Property 6.5. An n-qubit QCSA circuit produces $6972.7N$ meV power, where meV is the unit of measuring power and $N = n/4$.

Proof As an n-qubit QCSA has $6n$ quantum gates for the quantum ripple carry adder, 16 quantum gates for the QCC, and 9 quantum gates for the quantum MUX, as described in property 6.1, the power for an four-qubit QCSA circuit is $((6 \times 4 + 16 + 9) \times 142.3)$ meV = 6972.7 meV.

Therefore, the total power for an n-qubit QCSA circuit is $6972.7N$ meV where $N = n/4$.

Property 6.6. An n-qubit QCSA circuit can produce $14N$ garbage outputs, where $N = n/4$.

Proof The above statement is proved by mathematical induction.

In a four-qubit QCSA, the quantum ripple carry adder does not produce any garbage, but each QCC and quantum MUX produces ten and four garbage outputs, respectively, as shown in figure 6.3. Thus a four-bit QCSA produces $G_4 = G_{QCG} + G_{QMG} = 10 + 4 = 14$ garbage outputs. Thus the statement holds for base case $n = 4$.

Assume that the statement holds for $n = k$. Thus a k-qubit QCSA circuit produces $14K$ garbage outputs, where $K = k/4$.

A $(k + 1)$-qubit QCSA requires $(k + 1)$ QCSA blocks. As a result, the total number of garbage outputs produced for a $(k + 1)$-qubit QCSA is $14 \times (K + 1)$ where $K = k/4$. Thus the statement holds for $n = k + 1$.

Therefore, an n-qubit QCSA circuit can be realized with $14N$ garbage outputs, where $N = n/4$.

Example 6.3. When $n = 4$, see figure 6.5, the total garbage outputs of a four-qubit QCSA are as follows:

$(14 \times 1) = 14$, where $N = n/4 = 4/4 = 1$.

6.5 The quantum BCD adder

The BCD adder comprises three units, a four-qubit binary adder, an over-9-detection unit, and a correction unit. The first part is a binary adder which performs on two four-qubit BCD digits and a one-qubit carry input. The second part, the over-9-detector, recognizes if the result of the first part is more than 9 or not. Finally, in the third part, if the output of the detector overflow detection qubit (ODQ) is 1, the sum is added to by 6, otherwise do nothing. The four-qubit binary adder is the cascade of a four-qubit carry–propagate adder. The detection part is constructed using two AND gates and one OR gate. The correction unit adds 0 to the binary

A:	**0111**	**7**
B:	**0101**	**5**
A+B: 0	**1100**	**0 12 if the sum is greater than 9**
	0110	**6 then add 6 to the sum**
	1 0010	**1 2**
Sum:	**0010**	**BCD representation**
C_{out}:	**1**	**BCD overflow**

(a)

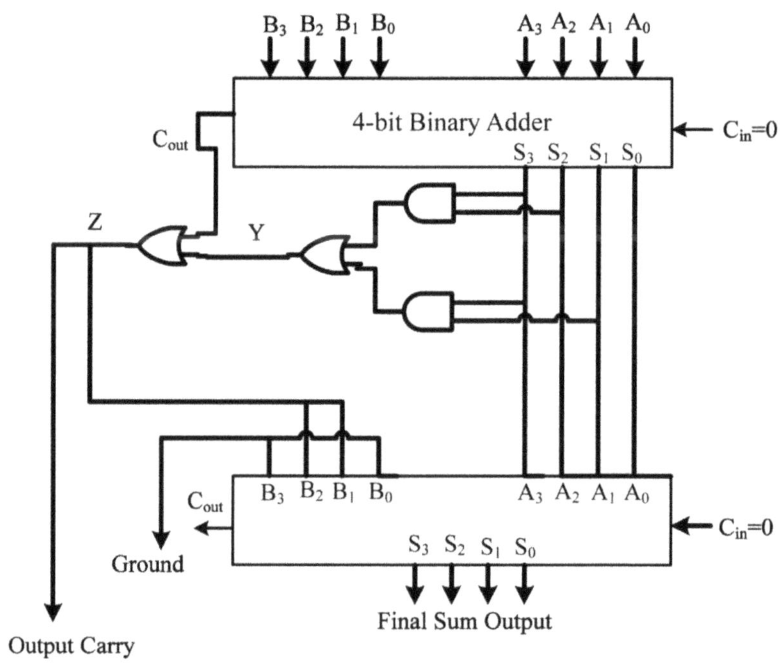

(b)

Figure 6.6. BCD addition. (a) BCD addition procedure. (b) Detailed architecture of a conventional or irreversible BCD adder.

number if the binary result is less than 10 and adds 6 to the binary result if it is more than 9.

Let A and B be two one-digit BCD numbers. The BCD addition procedure of these numbers is shown in figure 6.6(a) and the detailed architecture of a conventional or irreversible BCD adder is shown in figure 6.6(b).

Table 6.1. The binary sum of two BCD numbers.

BCD_1	BCD_2	Binary sum	
$A_3A_2A_1A_0$	$B_3B_2B_1B_0$	$Z_3Z_2Z_1Z_0$	
0 0 0 0	0 0 0 0	0 0 0 0	
.	.	.	
.	.	.	
.	.	.	
.	.	.	
1 0 0 1	1 0 0 1	1 0 0 1 0	(decimal 18, if $\text{carry}_{\text{in}} = 0$)
1 0 0 1	1 0 0 1	1 0 0 1 1	(decimal 19, if $\text{carry}_{\text{in}} = 1$)

Let $A_3A_2A_1A_0$ and $B_3B_2B_1B_0$ be two BCD numbers to be added and the resulting number is represented by $Z_3Z_2Z_1Z_0$. The carry out is represented by $C_4C_3C_2C_1$. C_4 is 1 when the resulting number is greater than binary 1111, i.e. decimal 15. As the single digit of the BCD number goes up to 9 (binary 1001), the maximum sum of the two one-digit BCD numbers is 19 (=9 + 9 + 1), shown in table 6.1.

In this section the quantum BCD adder is presented. Four quantum full-adders can be used to design a four-bit quantum full-adder (a quantum full-adder is presented in section 4.1.1). Two quantum Fredkin gates and one quantum Toffoli gate can be used as the overflow detection unit. Finally, one quantum Peres gate, one quantum full-adder, and one quantum Ex-OR gate can be used to construct the correction unit. The whole design of the one-digit quantum BCD adder is shown in figure 6.7.

Property 6.7. Let g_{pa} be the minimum number of garbage outputs for a quantum four-qubit parallel adder, g_{od} be the minimum number of garbage outputs produced by the overflow detection circuit, and g_{ocl} be the minimum number of garbage outputs for overflow correction logic. Let g_{BCD} be the number of garbage outputs for a quantum BCD adder, then $g_{\text{BCD}} \geqslant g_{\text{pa}} + g_{\text{od}} + g_{\text{ocl}}$, where $g_{\text{pa}} \geqslant 8$, $g_{\text{od}} \geqslant 0$, and $g_{\text{ocl}} \geqslant 2$.

Proof. A quantum BCD adder consists of a four-qubit quantum parallel adder, overflow detection logic, and overflow correction logic circuits. A four-bit parallel adder consists of four full-adders and, according to figure 6.7, a four-bit parallel adder will be realized by at least eight garbage outputs. Thus the minimum number of garbage outputs for a quantum four-bit parallel adder is $g_{\text{pa}} = 8$.

In the overflow detection logic circuit, the overflow expression $F = (Z_1 + Z_2)Z_3 \oplus C_4$ is realized and, according to figure 6.7, the quantum BCD overflow detection unit can be realized using at least zero garbage outputs. Thus the minimum number of garbage outputs for the overflow detection unit is $g_{\text{od}} \geqslant 0$.

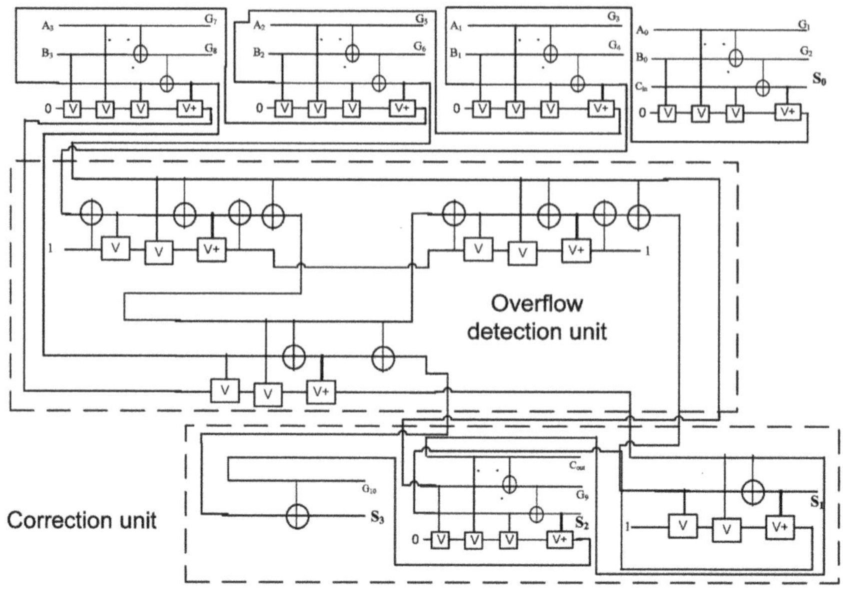

Figure 6.7. A one-qubit quantum BCD adder.

In the overflow correction logic, overflow F is propagated. According to the design presented in this section, overflow correction logic generates only two garbage outputs. Thus the minimum number of garbage outputs for overflow correction logic is $g_{\text{ocl}} \geqslant 2$.

As a result, the total number of garbage outputs for a quantum BCD adder is $g_{\text{BCD}} \geqslant g_{\text{pa}} + g_{\text{od}} + g_{\text{ocl}}$, where $g_{\text{pa}} \geqslant 8$, $g_{\text{od}} \geqslant 0$, and $g_{\text{ocl}} \geqslant 2$.

Property 6.8. Let gt_{pa} be the minimum number of quantum gates for a quantum four-bit parallel adder, gt_{od} be the minimum number of quantum gates required by the overflow detection circuit, and gt_{ocl} be the minimum number of quantum gates for the overflow correction logic circuit. Let gt_{BCD} be the total number of quantum gates for a quantum BCD adder, then

$$gt_{\text{BCD}} = gt_{\text{pa}} + gt_{\text{od}} + gt_{\text{ocl}}, \text{ where } g_{\text{pa}} = 24, \ g_{\text{od}} = 15, \text{ and } g_{\text{ocl}} = 11.$$

Proof A quantum BCD adder consists of four-qubit quantum parallel adder, overflow detection logic, and overflow correction logic circuits. A four-qubit parallel adder consists of four quantum full-adders and each quantum full-adder is realized by six quantum gates. Thus minimum number of gates for a quantum four-qubit parallel adder is $gt_{\text{pa}} = 4 \times 6 = 24$.

In the overflow detection logic, three blocks are used, each of which has five quantum gates. Thus the minimum number of gates for the overflow detection logic is $gt_{\text{od}} = 15$.

In the overflow correction logic, one quantum Peres gate, one quantum full-adder, and one quantum Ex-OR gate are used to construct the correction unit. The quantum Peres gate consists of four quantum gates, and one quantum full-adder consists of six quantum gates. Thus the minimum number of quantum gates for the overflow correction logic is $gt_{\text{ocl}} = 4 + 6 + 1 = 11$.

As a result, the total number of quantum gates for a quantum BCD adder is $gt_{\text{BCD}} = gt_{\text{pa}} + gt_{\text{od}} + gt_{\text{ocl}}$, where $g_{\text{pa}} = 24$, $g_{\text{od}} = 15$, and $g_{\text{ocl}} = 11$.

6.6 Summary

This chapter presents the design methodology of an n-qubit QCSA circuit with optimal delay and a quantum BCD adder. An efficient algorithm is shown for designing a compact QCSA. The BCD adder is designed using quantum full-adder circuits. The efficiency of the design is proved by presenting several properties. It is shown that the circuit has been constructed with the optimum number of quantum gates, delays, garbage outputs, quantum gate calculation complexity, area, and power. The adder is the basic arithmetic unit of a computer system and it can be used in diverse areas, such as complex quantum arithmetic circuits.

The design which has been shown in this chapter will help the reader to understand the next chapter better. In addition, the concept of designing a QCSA will help in designing quantum multiplier–accumulator circuits.

Critical thinking questions

1. How can the minimal propagation delay for a carry skip adder be calculated?
2. Which two methods are used to decrease the latency in quantum adders?
3. Construct the quantum comparison circuit and describe how it works.
4. Explain the properties of the quantum BCD adder.
5. Design an eight-qubit QCSA and explain its operation so that it may be implemented.
6. How many inputs are there in a quantum BCD adder?
7. Find the total quantum gate calculation complexity of a four-qubit QCSA circuit, when $n = 6$.
8. Calculate the total delay of a four-qubit QCSA circuit, when $n = 8$.

References

[1] Álvarez-Sánchez J J, Álvarez-Bravo J V and Nieto L M 2008 A quantum architecture for multiplying signed integers *J. Phys.: Conf. Ser.* 128 012013

[2] Babu H M H and Chowdhury A R 2005 Design of a reversible binary coded decimal adder by using reversible 4-bit parallel adder *18th Int. Conf. on VLSI Design and 4th Int. Conf. on Embedded Systems Design* pp 255–60

[3] Babu H M H and Chowdhury A R 2006 Design of a compact reversible binary coded decimal adder circuit *J. Syst. Archit.* **52** 272–82

[4] Balandin A A and Wang K L 1999 Implementation of quantum controlled-NOT gates using asymmetric semiconductor quantum dots *Quantum Computing and Quantum Communications* (Berlin: Springer) pp 460–7

[5] Biswas A K, Hasan M M, Chowdhury A R and Babu H M H 2008 Efficient approaches for designing reversible binary coded decimal adders *Microelectron. J.* **39** 1693–703

[6] Biswas A K, Hasan M M, Hasan M, Chowdhury A R and Babu H M H 2008 A novel approach to design BCD adder and carry skip BCD adder *21st Int. Conf. on VLSI Design (VLSID)* pp 566–71

[7] Bose A and Babu H M H 2015 Optimized designs of reversible fault tolerant BCD adder and fault tolerant reversible carry skip BCD adder *18th Int. Conf. on Computer and Information Technology (ICCIT)* pp 202–7

[8] Bose A, Babu H M H and Gupta S 2015 Design of compact reversible online testable ripple carry adder *IEEE Int. WIE Conf. on Electrical and Computer Engineering (WIECON-ECE)* pp 556–60

[9] Bruce J W, Thornton M A, Shivakumaraiah L, Kokate P S and Li X 2002 Efficient adder circuits based on a conservative reversible logic gate *IEEE Proc. Computer Society Annual Symp. on VLSI* (Piscataway, NJ: IEEE) pp 83–8

[10] Guyot A, Hochet B and Muller J-M 1987 A way to build efficient carry-skip adders *IEEE Trans. Comput.* **C-36** 1144–52

[11] Haghparast M 2011 Design and implementation of nanometric fault tolerant reversible BCD adder *Australian J. Basic Appl. Sci.* **5** 896–901

[12] Hamacher V C, Vranesic Z G and Zaky S G 2002 *Computer Organization* (New York: McGraw-Hill)

[13] Haque M U, Sworna Z T, Babu H M H and Biswas A K 2018 A fast FPGA-based bcd adder *Circuits Syst. Signal Process.* **37** 4384–408

[14] Hung W N N, Song X, Yang G, Yang J and Perkowski M 2006 Optimal synthesis of multiple output Boolean functions using a set of quantum gates by symbolic reachability analysis *IEEE Trans. Comput.-Aided Des. Integr. Circuits Syst.* **25** 1652–63

[15] Islam M *et al* 2010 Fault tolerant variable block carry skip logic (VBCSL) using parity preserving reversible gates arXiv:1009.3819

[16] Islam M S, Rahman M M, Begum Z and Hafiz M Z 2009 Fault tolerant reversible logic synthesis: carry look-ahead and carry-skip adders *Int. Conf. on Advances in Computational Tools for Engineering Applications* (Piscataway, NJ: IEEE) pp 396–401

[17] Lin Y S and Radhakrishnan D 2008 Delay efficient 32-bit carry-skip adder *VLSI Des.* **2008** 9

[18] Lisa N J and Babu H M H 2014 A compact realization of an *n*-bit quantum carry skip adder circuit with optimal delay *NASA/ESA Conf. on Adaptive Hardware and Systems (AHS)* (Piscataway, NJ: IEEE) pp 270–7

[19] Lisa N J and Babu H M H 2015 Design of a compact ternary parallel adder/subtractor circuit in quantum computing *IEEE Int. Symp. on Multiple-Valued Logic* pp 36–41

[20] Mano M M and Kime C R 2008 *Logic and Computer Design Fundamentals* (Upper Saddle River, NJ: Pearson Prentice-Hall)

[21] Mohammad M, Eshghi M, Haghparast M and Bahrololoom A 2008 Design and optimization of reversible BCD adder/subtractor circuit for quantum and nanotechnology based systems *World Appl. Sci. J.* **4** 787–92

[22] Mohammadi M, Haghparast M, Eshghi M and Navi K 2009 Minimization and optimization of reversible BCD-full adder/subtractor using genetic algorithm and don't care concept *Int. J. Quantum Inf.* **7** 969–89

[23] Nagendra C, Irwin M J and Owens R M 1996 Area-time-power tradeoffs in parallel adders *IEEE Trans. Circuits Syst. II* **43** 689–702

[24] Pang Y, Wang J and Wang S 2012 A 16-bit carry skip adder designed by reversible logic *5th Int. Conf. on Biomedical Engineering and Informatics* (Piscataway, NJ: IEEE) pp 1332–5

[25] Thapliyal H and Vinod A P 2006 Transistor realization of reversible TSG gate and reversible adder architectures *IEEE Asia Pacific Conf. on Circuits and Systems* (Piscataway, NJ: IEEE) pp 418–21

[26] Velagaleti S 2009 A novel high speed dynamic comparator with low power dissipation and low offset *Doctoral dissertation* National Institute of Technology, Rourkela.

[27] Viamontes G F, Markov I L and Hayes J P 2005 High-performance quantum circuit simulation *QuIDDPro* http://vlsicad.eecs.umich.edu/Quantum/qp/ (Accessed: 19 March 2020)

[28] Yao A C-C 1993 Quantum circuit complexity *Proc. IEEE 34th Annual Foundations of Computer Science* pp 352–61

IOP Publishing

Quantum Computing (Second Edition)
A pathway to quantum logic design
Hafiz Md Hasan Babu

Chapter 7

The quantum multiplier–accumulator

Learning objectives

- Describe the significance of the quantum multiplier–accumulator (QMAC) circuit.
- Construct a quantum multiplier circuit using a quantum AND operation and a quantum full-adder circuit.
- Develop an algorithm to explain multiplication.
- Construct an $n \times n$ quantum multiplier circuit.
- Discuss ways to reduce the garbage outputs and ancillary inputs of quantum circuits.

All quantum signal processing (QSP) algorithms use the quantum multiplier–accumulator (QMAC) operation for high-performance quantum processing systems. This operation is needed in filters, Fourier transforms, etc, since it eases the computation of convolution. A QMAC unit comprises of a quantum multiplier, quantum adder, and quantum accumulator. The quantum multiplier multiplies the inputs and gives the result to the quantum adder, which adds the quantum multiplier result to the previously accumulated result. A QMAC unit is used to perform the multiplication and quantum accumulator operations together to avoid unnecessary overhead on the processor in terms of processing time and the on-chip memory requirements.

High-speed multiplication has always been a fundamental requirement of high-performance processors and systems. In QSP applications, multiplication is one of the most utilized arithmetic operations. Improving multiplier design directly benefits the high-performance embedded processors and QSP applications used in consumer and industrial electronic products. Moreover, quantum information processing (QIP) is a high-impact research area in quantum information science for constructing a quantum computer. The main goal of QIP is to harness the fundamental laws of quantum mechanics to improve all aspects (e.g. acquisition, transmission, and

doi:10.1088/978-0-7503-5683-1ch7

processing) of information processing dramatically, as well as enhance the performance of quantum computers. There are several tasks for which a quantum computer will be useful. The one that is mentioned most frequently is that quantum computers will be able to read secret messages communicated over the Internet using the current technologies such as Rivest–Shamir–Adleman (RSA), Diffie–Hellman, and other cryptographic protocols that are based on the difficulty of number-theoretic problems such as factoring and discrete logarithms. In addition, quantum computers are also useful for scientists for conducting virtual experiments and searching in huge amounts of data.

7.1 The importance of a quantum multiplier–accumulator

The scope of the optimization of time complexity as well as the optimization of garbage outputs and constant inputs has been addressed, which has motivated the introduction of a completely new and efficient technique of multiplication for quantum multipliers with optimum time complexity. Time complexity reduction also minimizes the number of operations and hardware complexity. It is necessary to introduce an efficient and faster multiplication algorithm to optimize time complexity. Thus a new, faster, more efficient, and optimal tree-based multiplication technique which works like the breadth fast search (BFS) algorithm is introduced in this chapter.

To implement the quantum multiplier using the quantum multiplication technique, a new quantum full-adder (QFA) circuit and a new quantum ANDing circuit (QAC) are introduced with the fewest number of gates and less area, power, and gate level delay. In this chapter two algorithms for the construction of quantum partial product generation (PPG) and partial product addition (PPA) circuits are shown using QFA circuits and QACs that work in the manner of a heuristic algorithm, and divide and conquer algorithm, respectively, and a combination of them implements the quantum multiplier using the quantum multiplication technique and reduces the number of quantum gates, garbage outputs, and constant inputs, and the amount of area, power, and delay compared to the best known multipliers.

A large number of ancillary inputs and garbage outputs results in increased area, delay, power, and quantum cost. Thus it is important to design the quantum multiplier as well as the QMAC unit in such a way that they reduce the ancillary inputs and garbage outputs. For this purpose, in this chapter a new algorithm for designing the QMAC unit is described which is similar to the greedy algorithm. A QMAC unit is formed by combining quantum multiplier, full-adder, and quantum accumulator circuits. Thus the QMAC circuit has reduced ancillary inputs, garbage outputs, delay, area, power, quantum costs, and implementation complexities. Furthermore, this circuit is compact and efficient and requires a minimum number of quantum gates. In addition, this chapter presents a very novel and unparalleled design of a quantum multiplier as well as a QMAC unit that reduces the garbage outputs and ancillary inputs.

7.2 The quantum multiplication technique

The quantum multiplier is a very important device for quantum arithmetic logic units (ALUs). With advances in technology, researchers are trying to design quantum multipliers with high speed and low power consumption to make them suitable for various high-speed, low power, and compact VLSI implementations. To achieve high speed and low power consumption in a quantum multiplier, a tree-based multiplication technique is shown.

Property 7.1. A *tree* is a finite set T such that:

1. One node $r \in T$ is designated as the root.
2. The remaining nodes are partitioned into disjoint sets $T_1, T_2, \ldots, T_k$, where each of the sets is a tree and $k \geqslant 0$.

Property 7.2. In a tree, each node can have any number of child nodes.

Property 7.3. The *partial product addition tree* is a tree consisting of nodes in levels where the nodes in each level contain the products and their additions.

Property 7.4. A *node* of a tree contains the information or data. A node may be a child or parent node.

Property 7.5. The *level* of a node is the distance of that node from the root, where the distance between two vertically adjacent nodes is one.

Example 7.1. A tree, and its child and parent nodes, with the root and level are depicted in figure 7.1. In this figure node A is at the top of the tree and is called the root. Nodes which are just the next level of the root (B, C, D in figure 7.1) are the parent nodes and nodes at the bottom level are called the leaf or the child nodes of the previous parent node (I and J are the child nodes of the parent node D as shown in figure 7.1). To go to node E from root node A of the tree of figure 7.1, first it is

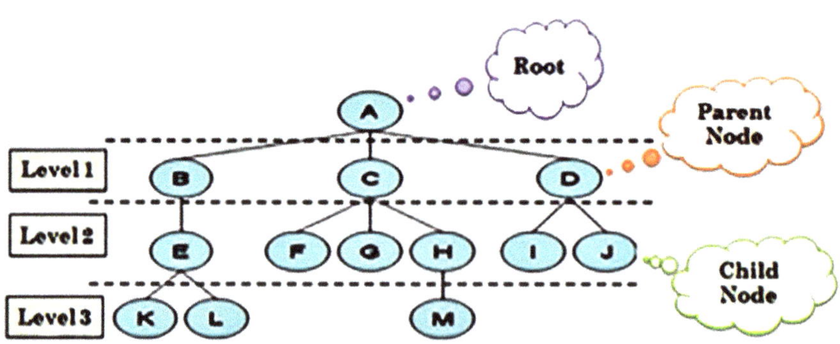

Figure 7.1. Examples of a tree, root, node, and level. Reproduced with permission from [4]. Copyright 2017 Springer Nature.

necessary to go to node B from root node A and then node E from node B. The distance from root node A to node B is 1, and from node B to node E is 1. Thus the total distance of node E from root node A is 2. Therefore, node E is at level 2.

Property 7.6. The set of qubits that is to be multiplied with other qubits is called the *multiplicand.*

Property 7.7. The qubits which are multiplied with the qubits of the multiplicand are called the *multiplier*.

Example 7.2. The multiplier and multiplicand qubits are depicted in figure 7.2. $|A7\rangle|A6\rangle|A5\rangle|A4\rangle|A3\rangle|A2\rangle|A1\rangle|A0\rangle$ is the set of the multiplicand qubits and $|B7\rangle|B6\rangle|B5\rangle|B4\rangle|B3\rangle|B2\rangle|B1\rangle|B0\rangle$ is the set of the multiplier qubits, as shown in figure 7.2.

Quantum logic preserves the properties of a linear combination of multiple states (superposition), whereas binary logic preserves the properties of only two states. Thus a multiplication algorithm based on the superposition states of quantum computing is shown. However, the multiplication technique is a generalized algorithm which is applicable for both quantum and classical computations.

The technique has two basic steps. First, partial products are generated in such a way that the copying of multiplicand qubits, multiplier qubits, and ANDing operations are performed in parallel, which results in the faster generation of partial products. Second, partial product additions are performed using a tree structure, where the tree has $((\log_2 n) - 1)$ levels. In the first layer of the tree, four successive rows of the partial product array are added by adders in parallel. In the following levels, the rate of increase of the number of adders will be $(n + 2 + i)$, where $i = 1, 2, \ldots, ((\log_2 n) - 2)$ and n is the number of qubits. This technique is

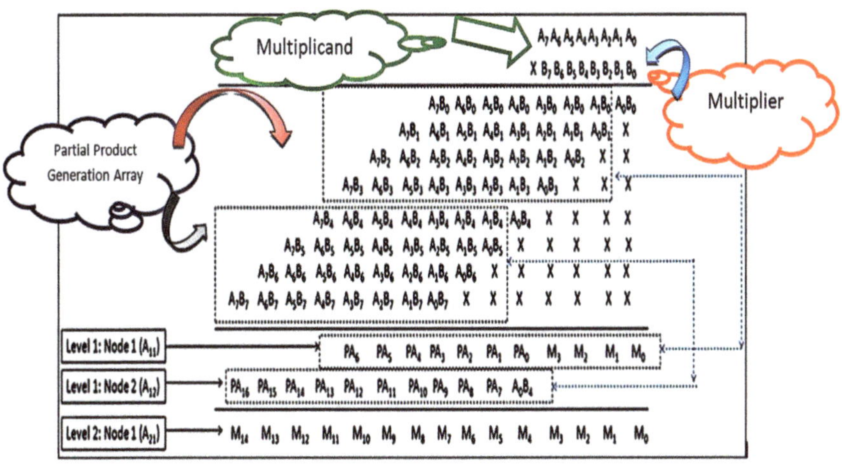

Figure 7.2. An example of the multiplication method. Reproduced with permission from [4]. Copyright 2017 Springer Nature.

1: START
2: Create a partial product generation array where ANDing and Copying operations are done in parallel to get a compact quantum partial product generation circuit.
3: Create a partial products addition tree with $((log_2\ n) - 1)$ levels.
4: Generate $n/4$ numbers of nodes at the first level of the partial product addition tree, where each node represents the addition of successive four rows of partial product array.
5: Define the next following levels of the tree where the addition operations are performed between two predecessor nodes in which the rate of the increase of the number of adders will be $(n{+}2{+}i)$ and $i = 1, 2, \ldots, ((log_2\ n) - 2)$.
6: Obtain the multiplication result in the $((log_2 n) - 1)^{th}$ level of the tree.
7: END

Algorithm 7.1. Multiplication technique.

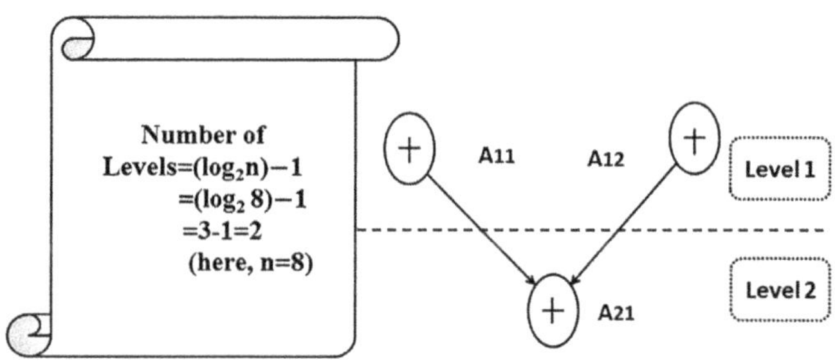

Figure 7.3. Partial product addition using the tree structure. Reproduced with permission from [4]. Copyright 2017 Springer Nature.

described by algorithm 7.1. The working principle of algorithm 7.1 is presented using example 7.3.

Example 7.3. The method is described using the following example.

To multiply two eight-qubit numbers, such as $\{|A7\rangle|A6\rangle|A5\rangle|A4\rangle|A3\rangle|A2\rangle|A1\rangle|A0\rangle\}$ (multiplicand qubits) and $\{|B7\rangle|B6\rangle|B5\rangle|B4\rangle|B3\rangle|B2\rangle|B1\rangle|B0\rangle\}$ (multiplier qubits), the partial product generation and partial product addition processes using the designed method for the given numbers are shown in figures 7.2 and 7.3, respectively. In figure 7.2 the partial products are generated using the partial product generation array and the symbol 'X' indicates the absence of the partial product term. In this figure A_iB_i indicates the partial product terms, PA_i is the result of the addition of the partial products, and M_i is the final multiplication result. Since in this example the number of qubits is $n = 8$, the partial product addition tree requires $((\log_2 n) - 1) = 2$ levels, as shown in figure 7.3. At the first level, the partial products are added in two different nodes A_{11} and A_{12}, and at the next level, there is a node A_{21} in which partial addition results are added to form the final result of the multiplication, as shown in figure 7.3, where the symbol '+' inside a node indicates the partial product addition operation.

Property 7.8. The time complexity of the multiplication approach is at least $O((3\log_2 n) + 1)$, where n is the number of multiplicand and multiplier qubits.

Proof. Property 7.8 is proved by mathematical induction. For an n-qubit multiplier and multiplicand, a total $\log_2 n^2$ of ANDing and copying operations are required in the partial product generation array. Thus the time for the ANDing and copying operations is $T(A) = O(\log_2 n^2)$.

To generate the partial product addition tree and to perform the addition of the partial products in the partial product addition tree, $\log_2 2n$ operations are required, when the multiplier and multiplicand are n qubits. Thus the total time required to generate the partial product addition tree and perform the partial product addition is

$$T(P) = O\left(\log_2 2n\right).$$

When $n = 1$, we have the one-qubit multiplier and multiplicand. Thus the time required for this multiplication is

$$T(1) = 1 = O\left(\log_2 n^2 + \log_2 2n\right) = O\left(\left(3\log_2 1\right) + 1\right).$$

Therefore, the statement holds for the base case $n = 1$.

Assume that the statement is true for $n = k$ and we have the k-qubit multiplier and multiplicand. Thus the time for this multiplication is $T(k) = O(\log_2 k + \log_2 2k) = O((3\log_2 k) + 1)$.

When $n = (k + 1)$, i.e. we have the $(k + 1)$-qubit multiplier and multiplicand, the total required time for this multiplication is

$$T(k+1) = O(\log_2(k+1)^2 + \log_2 2(k+1)) = O((3\log_2(k+1)) + 1).$$

Therefore, the statement is true for $n = (k + 1)$.

Thus, for all n, the time required for the multiplication process of an n-qubit multiplicand and multiplier is

$T(n) = T(A) + T(P) = O(\log_2 n) + O(\log_2 2n) = O(2\log_2 n + \log_2 n + \log_2 2) = O(3\log_2 n + 1)$.

Thus the time complexity of the multiplication approach is at least $O((3\log_2 n) + 1)$.

It is important to note that this multiplication algorithm is also suitable for quantum computation. In quantum computation the smallest particles are known as qubits, which are always in motion and can be superpositioned at any time. Therefore, the execution time of any quantum computation becomes random and probabilistic due to the superposition of the particles. Whenever the quantum particles are in the 'cat state' (a special pure quantum state, where the quantum particles are in a fixed equal superposition of all being $|0\rangle$ and $|1\rangle$, i.e. $|00...0\rangle + |11...1\rangle$), one can compute the execution time when performing any quantum operation, which in turns measures the time complexity of the quantum operation. As a result, in the case of quantum computation, the time complexity of the multiplication method will be similar when both the multiplicand qubits and multiplier qubits are in the 'cat state'.

Property 7.9. The multiplication algorithm holds multiple states of the superposition of quantum logic.

Proof. In the multiplication algorithm, the partial product addition is performed using a tree-based structure. Let T be the partial product addition tree of $|\phi\rangle$ where the qubit $|\phi\rangle$ is used to denote a multiplier or a multiplicand qubit. By applying an addition, the operator A_i on the ith qubit simply changes the superposition $\alpha|0\rangle_i + \beta|1\rangle_i$ (where α and β are the probabilities of being in the $|0\rangle$ and $|1\rangle$ states, respectively) of that qubit located at any arbitrary leaf (nodes at the bottom level of the tree) to a different superposition $\alpha'|0\rangle_i + \beta'|1\rangle_i$.

Thus the individual sum of the partial products can be represented by the partial product addition tree (T) using property 7.1 which holds the multiple states of the superposition of quantum logic.

A quantum multiplication circuit is designed using the multiplication method, which is shown in section 7.4.

7.3 Reduction of the garbage outputs and ancillary inputs of quantum circuits

Quantum computers of many qubits are extremely difficult to realize. This sets the major objective of optimizing the number of ancillary (constant) input qubits and the number of garbage outputs in quantum circuits. In other words, in the case of choosing between increasing the garbage outputs and increasing the number of gates in a quantum circuit implementation, the preference should be given to the design method that delivers the minimum amount of garbage outputs, as reducing garbage outputs and constant inputs in turn improves the logical complexity and quantum costs of the circuits. Sometimes in quantum circuits a huge number of ancillary (constant) inputs leads to an increase in the size (area) of the circuit as well as gate level delays and power consumption. This is, in particular, a problem in quantum computation, since in quantum circuits input lines are a highly limited resource (caused by the fact that the number of circuit lines corresponds to the number of qubits). Furthermore, if the numbers of constant inputs in the quantum circuits increase, it may decrease the reliability of the circuits. Thus keeping the number of constant inputs as small as possible is an important issue. As the optimization of the ancillary input bits and garbage outputs may produce a compact circuit in terms of the quantum cost, power, and delay, it is always the primary focus to optimize the number of ancillary input bits and garbage outputs when designing a circuit.

Different minimization algorithms and tree-based designs are given to decrease the number of ancillary inputs and garbage outputs. In addition to the algorithms, new quantum adder, quantum ANDing, and quantum multiplier circuits are used in order to reduce the ancillary inputs and garbage outputs in the final circuit.

This solution enables us to overcome the overheads of increasing gate level delay, quantum cost, area, and power due to decreasing the number of garbage outputs and constant inputs in quantum circuits.

7.4 The design of a quantum multiplier circuit

In this section a quantum multiplier circuit is designed using a multiplication algorithm. The techniques of fast multiplication as well as the quantum ANDing circuit and quantum full-adder, and the design of a compact quantum multiplier and quantum $n \times n$ multiplier circuit are presented in this subsection.

7.4.1 The quantum ANDing circuit

In this subsection a new 3×3 QAC is shown. The input vector (I_v) and the output vector (O_v) of the circuit are as follows:

$$|I_v\rangle = \{|A\rangle, |B\rangle, |C\rangle\}; \text{ and}$$
$$|O_v\rangle = \{|A\rangle, |AC \oplus B\rangle, |C\rangle\}.$$

Figure 7.4 shows the diagram of the 3×3 QAC and the circuit can also be realized by the decomposition of the quantum Toffoli gate. In the QAC, five basic quantum gates are used. Thus the quantum cost is 5. The output $|Q\rangle$ is the result of the ANDing of qubits $|A\rangle$ and $|C\rangle$. The truth table corresponding to the QAC is shown in table 7.1. It can be verified from the truth table that the input pattern corresponding to a particular output pattern can be uniquely determined. To perform the 'AND' operation using QAC, the $|B\rangle$ input qubit in figure 7.4 has to remain $|0\rangle$. Thus the QAC can be used to design the partial product generation circuit.

7.4.2 The quantum full-adder circuit

A quantum full-adder is a circuit that performs the arithmetic sum of three input qubits. It consists of three inputs and two outputs. If we take three input variables A,

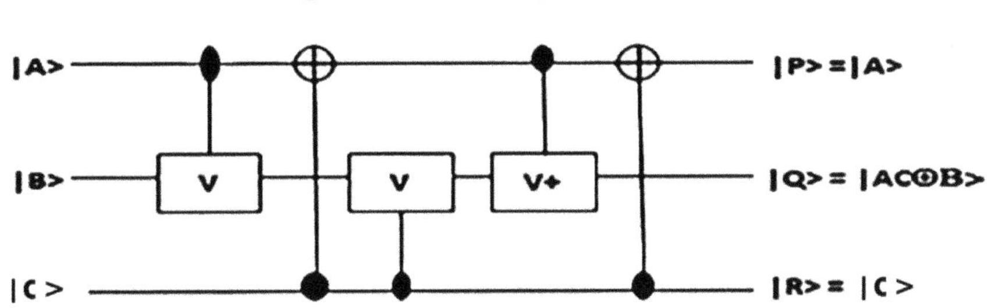

Figure 7.4. A quantum circuit for the ANDing operation. Reproduced with permission from [4]. Copyright 2017 Springer Nature.

Table 7.1. Reversibility of 3 × 3 QAC circuit.

Input			Output		
A	*B*	*C*	*P*	*Q*	*R*
0	0	0	0	0	0
0	0	1	0	0	1
0	1	0	0	1	0
0	1	1	0	1	1
1	0	0	1	0	0
1	0	1	1	1	1
1	1	0	1	1	0
1	1	1	1	0	1

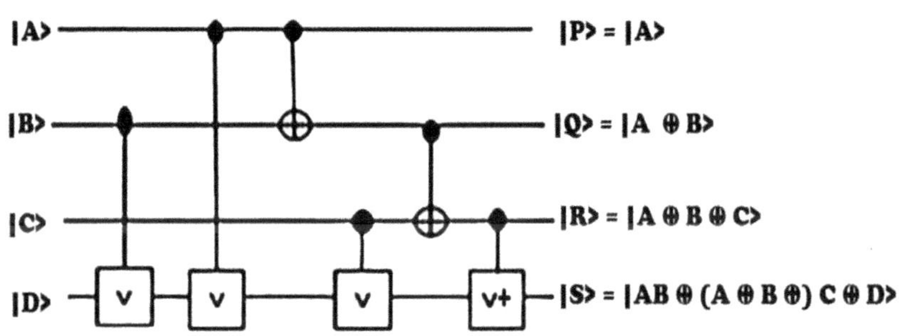

Figure 7.5. The quantum full-adder circuit, when $|D\rangle = |0\rangle$. Reproduced with permission from [4]. Copyright 2017 Springer Nature.

B, C_{in} and two output variables S (sum) and C_{out} (carry), then the following equations can be used to obtain the outputs:

$$S = A \oplus B \oplus C_{\text{in}}$$
$$C_{\text{out}} = AB \oplus BC_{\text{in}} \oplus C_{\text{in}}A.$$

Quantum full-adders are versatile and widely used building blocks in quantum arithmetic circuits. In this subsection a new 4 × 4 QFA circuit is shown.

The input vector I_{v} and the output vector O_{v} of the circuit are as follows:

$$|I_{\text{v}}\rangle = \{|A\rangle, |B\rangle, |C\rangle, |D\rangle\}; \text{ and}$$
$$|O_{\text{v}}\rangle = \{|A\rangle, |A \oplus B\rangle, |A \oplus B \oplus C\rangle, |AB \oplus (A \oplus B)C \oplus D\rangle\}.$$

Figure 7.5 shows the diagram of the 4 × 4 QFA circuit. In this figure, the output $|R\rangle$ is the S (sum) and $|S\rangle$ is the C_{out} (carry). The number of quantum gates of the QFA circuit is six and so the quantum cost is 6. The truth table corresponding to the QFA circuit is shown in table 7.2. It can be verified from the truth table that the input pattern corresponding to a particular output pattern can be uniquely

Table 7.2. Reversibility of 4 × 4 QFA circuit.

Input				Output			
A	*B*	*C*	*D*	*P*	*Q*	*R*	*S*
0	0	0	0	0	0	0	0
0	0	0	1	0	0	0	1
0	0	1	0	0	0	1	0
0	0	1	1	0	0	1	1
0	1	0	0	0	1	1	0
0	1	0	1	0	1	1	1
0	1	1	0	0	1	0	1
0	1	1	1	0	1	0	0
1	0	0	0	1	1	1	0
1	0	0	1	1	1	1	1
1	0	1	0	1	1	0	1
1	0	1	1	1	1	0	0
1	1	0	0	1	0	0	1
1	1	0	1	1	0	0	0
1	1	1	0	1	0	1	1
1	1	1	1	1	0	1	0

Input: |A>, |B>, |C_{in}> Output: |Sum>, |Carry>

START
Take an QFA circuit (|a>, |b>, |c>, |d>)(|p>, |q>, |r>, |s>)
Create a partial products addition tree with $((log_2\ n) - 1)$ levels.
Set, QFA.a = |A>; QFA.b = |B>; QFA.c = |C_{in}>; QFA.d = |0>;
Output |Sum> =QFA.r
END

Algorithm 7.2. Construction of the quantum full-adder.

determined. The QFA circuit is designed in such a way that it can be used efficiently as a quantum full-adder by setting the fourth input qubit as a constant zero $|0\rangle$. Algorithm 7.2 describes the design of QFA circuit as a quantum full-adder.

7.4.2.1 The quantum multiplier circuit

The design of the quantum multiplier is composed of two components: the quantum PPG circuit and a quantum PPA circuit. The quantum PPG circuit and quantum PPA circuit are described in the following sections. The number of constant input and garbage output lines has been reduced significantly by using quantum memory elements.

7.4.2.2 The quantum partial product generation circuit

The quantum PPG which is constructed using the QACs, takes the multiplicand and multiplier qubits as the input and produces the partial products as outputs. The multiplier qubits and the multiplicand qubits are fed to a component which produces partial products. The QAC has been used to construct this part.

Example 7.4. To construct a 4 × 4 quantum PPG circuit, sixteen QACs are needed to perform the ANDing operations for generating the partial products. In figure 7.6 the detailed quantum partial product generation circuit for a 4 × 4 quantum multiplier is shown. In this figure the qubit $|C_0\rangle$ is used as the constant input lines; and qubits $|X_i\rangle$ and $|Y_i\rangle$ are the input qubits of the multiplicand and multiplier, respectively. The partial products denoted by the qubits $|X_i Y_i\rangle$ are the outputs of the PPG circuit, where the symbol '…' is used to denote the single constant input line and also the single garbage output line, which are reused in the circuit wherever it is necessary.

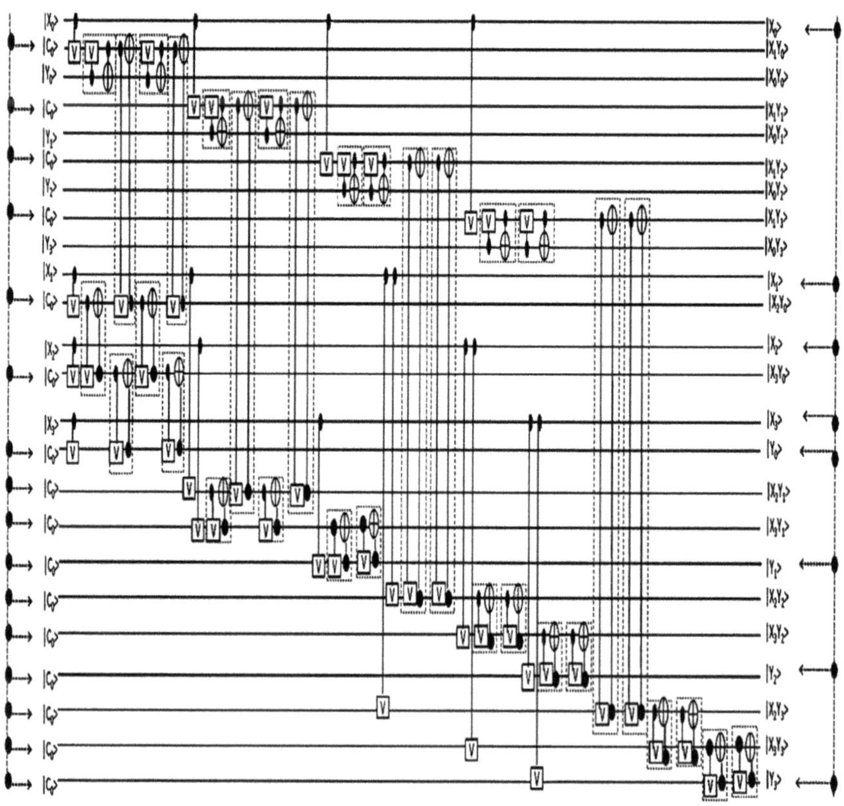

Figure 7.6. The quantum PPG circuit of a 4 × 4 multiplier. Reproduced with permission from [4]. Copyright 2017 Springer Nature.

7.4.2.3 The quantum partial product addition circuit

The quantum PPA circuit adds column-wise the partial products that have been generated by the quantum PPG circuit. The PPA circuit is constructed using the QFA circuits. The construction procedure of the quantum PPA circuit is as follows.

To construct a quantum PPA circuit a tree structure is followed, consisting of levels. A PPA tree is created with $((\log_2 n) - 1)$ levels. At the first level of the PPA tree, there are $n/4$ nodes, where each node represents the addition by a quantum adder f the successive four rows of a partial product array. In the following levels of the tree, the addition operations are performed between two predecessor nodes, where the rate of increase of the number of adders will be $(n + 2 + i)$, where $i = 1, 2, \ldots, ((\log_2 n) - 2)$ and n is the number of qubits. Finally, the multiplication result is obtained in the $((\log_2 n) - 1)$th level of the tree.

Example 7.5. To construct a 4 × 4 quantum PPA circuit, ten QFA circuits are needed to perform the addition operations for generating the PPA. In figure 7.7 the detailed quantum PPA circuit for a 4 × 4 multiplier is shown. In this figure the qubit $|C_0\rangle$ is used as the constant in the input lines. This PPA circuit takes the partial products of qubits ($|X_i Y_i\rangle$) as inputs and generates the result of addition denoted by

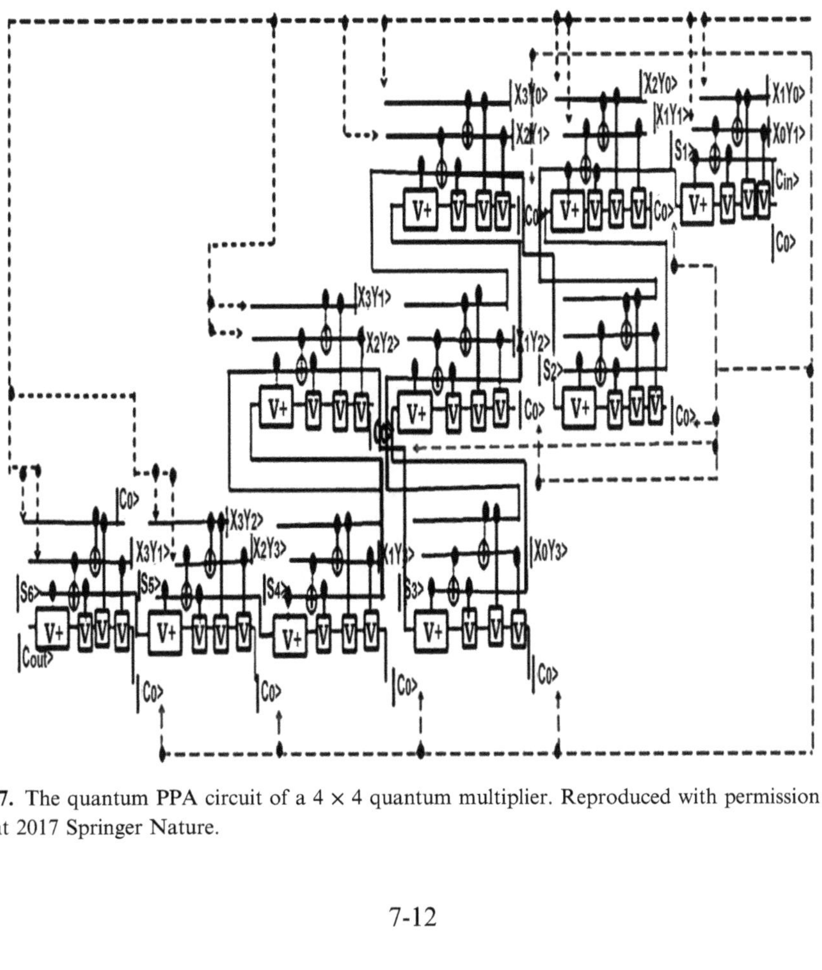

Figure 7.7. The quantum PPA circuit of a 4 × 4 quantum multiplier. Reproduced with permission from [4]. Copyright 2017 Springer Nature.

the qubit $|S_i\rangle$ as the output, where the symbol '…' is used to denote the single constant input line and also the single garbage output line, which are reused in the circuit wherever necessary.

7.4.3 The *n* × *n*-qubit quantum multiplier

A 4 × 4 quantum multiplier is constructed using QACs and QFA circuits. First, a quantum PPG circuit is constructed using a QAC circuit to generate partial products for a 4 × 4 quantum multiplier. The quantum PPA circuit is constructed for PPA using QFA circuits and obtains a 4 × 4 quantum multiplier. The generalized block diagrams for an $n \times n$ quantum PPG and quantum PPA circuits are shown in figures 7.8 and 7.9, respectively, where the qubit $|C\rangle$ is the constant input and the

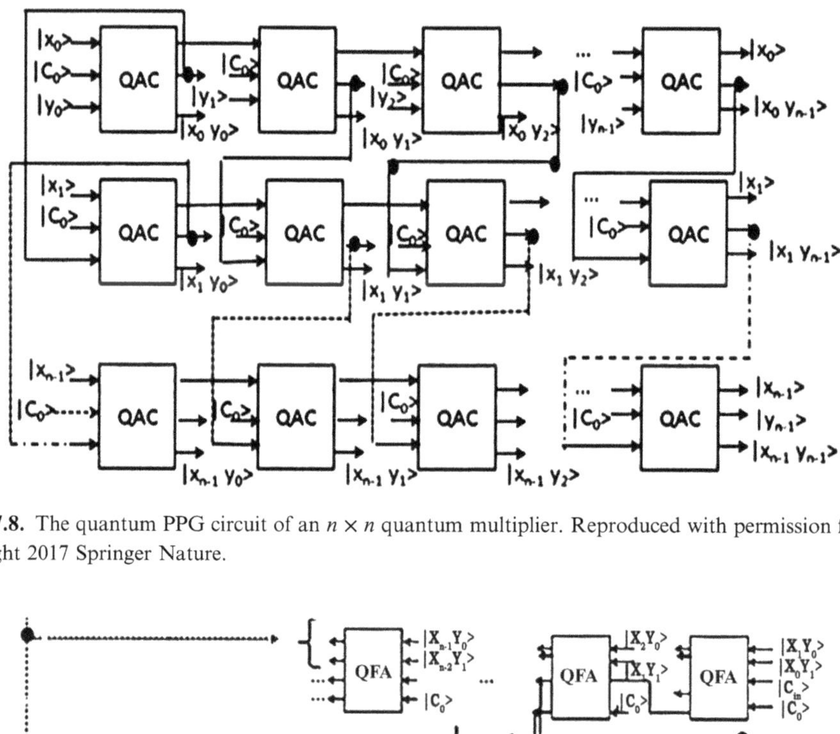

Figure 7.8. The quantum PPG circuit of an $n \times n$ quantum multiplier. Reproduced with permission from [4]. Copyright 2017 Springer Nature.

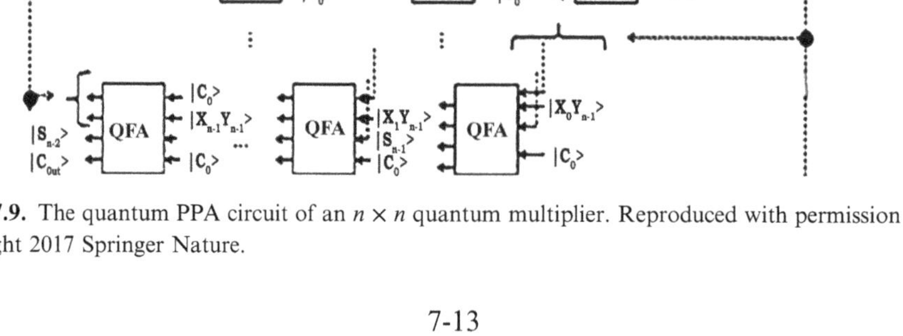

Figure 7.9. The quantum PPA circuit of an $n \times n$ quantum multiplier. Reproduced with permission from [4]. Copyright 2017 Springer Nature.

```
START
Divide the PPG-Array row wise into (n/4) parts
for i:=0 to n/4 do
    for j:=0 to (n-1) do
        Apply j number of QFA gates in j number column to get the addition of nodes of i^th layers
    end for
    Apply QFA gates to get the addition between layers
end for
END
```

Algorithm 7.3. Quantum partial product addition circuit.

qubit $|S\rangle$ stands for the sum of the qubits. In figure 7.8 each of the QAC blocks takes a multiplicand qubit $|X_i\rangle$ and a multiplier qubit $|Y_i\rangle$ and generates the product terms of qubits $|X_iY_i\rangle$ as outputs, which are carried out on the PPA circuits. The block diagram of the PPA circuit as shown in figure 7.9 is constructed by the QFA blocks where each QFA block takes two partial products of qubits $|X_iY_i\rangle$ and performs the addition of the products in which the sum ($|S\rangle$) is the output of the block. Thus the construction procedure of an $n \times n$ quantum multiplier circuit is as follows: at first an $n \times n$ quantum PPG circuit is generated and then the quantum PPA circuit of an $n \times n$ multiplier is constructed using algorithm 7.3.

Property 7.10. The depth of the design of the quantum multiplier circuit is $O(9n + 14\log_2 n)$, where n is the number of qubits of the multiplicand and multiplier.

Proof. The above statement is proved by the method of mathematical induction. A 2×2 PPG and a 2×2 PPA circuit are required for constructing a 2×2 quantum multiplier circuit using the design of the tree-based multiplier, where the total number of stages (depth) in a 2×2 PPG circuit is 20 and the number of stages (depth) in a 2×2 PPA circuit is 12.

When $n = 2$ for a 2×2 quantum multiplier circuit, the total depth of the 2×2 quantum multiplier circuit is

$$D_{2\times2} = D_{\mathrm{PPG}} + D_{\mathrm{PPA}}$$

$= 20 + 12 = 18 + 14 = 9n + 14\log_2 n$; [$\because n = 2$ here].

Thus the statement holds for the base case $n = 2$.

Assume that the statement also holds for $n = k$, which is a $k \times k$ quantum multiplier circuit. Thus $D_{k\times k} = 9k + 14\log_2 k$. Therefore, the depth of the $(k + 1) \times (k + 1)$ quantum multiplier circuit is

$D_{(k+1)\times(k+1)} = 9(k + 1) + 14\log_2(k + 1)$.

Thus the statement is also true for $n = (k + 1)$ and thus the depth of the design of the quantum multiplier circuit is $O(9n + 14\log_2 n)$.

It is important to note that the depth of the quantum multiplier circuit is exponential or linear, whereas the depth of the quantum multiplier is logarithmic.

7.5 Accumulator

A quantum accumulator is a quantum register used by the quantum central processing unit (CPU) of a quantum computer to temporarily store arithmetic and logic data. However, it is rarely used to refer to modern quantum CPUs because the term 'register' emerged at the turn of the millennium. Any quantum ALU operation can have one of the two operands stored in the quantum accumulator. This is only an example, but to add two qubits together, one would go into the quantum accumulator and the other into memory or a general-purpose quantum register. The two integers are used as the quantum ALU's input when it is put into action.

The two-qubit quantum accumulator is made up of two quantum D-flip flops, each of which is composed of four quantum NAND operations and one quantum AND operation. The inputs of the quantum AND operation are |LOAD⟩ and |CLK⟩, and the outputs of the quantum AND operation serve as the input for each quantum D-flip flop. The final outputs of a quantum accumulator circuit come from the first and second quantum D-flip flops, respectively. The circuit diagram of a quantum accumulator is shown in figure 7.10.

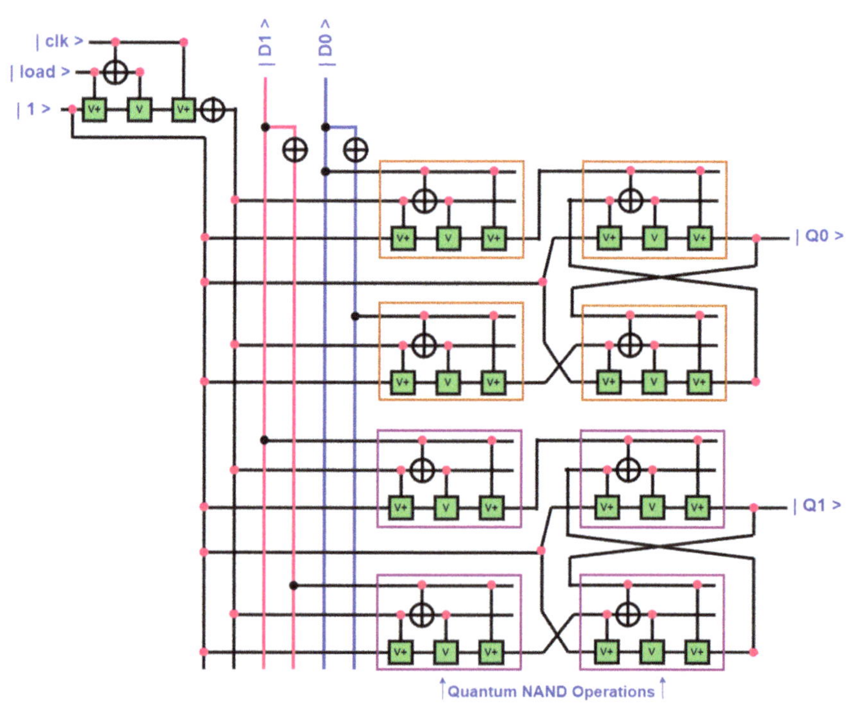

Figure 7.10. The quantum accumulator.

This type of quantum register is known as a quantum accumulator, which serves as a momentary storage space and stores a value used as a bridge in logical and mathematical processes. For example, while performing the operation '5 + 6 + 7', the quantum accumulator will initially store the value 5, then the value 11, and eventually the value 18. If the result of the quantum AND operation is $|1\rangle$, then the data will be stored in the quantum accumulator. There will be no data saved in the accumulator if the output of the quantum AND operation is $|0\rangle$.

7.6 Summary

This chapter presents the design methodology of a novel quantum multiplier–accumulator unit using quantum multiplier and quantum accumulator circuits. As the multiplier is the major component of a multiplier–accumulator unit, in this chapter a new design methodology of the quantum multiplier device is presented using a multiplication technique. The design methodology generates the partial products using the quantum ANDing circuit and performs the addition of partial products in parallel by using the quantum full-adders with reduced ancillary and garbage bits. The unique tree-based multiplication technique along with the use of quantum full-adders and quantum ANDing circuits reduces the number of ancillary and garbage bits in the quantum multiplier design. A quantum accumulator circuit is also shown which is more compact and efficient. The design methodology can be integrated as part of a quantum CPU, quantum signal processing, and quantum ALU, optimized in terms of ancillary inputs and garbage outputs.

Critical thinking questions

1. Give a short explanation of the characteristics of a quantum multiplier–accumulator.
2. Construct a 6×6 multiplier circuit using a quantum accumulator circuit and quantum adder circuits.
3. Is it possible to construct a 5×5 quantum accumulator circuit? If so, design it using the appropriate circuits.
4. Give a brief explanation of how the ancillary input and garbage output of a quantum circuit work.
5. Prove that the quantum superposition states of the multiplication algorithm may exist in numerous states.
6. Identify the multiplication strategy's temporal complexity, when n is 9.
7. Describe briefly the reduction of an ancillary input and a garbage output of a quantum circuit.
8. Is it possible to get rid of the garbage value from the circuit below? Explain in detail.

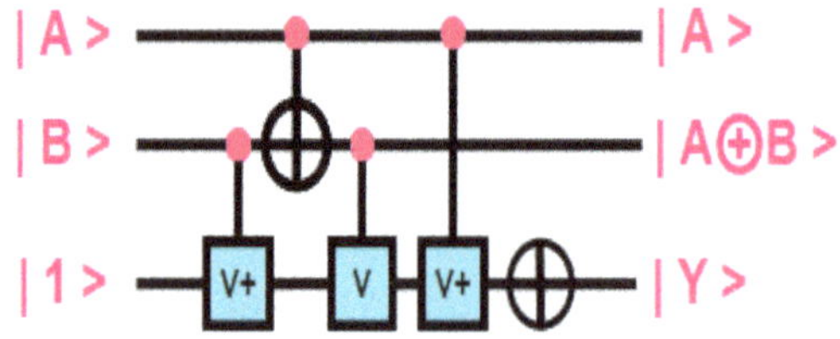

References

[1] Ömer B QCL—a programming language for quantum computers http://tph.tuwien.ac.at/~oemer/qcl.html (Accessed: 5 December 2018)

[2] Akbar E P A, Haghparast M and Navi K 2011 Novel design of a fast reversible Wallace sign multiplier circuit in nanotechnology *Microelectron. J.* **42** 973–81

[3] Álvarez-Sánchez J J, Álvarez-Bravo J V and Nieto L M 2008 A quantum architecture for multiplying signed integers *J. Phys.: Conf. Ser.* 128 012013

[4] Babu H M H 2017 Cost-efficient design of a quantum multiplier-accumulator unit *Quantum Inf. Process.* **16** 30

[5] Cai X-D, Wu D, Su Z-E, Chen M-C, Wang X-L, Li L, Liu N-L, Lu C-Y and Pan J-W 2015 Entanglement-based machine learning on a quantum computer *Phys. Rev. Lett.* **114** 110504

[6] Chakrabarti A and Sur-Kolay S 2008 Designing quantum adder circuits and evaluating their error performance *Int. Conf. on Electronic Design* (Piscataway, NJ: IEEE) pp 1–6

[7] Cormen T H 2001 *Introduction to Algorithms* ed T H Cormen, C E Leiserson, R L Rivest and C Stein (Cambridge, MA: MIT Press)

[8] Hung W N N, Song X, Yang G, Yang J and Perkowski M 2006 Optimal synthesis of multiple output Boolean functions using a set of quantum gates by symbolic reachability analysis *IEEE Trans. Comput.-Aided Des. Integr. Circuits Syst.* **25** 1652–63

[9] Kaye P *et al* 2007 *An Introduction to Quantum Computing* (Oxford: Oxford University Press)

[10] Kitaev A Y 1997 Quantum computations: algorithms and error correction *Russ. Math. Surv.* **52** 1191–249

[11] Kotiyal S, Thapliyal H and Ranganathan N 2014 Circuit for reversible quantum multiplier based on binary tree optimizing ancilla and garbage bits *27th Int. Conf. on VLSI Design and 13th Int. Conf. on Embedded Systems* (Piscataway, NJ: IEEE) pp 545–50

[12] Li D-F, Wang R-J, Zhang F-L, Deng F-H and Baagyere E 2015 Quantum information splitting of arbitrary two-qubit state by using four-qubit cluster state and Bell-state *Quantum Inf. Process.* **14** 1103–16

[13] Maynard C M and Pius E 2014 A quantum multiply-accumulator *Quantum Inf. Process.* **13** 1127–38

[14] Mogensen T Æ 2013 Garbage-free reversible constant multipliers for arbitrary integers *Int. Conf. on Reversible Computation* (Berlin: Springer) pp 70–83

[15] Nielsen M A and Chuang I 2002 *Quantum Computation and Quantum Information* (Leiden: Cambridge University Press)

[16] Ramya P R and Vani Y S 2013 Optimization and implementation of reversible BCD adder in terms of number of lines *Int. J. Reconfigurable Embedded Syst.* **2** 21–6

[17] Schubert M and Rana F 2006 Analysis of terahertz surface emitting quantum-cascade lasers *IEEE J. Quantum Electron.* **42** 257–65

[18] Viamontes G F, Markov I L and Hayes J P 2005 High-performance quantum circuit simulation *QuIDDPro* http://vlsicad.eecs.umich.edu/Quantum/qp/ (Accessed: 19 March 2020)

[19] Wille R, Soeken M and Drechsler R 2010 Reducing the number of lines in reversible circuits *Proc. 47th Design Automation Conf.* (New York: ACM) pp 647–52

[20] Yao A C-C 1993 Quantum circuit complexity *Proc. IEEE 34th Annual Foundations of Computer Science* pp 352–61

IOP Publishing

Quantum Computing (Second Edition)

A pathway to quantum logic design

Hafiz Md Hasan Babu

Chapter 8

The quantum divider

Learning objectives

- Discuss the traditional integer division algorithm.
- Construct a quantum divider circuit.
- Design an algorithm of the quantum integer division.
- Acquire knowledge about the components of the quantum divider circuit.
- Explain the importance of the quantum divider.
- Implement an *n*-qubit quantum divider circuit using its algorithm.
- Describe the technique for tree-based quantum division.

Among all the fundamental operations in a quantum computer, division is the most complex operation. Division invokes operations such as addition, multiplication, and subtraction. The quantum divider is an indispensable part of a quantum processor and comprises arithmetic sub-modules such as a quantum comparator, quantum adder, quantum multiplier, and quantum subtractor.

A computer processor is composed of a large number of units to execute instructions. Arithmetic units are important units in computer hardware because they have many uses in computer systems. Therefore, designing quantum circuits for arithmetic units is very important. One of these arithmetic units is the division unit. The division circuit is a somewhat complicated component in computer hardware.

8.1 Division algorithms

This section presents classical integer division algorithms as well as quantum algorithms for integer division.

8.1.1 Classical integer division algorithms

Integer division is a basic part of the elementary theory of computation. A number of classical integer division algorithms have some advantages and disadvantages. The textbook division algorithm for integer division described by Knuth is similar to

doi:10.1088/978-0-7503-5683-1ch8

conventional mathematical division, where both the dividend and the divisor must be two-word integers and it has a time complexity of $O(n^3)$. This division algorithm is more complex and slower than other division algorithms. Moller showed an improved integer division algorithm which solved the problem of dividing a two-word integer by a single-word integer using single-word approximate reciprocals. This method requires a lower multiplication time and omits the most significant half of the products. However, this technique requires a time complexity of $O(n^2)$ and still needs more operations to perform the division. The restoring division algorithm first operates on fixed-point fractional numbers and uses an extra register to save the partial numerators obtained by the multiplication, and then reuses the result when the result of the subtraction is negative. This method has a time complexity of $O(n^2)$.

8.1.2 Quantum integer division algorithms

There is no explicit quantum integer division algorithm. All the algorithms incorporate quantum division in the classical integer division algorithms. One of the most well-known algorithms in the field of quantum computation is Shor's factorization algorithm. This is a quantum algorithm that finds the factors of a composite integer in polynomial time with a great probability of success. Division is not incorporated in this factorization algorithm as well as other implementations of the modular exponentiation part of Shor's algorithm. The quantum modular exponentiation architecture is described for Shor's factoring algorithm, where division has been incorporated. It is currently one of the most renowned design architectures of integer division using Shor's quantum factoring algorithm. This quantum modular exponentiation circuit requires $(9n + 2)$ qubits and has a depth of $O(2000n^2)$, where n is the number of qubits to be factored. The time complexity of this method for quantum integer division is $O(72(\log_2 n)^3)$.

Alireza devised a design to implement the quantum divider for the division of two n-qubit fixed-point numbers in binary format using the restoring division algorithm. The design methodology reduces the number of ancillary qubits and eliminates the use of quantum arithmetic sub-modules such as quantum shift register, quantum adder, subtractor, etc. However this architecture requires a time complexity of $(5n/6)^2 + O(n^2)$ and a total $3n^3 + 6n^2 + n$ operations are required, where n is the number of qubits of the dividend and divisor.

All the above-mentioned quantum algorithms for integer division have a high time complexity and a high total number of operations to perform the division operation of two arrays of qubits representing integers. Therefore it is expedient to incorporate a new quantum division algorithm for integer division with minimal time complexity and number of operations.

8.2 The importance of the quantum divider

In this chapter the optimization of time complexity as well as the reduction of number of operations are addressed, which has motivated the introduction of a completely new and efficient technique of integer division with optimum time complexity. *Time complexity* is the amount of instructions that will be executed during the run time of that particular algorithm. Obviously, the greater the time complexity of the algorithm, the more operations the are, and the greater is the delay of simulation. A large time complexity makes the algorithm slower. Increasing the time complexity of an algorithm is likely to increase the number of operations. This increase in turn increases the cost and delay of the execution of operations. Algorithmic complexity is the property of an algorithm which is mapped to the amount of computational resources used by the algorithm. This factor is analogous to the engineering productivity for a repeating or continuous process. Thus if the complexity changes, the resources such as memory space, access time, cost, delay, area, and power of execution of the algorithm will also fluctuate proportionally. This will make the simulation of the algorithm more vulnerable, particularly in the case of large sets of data. The reduction of this time complexity minimizes the number of operations and hardware complexities as well. This optimization in turn makes the division operation faster. Thus it is expedient to reduce the time complexity of the division algorithm to make it faster and more efficient. Researchers are looking for an efficient and faster division algorithm to optimize time complexity. Therefore a new, faster, more efficient, and optimal tree-based division technique which works like the breadth fast search (BFS) algorithm is introduced in this chapter.

8.3 The tree-based quantum division technique

In this section a tree-based quantum division algorithm is described to perform the division operation of the integers, which is the first such ever presented in the literature. The quantum algorithms for integer division based on the classical division algorithms require a large time complexity and a greater number of operations to execute the division of the dividend by the divisor.

8.3.1 Definitions and properties of the division technique

In this subsection the terms divisor, dividend, quotient, remainder, partial numerator, node, leaf, and root, and the properties of the tree that are used frequently throughout this chapter are elaborated to establish the main purpose and the theme of this chapter.

Definition 8.1. The qubits representing the integer that is to be divided are called the *dividend.*

Definition 8.2. The qubits of the integer that divides the dividend are known as the *divisor.*

Definition 8.3. The *quotient* is the resulting qubits of the integer obtained by dividing the dividend using the divisor.

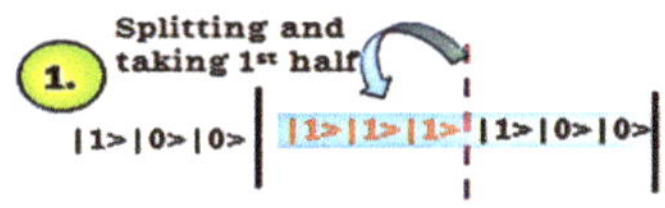

Figure 8.1. The divisor, dividend, remainder, quotient, and partial numerator.

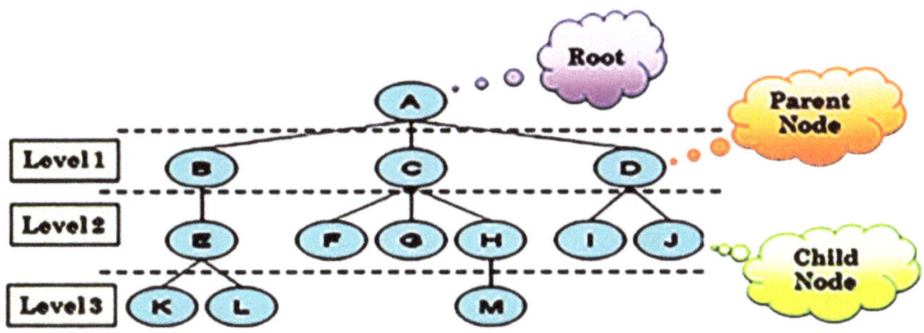

Figure 8.2. Example of a tree, root, node, and level.

Definition 8.4. In any step of the division process when the quotient qubits are multiplied with the divisor qubits, it generates *partial numerator* qubits that are subtracted from the original dividend qubits.

Definition 8.5. *Remainder* qubits representing integers are the result obtained after subtracting the current partial numerator qubits from the original divisor qubits.

Example 8.1. Suppose a qubit string of integer $\{|1\rangle|1\rangle|0\rangle|1\rangle|0\rangle\}$ is divided by $\{|1\rangle|0\rangle|1\rangle\}$, then first one is the dividend and the second qubit string is the divisor. The quotient is $\{|1\rangle|0\rangle|1\rangle\}$, the partial numerator is $\{|1\rangle|0\rangle|1\rangle\}$, and the remainder is $\{|1\rangle\}$. Figure 8.1 illustrates this scenario.

Property 8.1. A connected graph G is called a *tree* if:

1. The removal of any of its edges makes G disconnected.
2. The top most node $r \in G$ is designated as the root.
3. The remaining nodes are partitioned into disjoint sets $G_1, G_2, \ldots, G_k$, where each of the sets is a tree and $k \geqslant 0$.
4. G is connected and contains no closed path or cycles.

Definition 8.6. (Node) A *node* in a tree contains the information or data. The node may be a child or a parent node.

Definition 8.7. (Level) The *level* of a node is the unique path of that node from the root of the tree.

Example 8.2. A tree, its child and parent nodes, root, and level are depicted in figure 8.2.

8.3.2 The algorithm of the division technique

In the algorithm the tree structure approach is used for simplicity and generality as well as optimizing the time complexity through parallelism. A tree-based division algorithm makes the operation faster and it also simplifies the circuits.

The technique of division has five basic steps:

Step 1. First the midpoint qubit of the dividend is determined and the dividend is split using the quantum dividend splitter (QDS) circuit, as shown in figure 8.3. The first half will be taken after splitting, as shown in figure 8.1.

Step 2. Then the partial numerator qubits are generated in such a way that the copying and ANDing operations of the divisor and quotient qubits are done in parallel in the split dividend for faster operation. In this step, first an additional $|0\rangle$ is added at the beginning of the first half of step 1 to form the divisor qubits and then the quotient qubits are multiplied with the divisor qubits to obtain the results which are stored in the partial numerator qubit generation (PQG) array. Figure 8.3 illustrates the procedure.

Step 3. In this step the partial numerator additions are performed in a tree structure known as the partial numerator addition (PNA) tree, where the tree has $((\log_2 n) - 1)$ levels for numerator generation and addition. In the first level, four concurrent rows of the PQG array are combined by adders in parallel, and in the next levels (up to $((\log_2 n) - 1)$th) the results of the additions between the individual qubits of divisor and quotient are added together to form the partial numerator. According to example 8.3 we have $n = 4$. Thus only $((\log_2 n) - 1) = 1$ levels are needed to complete the addition of the results of the multiplication stored in the PQG array, as shown in figure 8.4.

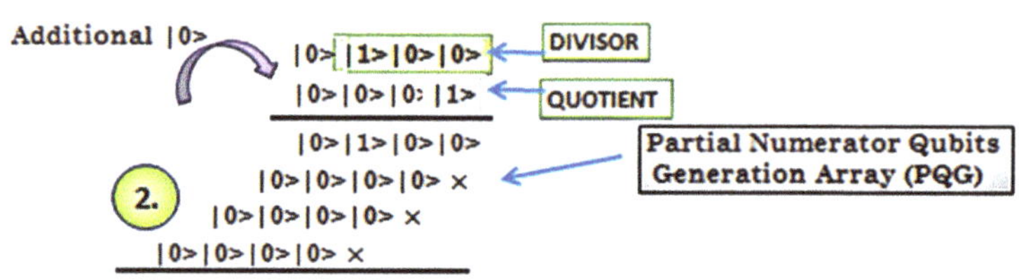

Figure 8.3. Step 2, multiplication of the divisor and quotient, and storing the result in the PQG array.

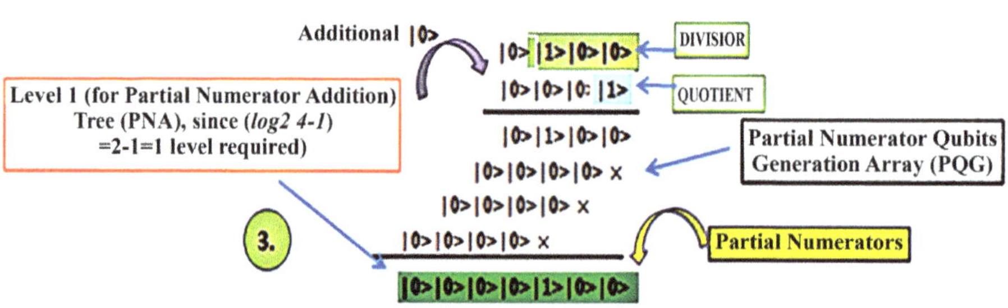

Figure 8.4. Step 3, the partial numerator obtained by addition in the PNA tree.

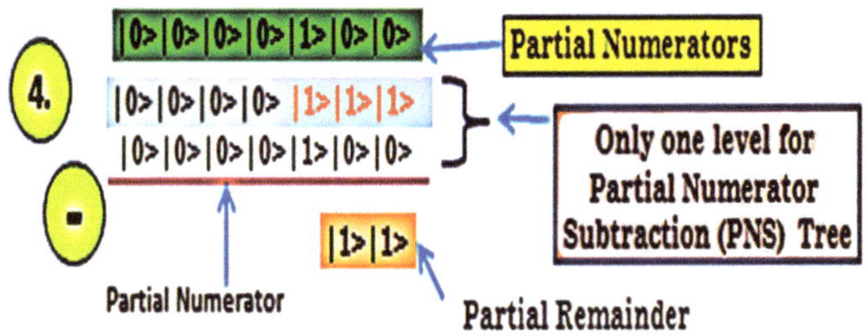

Figure 8.5. Step 4, subtraction of a partial numerator from the original dividend in the PNS tree.

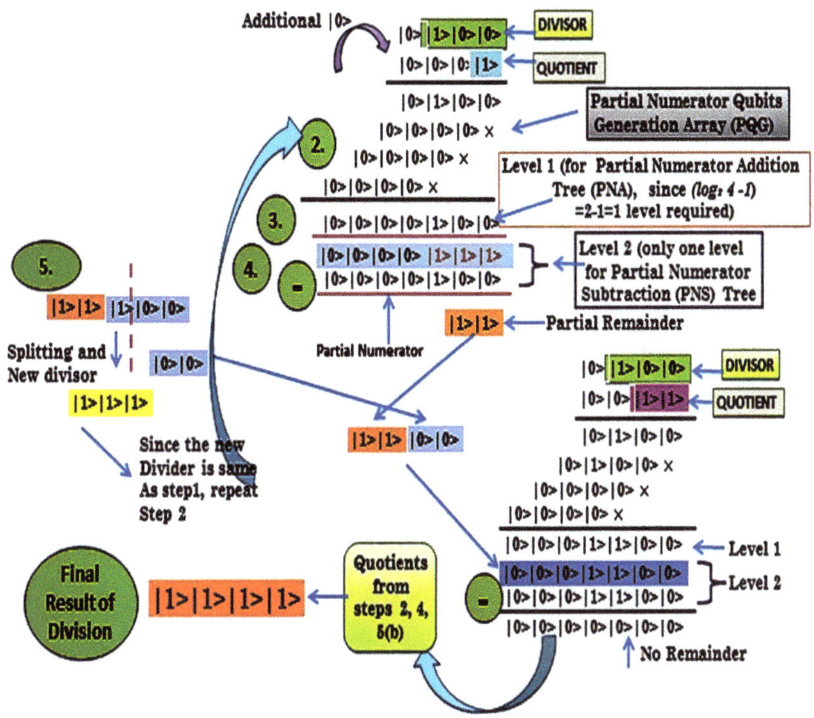

Figure 8.6. Step 5, repetition of the same process until there is a remainder.

Step 4. Now a partial numerator subtractor (PNS) tree with only one level is necessary for subtracting the partial numerator from the original dividend. Figure 8.5 illustrates the procedure.

Step 5. If there is no remainder after subtraction, then the division is complete, where the quotient qubits form the result. Otherwise, the remainders are added at the beginning positions of the rest of the qubits of the dividend before splitting the newly formed dividend. Repeat the same operation until a remainder exists. In example 8.3 there are remainders after step 4. Thus the same process is repeated, as shown in figure 8.6.

Input: Two qubit arrays *|Di [i]>* and *|Dd [j]>* for divisor and dividend, respectively.
Output: *(|Di [j]>) / (|Dd [i]>) i.e.,* the resulting qubits of the division algorithm.

```
1: START
2: (|Dd [i]>) = Array containing the qubits of the Dividend
3: (|Di [j]>) = Array containing the qubits of the Divisor
4: (|Qa [i]>) = Array that stores the Partial Numerator Addition Qubits
5: (|Qg [i]>) = Array that generates the Partial Numerator Qubits
6: (|Q [i]>) = Array that contains the Quotients
7: (|R [i]>) = Array that contains the Remainders
8: n = size of the dividend qubits array
9: m = size of the divisor qubits array
10: mid = ceil(n/2)
11: if mid >= m then
12:    Divide (Dd [i]>) by (Di [j]>), where 0 ≤ i ≤mid and 0 ≤ j ≤ m
13: end if
14: Generate the Partial Numerator Qubits Generation Array (PQG) using
    (|Qg[i]>), for copying and ANDingoperation of (Dd [i]>) & (|Q [i]>)
    which is to be done in parallel
15: Create the Partial Numerator Addition Tree (PNA) with (log2 n − 1)
    levels.
16: Generate (n/4) nodes at first level of the tree where each node repre-
    sents the addition of successive 4 rows of the PQG array.
17: Store the result of PNA tree in (|Qa [i]>)
18: Save the quotient in the (|Q [i]>)
19: Create the Partial Numerator Subtractor (PNS) Tree with 1 level
20: Subtract the (|Qa [i]>) from (|Dd [i]>)
21: Store the result in (|R [i]>)
22: if |R[i] >== |0 > for all qubit positions and all qubits are executed,
    then
23:    Result of the division would be (|Q[i] >)
24: else
25:    Add the values of (|R[i] >) in the beginning positions of the remain-
       ing qubits of (|Dd[i] >)
26: end if
27: loop
28:    Steps 10 to 26 for new qubits of the (|Dd[i] >) with the same divisor
29: end loop
30: print ("Division is not possible since length of divisor is larger than
    dividend");
31: END                                  102
```

Algorithm 8.1 Division technique.

Algorithm 8.1 describes the division technique while example 8.3 explains it.

Example 8.3. Let the divisor be a qubit string $\{|1\rangle|0\rangle|0\rangle\}$ and the dividend be $\{|1\rangle|1\rangle|1\rangle|1\rangle|0\rangle|0\rangle\}$. First, it is necessary to split the dividend according to algorithm 8.1. After splitting, a dividend consisting of the first three qubits, i.e. $\{|1\rangle|1\rangle|1\rangle\}$, is obtained. In the next step the partial numerator qubits (|1⟩, |0⟩, |0⟩, ...) are generated by the copying and ANDing operations of the divisor {|1⟩|0⟩|0⟩} and quotient qubit {|1⟩}. Then the PNA tree is generated and the partial numerator {|1⟩|0⟩|0⟩} is produced by the addition operations of the partial numerator qubits generated in the previous step. Then a PNS tree with a single level is generated and the partial numerator {|1⟩|0⟩|0⟩} is subtracted from the split dividend {|1⟩|1⟩|1⟩}. In the next step the remainder {|1⟩|1⟩} is added at the beginning of the rest of the dividend qubits {|1⟩|0⟩|0⟩} and the newly formed dividend becomes {|1⟩|1⟩|1⟩|0⟩|0⟩}. Now again, after splitting, the dividend becomes three qubits, i.e. {|1⟩|1⟩|1⟩}. Again the partial numerator produced by the PNA tree using quotient {|1⟩} is {|1⟩|0⟩|0⟩}, which is subtracted from the split dividend in the PNS tree and the remainder {|1⟩|1⟩} is added at the beginning of the rest of the dividend qubits and that becomes {|1⟩|1⟩|0⟩|0⟩}. No more splitting is needed and using the divisor {|1⟩|0⟩|0⟩} and quotient {|1⟩|1⟩}, a partial numerator {|1⟩|1⟩|0⟩|0⟩} is produced, which is then subtracted from the dividend {|1⟩|1⟩|0⟩|0⟩}; there is no remainder. Thus the quotient qubits form the result of the division, i.e. {|1⟩|1⟩|1⟩|1⟩}. This whole procedure is clarified in the figures 8.1 and figure 8.3–8.6.

Property 8.2. The time complexity of the division algorithm is at least $O(3n((\log_2 n) + 1) - n)$, where n is the number of qubits of the dividend.

Proof. Let n be the size of the dividend and m be the size of the divisor arrays of qubits, $n \geqslant m$. For a one-qubit division, the best known method requires a time complexity of $(5/6)^2 + O(1)$, whereas the best case time complexity of the division algorithm for one qubit is $O(2)$.

According to the algorithm, in the first step the dividend is split and it requires the best case time complexity of $T(P) = O(\log_2 n)$, where n is the number of qubits of the dividend.

In the second step a PNA tree is produced to perform the addition of the partial numerators. Assume that T is the PNA tree and $N(n)$ is the total number of levels of the tree. Thus, first the required time for the numerator generation and addition of the partial numerators is calculated using a tree structure by the method of mathematical induction. It is assumed that the time of numerator generation and addition operations using the tree is $O((\log_2 n) - 1)$.

When $n = 1$, T is a single level tree. Thus $N(1) = 1 = |\ (\log_2 1) - 1\ |$. Thus the statement holds for the base case which is $n = 1$.

Assume $N(i) = |\ (\log_2 i) - 1\ |$ for $0 \leqslant i < n$. A partial product addition tree consists of $(n - 1)$ levels and the root node. Thus,

$$N(n) = 2 \times N(n - 1) + 1 = \log_2(2n - 2) - 2 + 1$$

$$=(\log_2 n) + (\log_2 2) - (\log_2 2) - 1 = (\log_2 n) - 1.$$

Therefore in the best case, the time for the generation and the addition operation of the numerator using a PNA tree is $T(G) = O((\log_2 n) - 1)$.

In the third step it is necessary to access arrays. The array access time is linear and it is $O(1)$. According to the algorithm, three arrays have to be accessed: one for storing partial numerators, one for storing quotients, and the last one for storing partial remainders. Thus, in the best case, the total access time for three arrays is $T(A) = O(3)$.

In the fourth step the subtraction of the partial numerator from the partitioned dividend is performed which requires $T(S) = O(\log_2 n)$ time in the best case. Thus, in the best case, the total time complexity of the division algorithm is $T(N) = T(P) + T(G) + T(A) + T(S) = O(\log_2 n) + O((\log_2 n) - 1) + O(3) + O(\log_2 n) = O(3((\log_2 n) + 1) - 1)$.

Therefore the division algorithm requires a time complexity of $O(3((\log_2 n + 1) - 1)$ in the best case, where n is the number of dividend qubits. Thus the division algorithm requires a time complexity of $O(3n((\log_2 n) + 1) - n)$ for the n times iterations of the division process.

8.4 The design of a quantum divider circuit

Quantum divider design techniques are a very important topic at present. All the quantum divider circuits are designed using naive approaches. The designs of these quantum divider circuits require high area, delay, power, and quantum costs. As these designs produce quantum divider circuits with higher depths, the design techniques are not efficient enough. Thus a tree-based design algorithm is introduced. The splitting operation of the dividend is first performed in the root node of a tree. The partial numerator generator and addition circuits are considered as the parent nodes of the tree. In the last level of the tree there will be leaf nodes for the subtractors, which are the child nodes of the numerator generator and addition nodes.

8.4.1 A technique to minimize the number of ancillary inputs in the quantum circuit realization

In a quantum circuit an important parameter to measure the performance is the number of qubits in the circuit required for it to operate. It is expected that the number of qubits is to be optimized during the realization of quantum circuits. The number of input lines of the circuit is the indicator of required qubits as the input lines are limited resources in quantum circuits. Extra ancillary input lines help to increase the number of input lines. Therefore, in this subsection a merging technique is shown to optimize the number of ancillary inputs in large quantum circuits. This procedure works like the merge join algorithm. First it is necessary to divide the initial quantum circuit into sub-modules, then find the number of ancillary input lines in those sub-modules. Now combine the sub-modules and form the original structure up to the $m(2^n)$th ancillary input lines, where n is the number of qubits of each operand and m is the number of ancillary inputs in each sub-module.

Finally, calculate whether the number of quantum gates in the ancillary lines of the sub-modules are even or odd. If the value is odd, combine the quantum gates of

---

Input: (2^{n+1}) numbers of sub-modules of quantum circuits
Output: n-qubit circuit with $m(2^n+2)$ ancillary input lines

```
1: START
2: m=the number of ancillary inputs in each sub-module
3: p= the number of quantum gates in the ancillary line of the sub-
   module
4: c= the number of ancillary inputs in large circuit (initially 0)
5: for each sub-module do
6:    combine them together without explicit change
7:    c = c + m
8:    if (c == (m(2^n + 1))) then
9:       Step 3
10:   else
11:      Step 5
12:   end if
13:   if (p%2 == 0) then
14:      combine the 1^st p/2 quantum gates with quantum gates in the
         m(2^n+1)^th ancillary input line
15:      combine the rest p/2 along with the p gates of another sub-module
         with the p gates of the m(2^n+2)^th ancillary input line
16:   else
17:      combine the p gates with gates in the m(2^n+1)^th ancillary input
         line
18:      combine p gates of another sub-module with the p gates of the
         m(2^n+2)^th ancillary input line
19:   end if
20: end for
21: END
```

Algorithm 8.2. Reduction of the number of ancillary input lines in the design of the quantum divider circuit.

the ancillary input lines of the sub-modules with the quantum gates in each of the $m(2^n + 1)$th and $m(2^n + 2)$th ancillary input lines of the initial circuit and then combine the first half of the quantum gates in the ancillary input lines of the odd sub-modules with quantum gates in the $m(2^n + 1)$th ancillary input line of the initial circuit. Then combine the other half of the quantum gates in the ancillary input lines of the other half of the odd sub-modules with the gates in the ancillary input lines of the even sub-modules and the gates of the $m(2^n + 2)$th ancillary input line of the initial circuit. Algorithm 8.2 gives the procedure of the minimization of the number of ancillary input lines and example 8.4 describes the whole scenario.

Example 8.4. Consider the sub-modules of the quantum ANDing circuit (QAC) presented in figure 8.7. Suppose we need to combine eight ($2^{n+1} = 2^{2+} = 8$) of these sub-modules (the QACs) of the quantum divider circuit to form a two-qubit

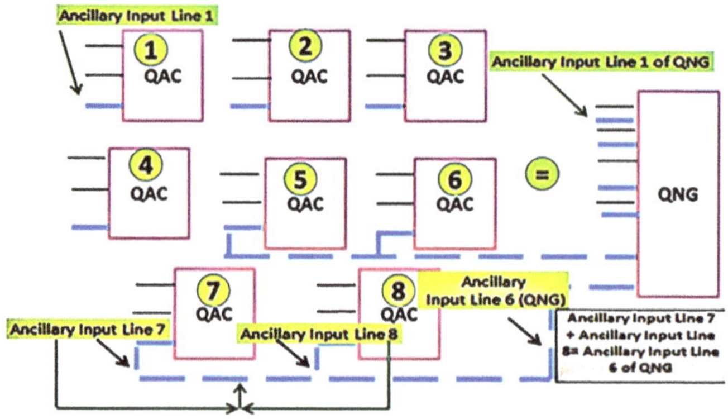

Figure 8.7. The quantum circuit for the ANDing operation.

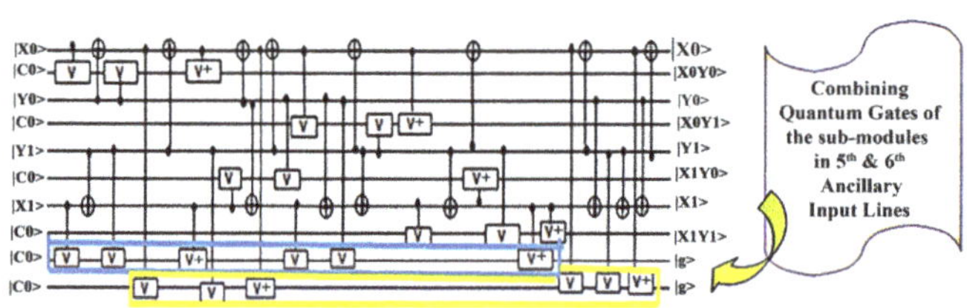

Figure 8.8. Block diagram to illustrate algorithm 8.2 to reduce ancillary inputs in a two-qubit QNG circuit.

quantum numerator generator circuit, as shown in figure 8.8, where the number of ancillary inputs in each sub-module (the QAC) is 1 ($m = 1$) and the number of quantum gates in the ancillary line of the sub-module is 3. Thus according to algorithm 8.2, the first four ($2^n = 2^2 = 4$) sub-modules (the QACs) are combined without joining the ancillary input lines. Thus the QNG circuit will have four ancillary input lines at its input from the four QACs.

Combine three quantum gates on the ancillary input line of sub-module 7 (ancillary input line 7) with the three quantum gates on the ancillary input line of sub-module 8 (ancillary input line 8) and place them on the sixth ($m(2^n + 2)$ th = 6th) ancillary input line of the initial QNG circuit. Figure 8.8 provides the block diagram to show how the sub-modules are combined to reduce the number of ancillary inputs according to algorithm 8.2 to form a two-qubit QNG circuit. In this figure the ancillary input line 7 and ancillary input line 8 are joined together and form the ancillary input line 6 of the QNG circuit. Similarly, ancillary input line 5 and ancillary input line 6 are joined together and finally form the QNG circuit with reduced ancillary inputs. The algorithm requires only six ancillary inputs instead of eight, as shown in figure 8.8. A detail of this implementation at the gate level is illustrated in figure 8.9.

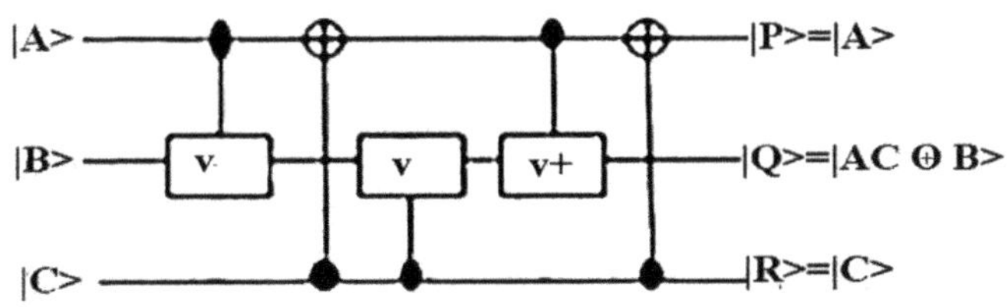

Figure 8.9. A two-qubit QNG circuit.

Table 8.1. Reversibility of a 3 × 3 QAC.

Input			Output		
A	*B*	*C*	*P*	*Q*	*R*
0	0	0	0	0	0
0	0	1	0	0	1
0	1	0	0	1	0
0	1	1	0	1	1
1	0	0	1	0	0
1	0	1	1	1	1
1	1	0	1	1	0
1	1	1	1	0	1

8.4.2 The components of the quantum divider circuit

The main components of the quantum divider circuit are the QAC, quantum full-subtractor circuit, QNG circuit, quantum numerator addition (QNA) circuit, and QDS circuit. These circuit components of the quantum divider are elaborated in the following.

8.4.2.1 The quantum ANDing circuit

A new 3 × 3 QAC which performs the ANDing operation is presented in this subsection. The input vector, I_v, and the output vector, O_v, of the circuit are as follows:

$$|I_v\rangle = \{|A\rangle, |B\rangle, |C\rangle\}; \text{ and}$$
$$|O_v\rangle = \{|A\rangle, |AC \oplus B\rangle, |C\rangle\}.$$

Here the input $|B\rangle$ is the target qubit and $|A\rangle$ and $|C\rangle$ are two control qubits. The QAC is constructed by two controlled-V, one controlled-V^+, and two CNOT gates, as shown in figure 8.7. Hence the quantum cost of the QAC is 5. The corresponding truth table of the circuit is shown in table 8.1. To perform the AND operation using the QAC the $|B\rangle$ input qubit in figure 8.7 has to remain $|0\rangle$.

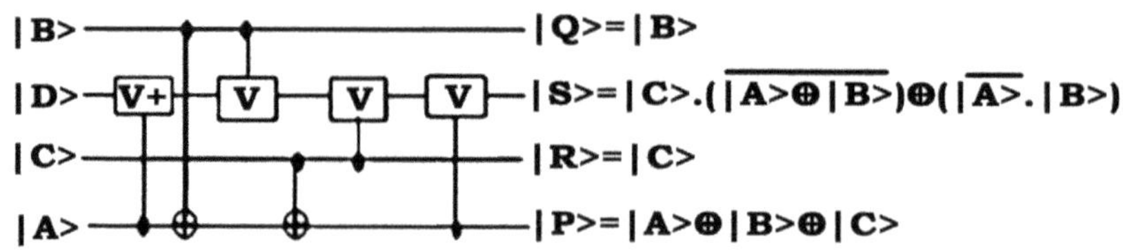

Figure 8.10. The quantum full-subtractor, when $|D\rangle = |0\rangle$.

8.4.2.2 The quantum full-subtractor circuit

A QFS is a quantum circuit that performs the arithmetic subtraction of three input qubits. It consists of three inputs and two outputs. If we take three input variables as $|A\rangle$, $|B\rangle$, and $|C\rangle$, and two output variables $|D\rangle$ (difference) and $|B_{\text{out}}\rangle$ (borrow), then the following equations can be used to obtain the outputs:

$$|D\rangle = |A\rangle \oplus |B\rangle \oplus |C\rangle$$

$$|B_{\text{out}}\rangle = (|C\rangle.\,(\overline{|A \oplus B\rangle})) \oplus (\overline{|A}.\,B\rangle).$$

In this subsection, a new 4 × 4 QFS circuit is shown. The input vector (I_v) and the output vector (O_v) of the circuit are as follows:

$|I_v\rangle = \{|B\rangle, |D\rangle, |C\rangle, |A\rangle\}$; and

$|O_v\rangle = \{|B\rangle, (|C\rangle.(\overline{|A \oplus B\rangle}))\oplus (\overline{|A}.B\rangle), |C\rangle, |A\rangle\oplus|B\rangle\oplus|C\rangle\}$.

Here the input $|D\rangle$ is the target qubit and $|A\rangle$, $|B\rangle$, and $|C\rangle$ are control qubits. The QFS circuit is constructed by three controlled-V, one controlled-V^+, and two CNOT gates, as shown in figure 8.10. Hence the quantum cost of the QFS circuit is 6. The corresponding truth table of the circuit is shown in table 8.2. Input $|A\rangle$ is used to take the role of the borrow in and output $|S\rangle$ is the borrow out, as shown in figure 8.10. Output $|P\rangle$ is the difference of the inputs $|B\rangle$, $|C\rangle$, and borrow in $|A\rangle$. The QFS circuit is designed in such a way that it can act as a full-subtractor by setting $|D\rangle = |0\rangle$.

8.4.2.3 The quantum dividend splitter circuit

A 4 × 4 QDS circuit is shown that takes the string of dividend qubits and divides it into two parts which are carried out to the QFS circuit. The input vector I_v and the output vector O_v of the circuit are as follows:

$$|I_v\rangle = \{|Dd[i/2]\rangle,\ Di[j/2]\rangle,\ |0\rangle,\ |0\rangle\};\ \text{and}$$

$$|O_v\rangle = \{|Dd[i/2]\rangle,\ \overline{(|Dd[i/2]\rangle \oplus |Di[j/2]\rangle)},$$
$$\overline{(|Di[j/2]\rangle} \cdot |Dd[i/2]\rangle),$$
$$\overline{(|Dd[i/2]\rangle} \cdot |Di[j/2]\rangle)\}.$$

In figure 8.11 the last input is the target qubit ($|0\rangle$) and other inputs are control qubits. The QDS circuit is constructed by two controlled-V, one controlled-V^+, one

Table 8.2. Reversibility of a 4 × 4 QFS circuit.

Input				Output			
A	*B*	*C*	*D*	*P*	*Q*	*R*	*S*
0	0	0	0	0	0	0	0
0	0	0	1	0	0	0	1
0	0	1	0	1	0	1	1
0	0	1	1	1	0	1	0
0	1	0	0	1	1	0	1
0	1	0	1	1	1	0	0
0	1	1	0	0	1	1	1
0	1	1	1	0	1	1	0
1	0	0	0	1	0	0	0
1	0	0	1	1	0	0	1
1	0	1	0	0	0	1	0
1	1	0	0	0	1	0	0
1	1	0	1	0	1	0	1
1	1	1	0	1	1	1	1
1	1	1	1	1	1	1	0

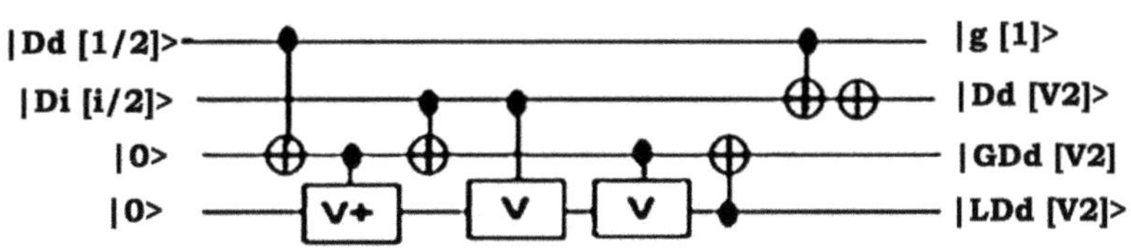

Figure 8.11. The QDS circuit.

NOT, and four CNOT gates, as shown in figure 8.11. Hence the quantum cost of the QDS circuit is 8. The main purpose of the QDS circuit is to separate the set of qubits of the dividend. If the dividend consists of four qubits, the midpoint qubit (second qubit) passes as ($|Dd[i/2]\rangle$) output. Then the inputs, the first and third input qubits, are Ex-ORed, which goes out as outputs $|GDd[i/2]\rangle$ and $|LDd[i/2]\rangle$, respectively, as shown in figure 8.11.

8.4.2.4 The quantum numerator generator circuit

The QNG circuit takes the divisor and quotient as the input and produces partial numerators as outputs by performing all possible ANDing operations using the divisor and quotient qubits. The divisor qubits and the quotient qubits are fed to this component which produces the product terms of the qubits known as partial numerators. The QAC has been used to construct this part. If n-qubit divisors and n-qubit quotients are used, the QNG circuit requires (2^{n+1}) QACs and $m(2^n + n)$ ancillary input lines, where m is the number of ancillary qubits in each QAC. Thus, including the divisor and quotient qubits, a total of $2n + m(2^n + n)$ qubits are required to construct the n-qubit QNG circuit.

Example 8.5. To construct a two-qubit QNG circuit, $(2^{n\ 1}) = 8$ QACs and $m(2^n + n) = 6$ ancillary input lines are required, where m is the number of ancillary qubits in each QAC (here, $m = 1$). In figure 8.8 the detailed two-qubit QNG circuit is shown. From figure 8.8 it is clear that the two-qubit QNG circuit needs $2n + m\ (2^n + n) = 4 + 6 = 10$ qubits according to the ancillary input reduction, as shown in algorithm 8.2.

8.4.2.5 The quantum numerator addition circuit

The QNA circuit adds column-wise the partial numerators that have been generated by the QNG circuit. The quantum full-adder (QFA) is used to construct the QNA circuit. For an n-qubit QNA circuit, $(5n/2)$ QFAs are needed. Thus for that purpose $m(2^n + n)$ ancillary input lines are needed, where m is the number of ancillary qubits in each QFA. If the number of QFAs required to construct the QNA circuit is greater than five (when $n \geq 4$), then it is necessary to reduce the ancillary input lines by combining these lines of the quantum gates according to the reduction algorithm.

Example 8.6. To construct a two-qubit QNA circuit, $(5n/2) = 5$ QFA circuits are required to perform the addition operations of the partial numerators generated by QNG. Since the number of QFAs is 5, merging of the quantum gates is not required to reduce the number of ancillary input lines. Thus the five ancillary input lines are required. In figure 8.12 the detailed quantum numerator addition circuit (QNA) for a two-qubit divider is shown.

8.4.2.6 The n*-qubit quantum divider*

The tree-based design structure of the quantum divider circuit reduces the depth of the circuit. The design also minimizes the number of ancillary inputs. In this subsection, details of the design of a two-qubit quantum divider circuit are presented

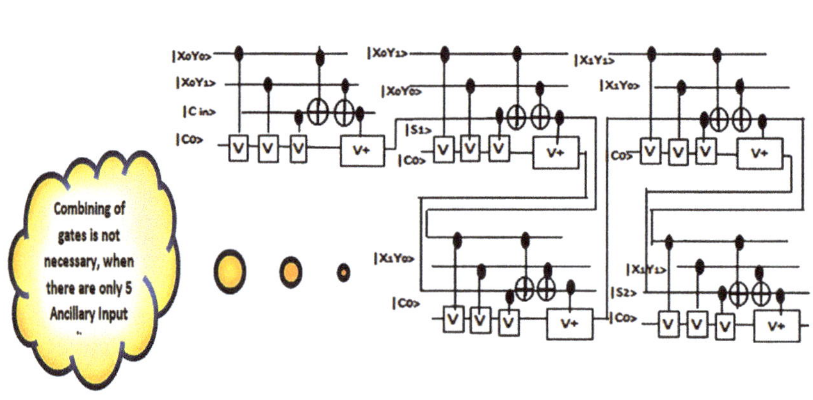

Figure 8.12. The two-qubit QNA circuit.

Input: n-qubit dividend $\{|Z_1>, |Z_2>, \ldots, |Z_n>\}$ and n-qubit divisor $\{|X_1>, |X_2>, \ldots, |X_n>\}$
Output: Result (Quotient Qubits) of n-qubit division: $\{|Z_1>, |Z_2>, \ldots, |Z_n>\}/\{|X_1>, |X_2>, \ldots, |X_n>\}$

START
for ($log_2 n$ times) **do**
    Split the n-qubit dividend into (log_2 n) numbers of 2-qubit divisor by the QDS circuit
    Each time put the qubits to a separate QFS circuit
    Call the ***Micro Division*** procedure
end for
Micro Division procedure
Put the 2-qubit divisor and 2-qubit quotient to the circuit having *(n/2)* numbers of 2-qubit QNG circuits
Send the output of the QNG circuit to *(n/2)* numbers of 2-qubit QNA circuit
Subtract the output of QNA from dividend qubits in QFS
Apply *2* QFS circuits to subtract the qubit of each position of the dividend
if the output of each QFS is ($|0>$) in each QFS **then**
    Take the quotient as result of the division
else
    Add the remainders as beginning of the other split and continue the same process again until no remainder ($|0>$) state is obtained
end if
END

Algorithm 8.3. Design of an n-qubit quantum divider circuit.

using algorithm 8.3, as shown in figure 8.13. In addition, a block diagram of an n-qubit quantum divider circuit using the two-qubit quantum divider circuit is presented using algorithms 8.2 and 8.3, as shown in figure 8.14.

The algorithm for the tree-based architecture of an n-qubit quantum divider circuit works recursively. This algorithm first creates the design of the two-qubit divider circuit using a procedure of micro-division and then it is called recursively to generate an n-qubit quantum divider circuit, where in the procedure of micro-division of the algorithm 8.3, the QDS first splits the dividend and sends it to the subtractor for the subtraction operation. Then, the QNG circuit generates the partial numerators that are added using the QNA circuit and the outputs of the QNA circuits are carried to the QDS circuit for the subtraction operation. In figure 8.14, a block diagram of an n-qubit quantum divider applying algorithm 8.3 is presented. The n-qubit quantum divider circuit consists of ($n/2$) two-qubit QNG circuits, ($n/2$) two-qubit QNA circuits, n QFS circuits, and ($\log_2 n$) QDS circuits.

The number of quantum wires can be minimized using the quantum memory elements in a quantum circuit. As a consequence the unwanted outputs (garbage outputs) can be reduced using quantum memory elements.

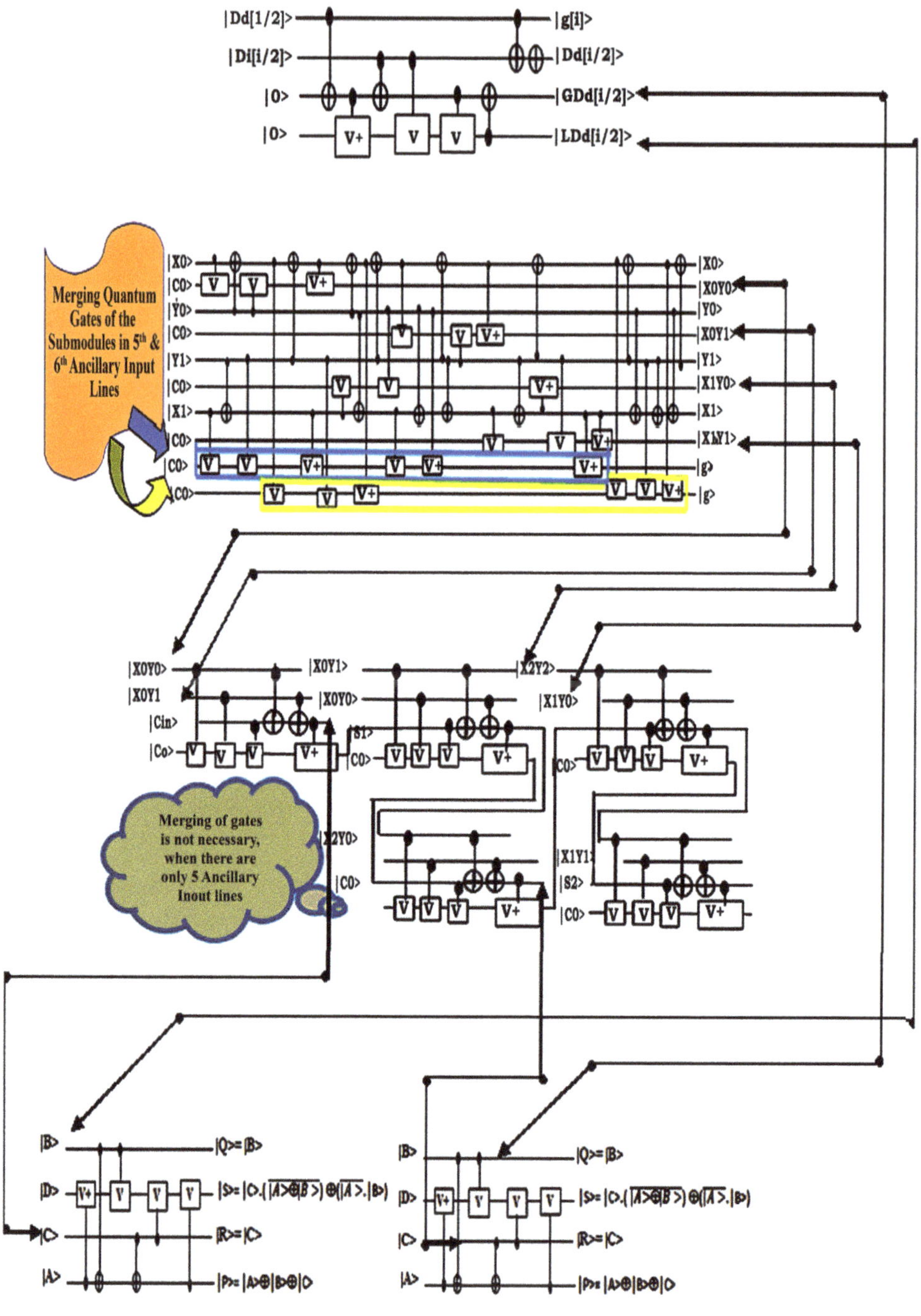

Figure 8.13. The two-qubit quantum divider circuit.

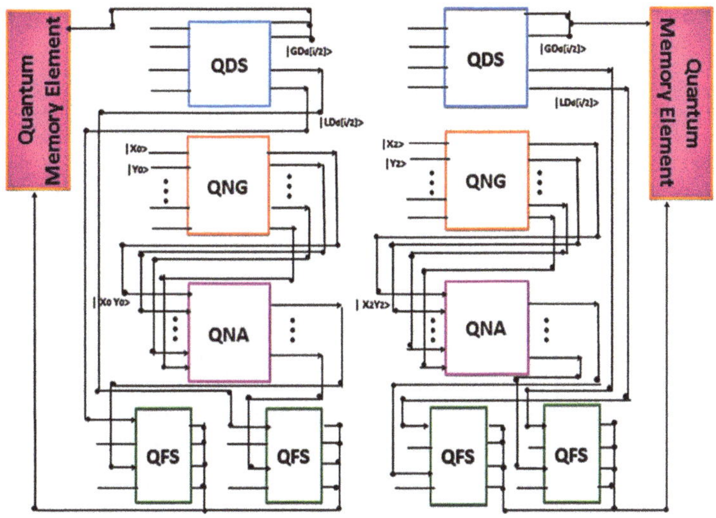

Figure 8.14. The block diagram of the four-qubit quantum divider circuit.

Here, this concept of the quantum memory element is used to reduce the number of garbage outputs. There is a single garbage output line that collects garbage from different locations and stores it in the memory locations for further reuse and thus shrinks the number of garbage output lines in the circuit design, as shown in figure 8.14.

Example 8.7. A four-qubit quantum divider circuit is designed using a two-qubit quantum divider, as shown in figure 8.13, by applying algorithm 8.3. According to the design procedure of algorithm 8.3, a four-qubit quantum divider circuit requires two ($\log_2 n = \log_2 4 = 2$) QDS circuits, two ($n/2 = 4/2 = 2$) two-qubit QNG circuits, two ($n/2 = 4/2 = 2$) two-qubit QNA circuits, and four QFS circuits. First the QDS splits the dividend in half until it is a two-qubit dividend. Then this two-qubit goes to the QFS where the partial numerator qubit is subtracted from the dividend qubit, which is obtained by the QDS and the recursive procedure of micro-division of algorithm 8.3 (two-qubit divider design). If the result of the subtraction is $|0\rangle$, then the quotient is the result of division of the four-qubit quantum divider circuit, as constructed by algorithm 8.3, otherwise the remainder qubits are added at the beginning positions of the initial dividend to form a new dividend which is split further. To obtain a four-qubit quantum divider circuit, the procedure of micro-division of algorithm 8.3 continues until the remainder exists. The four-qubit quantum divider circuit is shown in figure 8.15.

Property 8.3. The depth of the design of the quantum divider circuit is at least $O(32n + 8 \log_2 n)$, where n is the number of qubits of the dividend.

Proof. The above-mentioned statement is proved by the method of mathematical induction. A QDS circuit, a two-qubit QNG circuit, a two-qubit QNA circuit, and

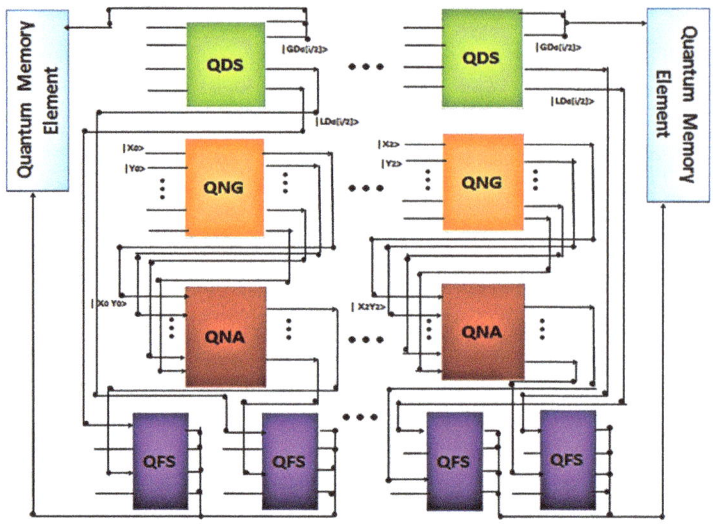

Figure 8.15. The block diagram of an n-qubit quantum divider circuit.

two QFS circuits are required for constructing a two-qubit quantum divider circuit using the design method of the tree-based divider in the best case where there is no remainder after the first iteration. The total number of stages in a QDS circuit is eight and the numbers of stages in a two-qubit QNG circuit, a two-qubit QNA circuit, and a QFS circuit are 34, 18, and 6, respectively.

When $n = 2$, i.e. for a two-qubit quantum divider circuit, the total depth of the two-qubit quantum divider circuit in the best case is

$$D_{\text{two-qubit}} = D_{\text{QDS}} + D_{\text{QNG}} + D_{\text{QNA}} + 2 \times D_{\text{QFS}}$$
$$= 8 + 34 + 18 + 2 \times 6 = 64 + 8 = 32n + 8\log_2 n; \ [\because n = 2 \text{ here}].$$

Thus the statement holds for the base case $n = 2$.

Assume that the statement also holds for $n = k$, which is a k-qubit quantum divider circuit.

Thus $D_{k\text{-qubit}} = 32k + 8\log_2 k$.

Now in the best case, $(k + 1)/2$ two-qubit QNG circuits, $(k + 1)/2$ two-qubit QNA circuits, $(k + 1)$ QFS circuits, and $\log_2(k + 1)$ QDS circuits are needed for the $n = (k + 1)$-qubit quantum divider circuit.

Thus the depth of the $(k + 1)$-qubit quantum divider circuit is

$$D_{(k+1)\text{-qubit}} = \log_2(k + 1) \times D_{\text{QDS}} + ((k + 1)/2) \times D_{\text{QNG}} + ((k + 1)/2) \times D_{\text{QNA}}$$
$$+ (k + 1) \times D_{\text{QFS}} = 8\log_2(k + 1) + 17(k + 1) + 9(k + 1) + 6(k + 1)$$
$$= 32(k + 1) + 8\log_2(k + 1).$$

Therefore, the statement is also true for $n = (k + 1)$.

Thus, the depth of the design of the quantum divider circuit is at least $O(32n + 8\log_2 n)$. □

8.5 Summary

In this chapter a tree-based quantum integer division algorithm is developed. This division technique is a paragon in the emerging technologies in computer aided design, since it has reduced time complexity to a large extent. The technique reduces the number of ancillary input lines in the quantum circuit, which in turn enables the reduction of the required qubits for the circuit to operate. In addition to this, an algorithm is introduced using a tree structure that is applicable for designing compact quantum divider circuits with optimum depth. The quantum divider design algorithm with optimized depth is an epoch-making achievement in the field of quantum computation.

Critical thinking questions

1. Explain the quantum division algorithm.
2. Summarize the characteristics of the quantum divider in details.
3. What factors are allowed to reduce the number of ancillary inputs in the quantum circuit realization? Explain in detail.
4. Write the components of the quantum divider circuit and explain each of the components in detail.
5. For a two-qubit quantum divider circuit, what will be the total depth of the two-qubit quantum divider circuit in the best case when $n = 4$?

References

[1] Abdessaied N, Wille R, Soeken M and Drechsler R 2013 Reducing the depth of quantum circuits using additional circuit lines *Int. Conf. on Reversible Computation* (Berlin: Springer) pp 221–33

[2] Aggarwal N, Asooja K, Verma S S and Negi S 2009 An improvement in the restoring division algorithm (needy restoring division algorithm) *2nd IEEE Int. Conf. on Computer Science and Information Technology* (Piscataway, NJ: IEEE) pp 246–9

[3] Barenco A, Bennett C H, Cleve R, DiVincenzo D P, Margolus N, Shor P, Sleator T, Smolin J A and Weinfurter H 1995 Elementary gates for quantum computation *Phys. Rev.* A **52** 3457

[4] Bera D, Green F and Homer S 2007 SIGACT News Complexity Theory Column 55 https://mathcs.clarku.edu/~fgreen/papers/constQDepthSurvey.pdf

[5] Cheng K-W and Tseng C-C 2002 Quantum full adder and subtractor *Electron. Lett.* **38** 1343–4

[6] Cormen T H 2001 *Introduction to Algorithms* ed T H Cormen, C E Leiserson, R L Rivest and C Stein (Cambridge, MA: MIT Press)

[7] Cuccaro S A, Draper T G, Kutin S A and Moulton D P 2004 A new quantum ripple-carry addition circuit arXiv: quant-ph/0410.184

[8] Deutsch D E 1989 Quantum computational networks *Proc. R. Soc. Lond.* A **425** 73–90

[9] Haghparast M and Navi K 2007 A novel reversible full adder circuit for nanotechnology based systems *J. Appl. Sci.* **7** 3995–4000

[10] Kaye P *et al* 2007 *An Introduction to Quantum Computing* (Oxford: Oxford University Press)

[11] Khosropour A, Aghababa H and Forouzandeh B 2011 Quantum division circuit based on restoring division algorithm *Eighth Int. Conf. on Information Technology: New Generations* (Piscataway, NJ: IEEE) pp 1037–40

[12] Knuth D E 2007 Seminumerical algorithms *The Art of Computer Programming* vol 2 3rd edn (Reading, MA: Addison-Wesley)

[13] Moller N and Granlund T 2011 Improved division by invariant integers *IEEE Trans. Comput.* **60** 165–75

[14] Nielsen M A and Chuang I L 2000 Quantum computation *Quantum Information* (Cambridge: Cambridge University Press)

[15] Pavlidis A and Gizopoulos D 2012 Fast quantum modular exponentiation architecture for Shor's factorization algorithm arXiv: 1207.0511

[16] Ramya P R and Vani Y S 2013 Optimization and implementation of reversible BCD adder in terms of number of lines *Int. J. Reconfigurable Embedded Syst.* **2** 21–6

[17] Sultana S and Radecka K 2014 Reversible architecture of computer arithmetic *Int. J. Comput. Appl.* **93** 6

[18] Yao A C-C 1993 Quantum circuit complexity *Proc. 34th IEEE Annual Foundations of Computer Science* (Piscataway, NJ: IEEE) pp 352–61

Chapter 9

The quantum BCD priority encoder

Learning objectives

- Define the quantum encoder.
- Discover the attributes of an encoder.
- Design a quantum binary coded decimal (BCD) priority encoder.
- Construct a 2*n*-to-*n* quantum BCD priority encoder circuit.
- Develop an algorithm of the quantum BCD priority encoder circuit.
- Analyze the characteristics of the encoder circuit.

In quantum circuit design, energy dissipation has become a crucial factor that needs to be taken into consideration. Since irreversible computing is one of the most significant causes of energy dissipation, designing quantum circuits using quantum gates is an efficient way to reduce the energy dissipation of the circuit. Although binary coded decimal (BCD) encoders are one of the most common types of unit circuits, they are all designed in an irreversible way.

The encoder is an essential element in many quantum devices. The BCD encoder is one of the most notable and common encoders in quantum systems. However, the conventional BCD encoder has certain faults in addition to energy dissipation, such as confused output codes when two or more inputs are valid. Considering these drawbacks, it is obviously necessary to focus on identifying a useful quantum BCD priority encoder circuit which not only resolves the problem of confused codes, but also dissipates little energy.

In this chapter the design of a quantum BCD priority encoder circuit that consists of quantum gates is analyzed.

9.1 The properties of a quantum encoder

Unlike a quantum multiplexer that selects one individual data input line and then sends these data to a single output line or switch, a *quantum encoder* takes all its data

doi:10.1088/978-0-7503-5683-1ch9

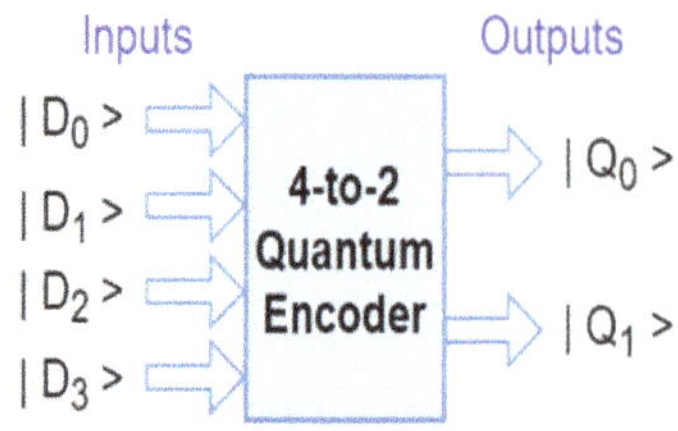

|D3>	|D2>	|D1>	|D0>	|Q1>	|Q0>
|0>	|0>	|0>	|1>	|0>	|0>
|0>	|0>	|1>	|0>	|0>	|1>
|0>	|1>	|0>	|0>	|1>	|0>
|1>	|0>	|0>	|0>	|1>	|1>
|0>	|0>	|0>	|0>	|x>	|x>

Figure 9.1. The block diagram of the 4-to-2 quantum encoder circuit and its truth table.

inputs one at a time and then converts them into a single encoded output. Thus a quantum encoder is a multi-input quantum logic circuit that converts the logic level 1 data at its inputs into an equivalent code at its output.

Generally, quantum encoders produce outputs of two-qubit, three-qubit, or four-qubit codes, depending on the number of data input lines. An n-qubit quantum encoder has 2^n input lines and n-qubit output lines. A 4-to-2 quantum encoder circuit is shown in figure 9.1.

The output lines of a quantum encoder generate the input line whose value is equal to 1 and are available to encode either a decimal or hexa-decimal input pattern to, typically, a binary or BCD output code.

One of the main disadvantages of standard quantum encoders is that they can generate the wrong output code when there is more than one input present at logic level 1. For example, if inputs D_1 and D_2 are both HIGH at logic 1 at the same time, the resulting output is neither at 01 or at 10, but will be at 11, which is an output that is different to the actual input present. Also, an output code of all logic 0s can be generated when all of its inputs are at 0 OR when input D_0 is equal to one.

One simple way to overcome this problem is to 'prioritize' the level of each input pin and thus, if there was more than one input at logic level 1, the actual output code would only correspond to the input with the highest designated priority. This type of quantum encoder is known commonly as a *priority encoder* or P-encoder for short.

The priority encoder solves the problems mentioned above by allocating a priority level to each input. The priority encoder's output corresponds to the currently active input which has the highest priority. Thus when an input with a higher priority is present, all other inputs with a lower priority will be ignored.

The priority encoder comes in many different forms with an example of an eight-input priority encoder along with its truth table shown in figure 9.2.

9.2 The design of a quantum BCD priority encoder circuit

In this section a quantum 2^n-to-n BCD priority encoder circuit is presented. The optimized 2^n-to-n BCD priority encoder circuit is designed using quantum gates. The working principle of an 8-to-3 quantum priority encoder is shown in figure 9.2.

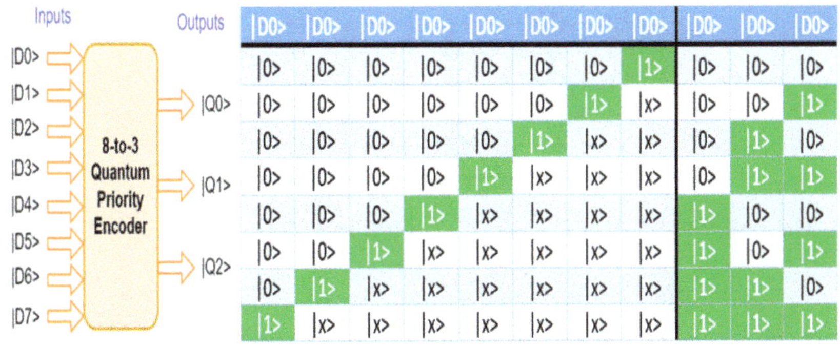

Figure 9.2. The block diagram of a 8-to-3 quantum priority encoder and its truth table.

Table 9.1. The truth table of a quantum BCD priority encoder.

Input								Output		
I_7	I_6	I_5	I_4	I_3	I_2	I_1	I_0	A	B	C
1	X	X	X	X	X	X	X	1	1	1
0	1	X	X	X	X	X	X	1	1	0
0	0	1	X	X	X	X	X	1	0	1
0	0	0	1	X	X	X	X	1	0	0
0	0	0	0	1	X	X	X	0	1	1
0	0	0	0	0	1	X	X	0	1	0
0	0	0	0	0	0	1	X	0	0	1
0	0	0	0	0	0	0	1	0	0	0

9.2.1 The quantum BCD priority encoder circuit

A traditional 8-to-3 BCD encoder has eight inputs which are I_0 to I_7, and three outputs which are A, B, and C. All the inputs and outputs are required to be binary. The function of a traditional BCD encoder can be represented as

$$A = I_7 + I_6 + I_5 + I_4$$
$$B = I_7 + I_6 + I_3 + I_2$$
$$C = I_7 + I_5 + I_3 + I_1.$$

Different to traditional BCD encoders, the function of the BCD priority encoder has priority levels. To be more precise, I_7 has the highest priority level and the other inputs' priority levels decrease in turn, so that I_0 has the lowest. For example, if I_5 and I_3 are both 1 and the rest of the inputs are 0, then since I_5 is in a higher priority level, the encoder will ignore I_3 and regard it as the situation where only I_5 is 1 and the rest of the inputs are all 0. The truth table of a BCD priority encoder is shown in table 9.1. 'X' represents uncertain values of inputs.

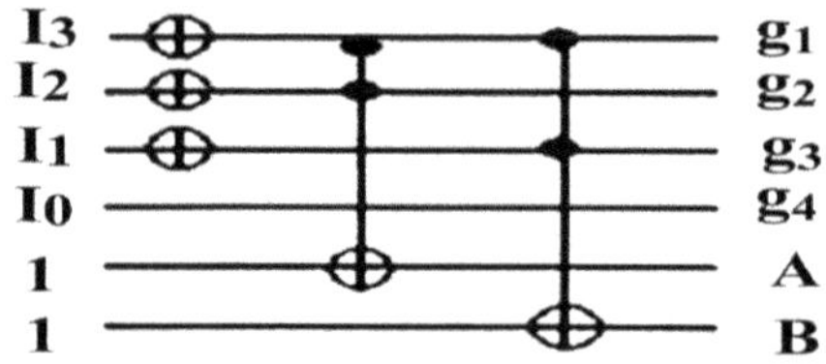

Figure 9.3. The quantum 4-to-2 BCD priority encoder circuit. Reproduced with permission from [6]. Copyright 2014 IEEE.

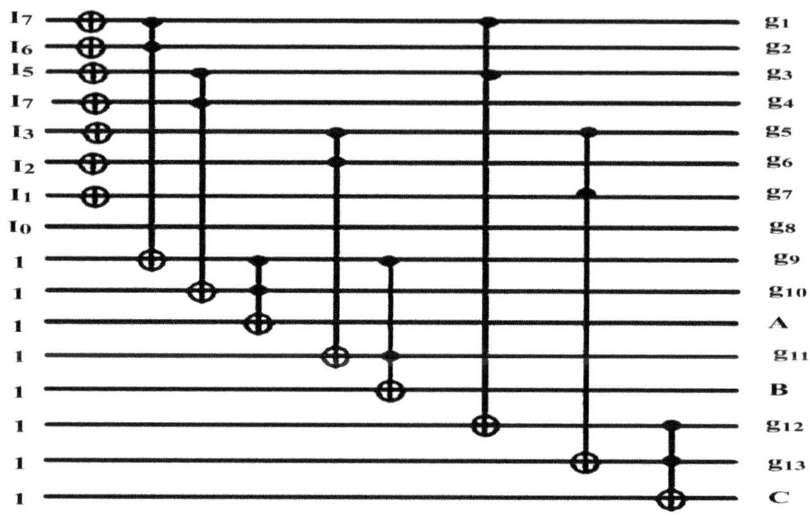

Figure 9.4. The quantum 8-to-3 BCD priority encoder circuit. Reproduced with permission from [6]. Copyright 2014 IEEE.

In order to design quantum BCD priority encoder circuit, irreversible logic gates such as AND and OR gates are not allowed in the circuit. Thus the functions of a traditional priority BCD encoder are transformed to a form which can be converted conveniently into a circuit diagram. An 8-to-3 quantum BCD priority encoder is shown in figure 9.3, where quantum NOT gates in all inputs are utilized to obtain the inverse. In the second step the OR function is completed using quantum gates. Similarly, figure 9.2 presents a 4-to-2 quantum BCD priority encoder circuit.

For example, consider the function of A. The inputs will be I_7', I_6', I_5', and I_4' at the first step, then the inputs I_5', I_4' and the additional input 1 are connected with quantum gates. Hence it produces $I_5 + I_4$. In the same way we can obtain $I_7 + I_6$. Finally, by cascading $I_5 + I_4$, $I_7 + I_6$, and the additional input line 1, $I_7 + I_6 + I_5 + I_4$ will be obtained, which is illustrated in figure 9.4. The generalized algorithm to construct a quantum 2^n-to-n BCD priority encoder circuit is presented in algorithm 9.1.

INPUT: 2^n-bit operand I ($I_{2n}, I_{2n-1}, \ldots, I_1$)
OUTPUT: n-bit operand A ($A_n, A_{n-1}, \ldots, A_1$)

START
for i=1 **to** 2^n-1 **do**
  Apply (2^n-1) number of quantum NOT gates at each input line to get the inverse input
end for
for i=1 **to** $[2(\frac{2^n}{2} - 1) + 2^{n-2}(n-2)]$ **do**
  if (i=1 or i = $[2(\frac{2^n}{2} - 1) + 2^{n-2}(n-2)]$) **then**
    Apply $(\frac{2^n}{2} - 1)$ number of quantum Toffoli gates to get the output (A_n or A_1)
  else
    Apply $2^{n-2}(n-2)$ number of quantum Toffoli gates to get the rest of the outputs ($A_{n-1}, A_{n-2}, A_{n-3}, \ldots, A_2$)
  end if
end for
END

Algorithm 9.1. Constructing a 2^n-to-n quantum BCD priority encoder circuit.

9.2.2 Analysis of the properties of the encoder circuit

In this subsection the properties of a quantum 2^n-to-n BCD priority encoder circuit are presented.

Property 9.1. A quantum 2^n-to-n BCD priority encoder circuit ($n \geqslant 2$) can be realized with $[5\,[2(\frac{2^n}{2} - 1) + 2^{n-2}(n-2)] + (2^n - 1)]$ quantum gates, where n is the number of output bits.

Proof. The property 9.1 is proved by mathematical induction. A 4-to-2 BCD encoder circuit consists of NOT quantum gates and quantum Toffoli gates, and is shown in figure 9.3. Thus the total number of quantum gates (NOQG) required to construct 4-to-2 BCD priority encoder circuit is

$$
\begin{aligned}
\text{NOQG}_{\text{4-to-2}} &= \text{NOQG}_{\text{NOT}} + 5 \times \text{NOQG}_{\text{Toffoli}} \\
&= (2^2 - 1) + 5 \times [2(\tfrac{2^2}{2} - 1) + 2^{2-2}(2-2)] \\
&= (4 - 1) + 5 \times [2(2-1) + 2 \times 0)] \\
&= 3 + 5 \times 2 \\
&= 13 \\
&= (4 - 1) + 5 \times [2(2-1) + 2 \times 0)].
\end{aligned}
$$

Thus the statement holds for base case $n = 2$.

Assume that the statement holds for $n = k$. Thus a quantum 2^k-to-k BCD priority encoder circuit can be realized with $[5[2(\frac{2^k}{2} - 1) + 2^{k-2}(k-2)] + (2^k - 1)]$ quantum gates.

A 2^{k+1}-to-$(k + 1)$ BCD priority encoder circuit requires $(2^{k+1} - 1)$ quantum NOT gates and $5\,[2(\frac{2^{k+1}}{2} - 1) + 2^{k+1-2}\,(k + 1 - 2)]$ quantum Toffoli gates. As a result, the total number of quantum gates required to construct a 2^{k+1}-to-$(k + 1)$ BCD priority encoder circuit is $(2^{k+1} - 1) + 5\,[2(\frac{2^{k+1}}{2} - 1) + 2^{k+1-2}\,(k + 1 - 2)]$.

Thus the statement holds for $n = k + 1$.

Therefore, a quantum 2^n-to-n BCD priority encoder circuit can be realized with $[5\,[2(\frac{2^n}{2} - 1) + 2^{n-2}\,(n - 2)] + (2^n - 1)]$ quantum gates.

Property 9.2. A quantum 2^n-to-n BCD priority encoder circuit produces $[5\,[2(\frac{2^n}{2} - 1) + 2^{n-2}\,(n - 2)] + 1]\Delta$ quantum delay, where Δ is the unit delay and n is the number of output bits.

Proof. A quantum BCD priority encoder circuit consists of quantum Toffoli gates and quantum NOT gates. A quantum Toffoli gate require 5Δ delay and a quantum NOT gate requires 1Δ delay. Thus 2^n number of quantum Toffoli gates require $5\,[2(\frac{2^n}{2} - 1) + 2^{n-2}\,(n - 2)]\Delta$ delay. Therefore, the total delay of a quantum 2^n-to-n BCD priority encoder circuit is $[5\,[2(\frac{2^n}{2} - 1) + 2^{n-2}\,(n - 2)] + 1]\Delta$, where $2^n - 1$ is the number of NOT gates at the same level.

Example 9.1. When $n = 3$ in figure 9.4, the total delay of an 8-to-3 BCD priority encoder is $[5\,[2(\frac{2^3}{2} - 1) + 2^{3-2}\,(3 - 2)] + 1] = 41\Delta$.

Property 9.3. A quantum 2^n-to-n BCD priority encoder circuit requires $[(3\Omega + 2\sigma)\,[2(\frac{2^n}{2} - 1) + 2^{n-2}\,(n - 2)] + (2^n - 1)\;]$ quantum gate calculation complexity, where n is the number of output bits, σ is the CNOT gate calculation complexity, and Ω is the controlled-V or controlled-V^+ gate calculation complexity.

Proof. From property 9.1 it is found that a quantum BCD priority encoder circuit consists of quantum Toffoli gates and quantum NOT gates. A quantum Toffoli gate has 2σ CNOT quantum gate calculation complexity and 3Ω controlled-V (controlled-V^+) quantum gate calculation complexity. Thus the total quantum gate calculation complexity for a 2^n-to-n quantum BCD priority encoder circuit is $2\,[2(\frac{2^n}{2} - 1) + 2^{n-2}\,(n - 2)]\,\sigma$ for $2[2(\frac{2^n}{2} - 1) + 2^{n-2}\,(n - 2)]$ CNOT gates, $3[2(\frac{2^n}{2} - 1) + 2^{n-2}\,(n - 2)]\Omega$ for $3[2(\frac{2^n}{2} - 1) + 2^{n-2}\,(n - 2)]$ controlled-V (controlled-V^+) gates, and $(2^n - 1)\rho$ for $(2^n - 1)$ NOT gates, i.e. $[(3\Omega + 2\sigma)\,[2(\frac{2^n}{2} - 1) + 2^{n-2}\,(n - 2)] + \rho(2^n - 1)]$ quantum gate calculation complexity.

Example 9.2. When $n = 3$ in figure 9.4 the total quantum gate calculation complexity of an 8-to-3 quantum BCD priority encoder circuit is $(3 + 2)\,[2(\frac{2^3}{2} - 1) + 2^{3-2}\,(3 - 2)] + (2^3 - 1)] = 7 + 16\sigma + 24\Omega$.

Property 9.4. A quantum 2^n-to-n BCD priority encoder circuit requires $[250\ [2(\frac{2^n}{2} - 1) + 2^{n-2}(n - 2)] + 50\ (2^n - 1)]$ Å area, where n is the number of output bits and Å is the unit of measuring area.

Proof. From property 9.1 it is found that a quantum BCD priority encoder circuit consists of quantum Toffoli gates and quantum NOT gates. A quantum Toffoli gate has five quantum gates. Thus a quantum 2^n-to-n BCD priority encoder circuit requires $5[2(\frac{2^n}{2} - 1) + 2^{n-2}(n - 2)]$ quantum Toffoli gates. Moreover, a quantum 2^n-to-n BCD priority encoder requires an extra $(2^n - 1)$ quantum NOT gates to obtain inverted input. Therefore, the total area for a quantum 2^n-to-n BCD priority encoder circuit is $((5\ [2(\frac{2^n}{2} - 1) + 2^{n-2}(n - 2)] \times 50) + (2^n - 1) \times 50)$ Å $= (250\ [2(\frac{2^n}{2} - 1) + 2^{n-2}(n - 2)] + (2^n - 1) \times 50)$ Å.

Property 9.5. A quantum 2^n-to-n BCD priority encoder circuit requires $142.3 \times [5[2(\frac{2^n}{2} - 1) + 2^{n-2}(n - 2)] + (2^n - 1)]$ meV power, where n is the number of output bits and meV is the unit of measuring power.

Proof. From property 9.1 it is found that a quantum BCD priority encoder circuit consists of quantum Toffoli gates and quantum NOT gates. A quantum Toffoli gate has five quantum gates. Thus a quantum 2^n-to-n BCD priority encoder circuit requires $5[2(\frac{2^n}{2} - 1) + 2^{n-2}(n - 2)]$ quantum Toffoli gates. Moreover, a quantum 2^n-to-n BCD priority encoder requires an extra $(2^n - 1)$ quantum NOT gates to obtain inverted input. Therefore, the total power for a quantum 2^n-to-n BCD priority encoder circuit is $142.3 \times [5[2(\frac{2^n}{2} - 1) + 2^{n-2}(n - 2)] + (2^n - 1)]$ meV, where meV is the unit of measuring power.

Property 9.6. A quantum 2^n-to-n BCD priority encoder circuit produces $[2^n + [[2(\frac{2^n}{2} - 1) + 2^{n-2}(n - 2)] - n]]$ garbage outputs, where n is the number of output bits.

Proof. The above statement is proved by mathematical induction.

In a quantum realization of 4-to-2 BCD encoder circuit, each input line produces $2^2 = 4$ garbage outputs, whereas quantum Toffoli gate produces $[2(\frac{2^2}{2} - 1) + 2^{2-2}(2 - 2)] - 2 = [2 \times 1 + 1 \times 0] - 2 = 0$ garbage outputs, as shown in figure 9.3. Thus a 4-to-2 BCD produces $G_{4\text{-to-}2} = G_{\text{inputlines}} + G_{\text{Toffoli}} = 2^2 + [2(\frac{2^2}{2} - 1) + 2^{2-2}(2 - 2) - 2] = 4 + [2 + 0 - 2] = 4$ garbage outputs.

Thus the statement holds for base case $n = 2$.

Assume that the statement holds for $n = k$. Thus a quantum 2^k-to-k BCD priority encoder circuit produces $[2^k + [[2(\frac{2^k}{2} - 1) + 2^{k-2}(k - 2)] - k]]$ garbage outputs.

A 2^{k+1}-to-$(k+1)$ BCD priority encoder circuit produces 2^{k+1} garbage outputs from input lines and $[[2(\frac{2}{2}^{k+1} - 1) + 2^{k+1-2}(k+1-2)] - (k+1)]$ garbage outputs from quantum Toffoli gates. As a result, the total number of garbage outputs produced for an 2^{k+1}-to-$(k+1)$ BCD priority encoder circuit is $[2^{k+1} + [2(\frac{2}{2}^{k+1} - 1) + 2^{k+1-2}(k+1-2) - (k+1)]]$.

Thus the statement holds for $n = k + 1$.

Therefore a quantum 2^n-to-n BCD priority encoder circuit can be realized with $[2^n + [[2(\frac{2}{2}^n - 1) + 2^{n-2}(n-2)] - n]]$ garbage outputs.

Example 9.3. When $n = 3$ in figure 9.4 the total garbage outputs of an 8-to-3 BCD priority encoder are $2^3 + [2(\frac{2}{2}^3 - 1) + 2^{3-2}(3-2)] - 3 = 8 + 8 - 3 = 13$.

Property 9.7. A quantum 2^n-to-n BCD priority encoder circuit requires $[2(\frac{2}{2}^n - 1) + 2^{n-2}(n-2)]$ constant inputs.

Proof. From property 9.1 it is found that a quantum BCD priority encoder circuit consists of quantum Toffoli gates and quantum NOT gates. Quantum Toffoli gates are used to perform the OR operation with one constant input. As a quantum 2^n-to-n quantum BCD priority encoder circuit requires $[2(\frac{2}{2}^n - 1) + 2^{n-2}(n-2)]$ quantum Toffoli gates, the total number of constant inputs is $[2(\frac{2}{2}^n - 1) + 2^{n-2}(n-2)]$.

Example 9.4. When $n = 3$ in figure 9.4, the total constant inputs of an 8-to-3 BCD priority encoder are $[2(\frac{2}{2}^3 - 1) + 2^{3-2}(3-2)] = 6 + 2 = 8$.

9.3 Summary

This chapter presents the design methodology of a 2^n-to-n quantum BCD priority encoder circuit, where n is the number of output bits. An algorithm has been presented for designing a compact 2^n-to-n quantum BCD priority encoder circuit. The circuit has been constructed with the optimum number of quantum gates, garbage outputs, constant inputs, delays, quantum gate calculation complexity, area, and power. The efficiency of the design has been proved by presenting several properties. The BCD encoder designed using quantum logic has the obvious advantage of low power dissipation, so it can be applied in wireless sensors, quantum signal processing, parallel circuits, etc.

Critical thinking questions

1. Outline the characteristics of the quantum encoder.
2. Describe briefly the characteristics of the quantum BCD priority encoder.
3. Solve the following problems using the circuit given below.

i. Find the overall latency, when $n = 5$.
ii. Find the overall complexity of the quantum gate calculations, when $n = 5$.
iii. Find the total garbage outputs, when $n = 5$.
iv. Calculate the total number of constant inputs, when $n = 5$.

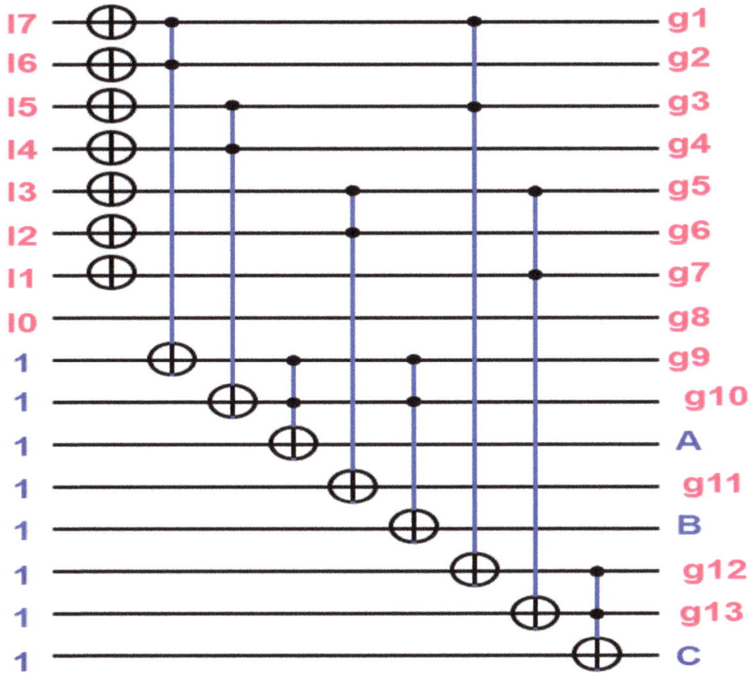

References

[1] Aspencore Priority encoder *Electronics Tutorial* https://www.electronics-tutorials.ws/combination/comb_4.html (Accessed: 4 December 2018)

[2] Abdel-Hafeez S and Harb S 2006 A VLSI high-performance priority encoder using standard CMOS library *IEEE Trans. Circuits Syst. II: Express Briefs* **53** 597–601

[3] Cleve R and Watrous J 2000 Fast parallel circuits for the quantum Fourier transform arXiv: quant-ph/0006004

[4] Delgado-Frias J G and Nyathi J 1998 A VLSI high-performance encoder with priority lookahead *Proc. 8th Great Lakes Symp. VLSI* (Piscataway, NJ: IEEE) pp 59–64

[5] Fsaifes I, Lepers C, Obaton A-F and Gallion P 2006 DS-OCDMA encoder/decoder performance analysis using optical low-coherence reflectometry *J. Lightwave Technol.* **24** 3121

[6] Lisa N J and Babu H M H 2014 Minimization of a reversible quantum 2*n*-to-*n* BCD priority encoder *IEEE/ACM Int. Symp. on Nanoscale Architectures (NANOARCH)* (Piscataway, NJ: IEEE) pp 77–82

[7] Voyiatzis I, Gizopoulos D and Paschalis A 2005 Accumulator-based test generation for robust sequential fault testing in DSP cores in near-optimal time *IEEE Trans. Very Large Scale Integr. Syst.* **13** 1079–86

[8] Wang J-C, Pang Y and Xia Y 2012 A BCD priority encoder designed by reversible logic *Wavelet Active Media Technology and Information Processing (ICWAMTIP), 2012 Int. Conf. on* (Piscataway, NJ: IEEE) pp 318–21

[9] Wang W, Swamy M N S and Ahmad M O 2002 A new architecture of RRNS error-correcting QC encoder/decoder and its FPGA implementation *IEEE Int. Symp. on Circuits and Systems* 5 (Piscataway, NJ: IEEE) p V

[10] Yunru L, Yeqing Y, Shaojie C and Lijun C 2011 Reversible watermarking algorithm for wireless sensor network using method of multi-node negotiation *Int. Conf. on E-Business and E-Government* (Piscataway, NJ: IEEE) pp 1–4

IOP Publishing

Quantum Computing (Second Edition)
A pathway to quantum logic design
Hafiz Md Hasan Babu

Chapter 10

The quantum decoder

Learning objectives

- Define a quantum decoder.
- Construct an n-to-$2n$ quantum decoder circuit.
- Discuss the characteristics of a quantum decoder.
- Describe the algorithm of an n-to-$2n$ quantum decoder.
- Discover how to construct a quantum decoder circuit.
- Explain the characteristics of a quantum decoder circuit.

Multiple valued logic is a promising approach to reduce the width of quantum circuits. Moreover, quaternary logic is considered a good choice for future quantum computing technology and hence it is very suitable for the encoded realization of binary logic functions through its grouping of two qubits together into quaternary values. The quaternary decoder is an essential unit of quaternary quantum systems.

A quantum decoder is a fundamental building block in many computing systems, which can take the form of a multiple-input, multiple-output logic circuit that converts coded inputs into coded outputs, where the input and output codes are different. Quantum decoders have been used in the memory system and input/output (I/O) of microprocessors.

In this chapter a useful quantum decoder circuit is designed. The quantum decoder circuit that consists of a quantum 2-to-4 decoder circuit and quantum Fredkin gates is analyzed.

10.1 The characteristics of a quantum decoder

The term 'decoder' means to translate or decode coded information from one format into another. Thus a quantum decoder transforms a set of quantum input signals into an equivalent output.

A quantum decoder is a quantum logic device that has inputs of two-qubit, three-qubit, or four-qubit codes depending on the number of data input lines. Thus a

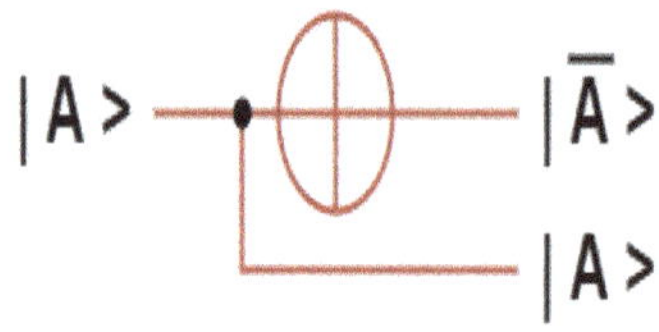

Figure 10.1. A quantum inverter circuit.

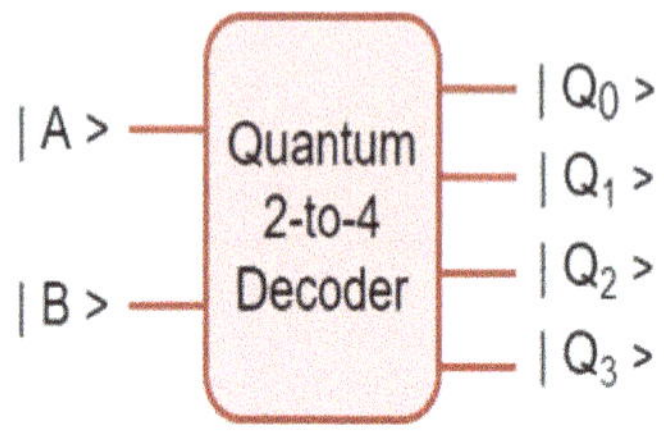

|A>	|B>	|Q0>	|Q1>	|Q2>	|Q3>
|0>	|0>	|1>	|0>	|0>	|0>
|0>	|1>	|0>	|1>	|0>	|0>
|1>	|0>	|0>	|0>	|1>	|0>
|1>	|1>	|0>	|0>	|0>	|1>

Figure 10.2. Block diagram of a quantum 2-to-4 decoder circuit and truth table.

quantum decoder that has a set of two or more qubits will be defined as having an n-qubit code, and therefore it has 2^n possible values. A quantum decoder generally decodes a quantum value.

If a quantum decoder receives n inputs, it activates only one of its 2^n outputs based on that input, where all other outputs are deactivated.

For example, an quantum inverter (quantum NOT gate) can be classified as a 1-to-2 quantum decoder as one input and two outputs (2^1) are possible, because with an input $|A\rangle$ it can produce two outputs $|A\rangle$ and $|\overline{A}\rangle$ (not-$|A\rangle$) as shown in figure 10.1.

A standard combinational logic decoder is an n-to-m decoder, where $m \leqslant 2^n$, and whose output Q is dependent only on its present input states. In other words, a quantum decoder looks at its current inputs, determines which binary code or binary number is present at its inputs, and selects the appropriate output which corresponds to that binary input.

The block diagram of a quantum 2-to-4 decoder circuit with its truth table is shown in figure 10.2. The two-qubit inputs labeled A and B are decoded into one of four outputs, where each output represents one of the miniterms of the two input variables (each output = a miniterm).

The inputs A and B determine which output line from Q_0 to Q_3 is HIGH with logic level 1 while the remaining outputs are LOW with logic 0. Thus only one output can be active (HIGH) at a time. Therefore, the output line with HIGH identifies its correspondence binary code located at the input. In other words it decodes the binary input.

Some quantum decoders have an additional input pin labeled 'enable' that controls the outputs from the device. This extra input allows the quantum decoder outputs to be turned ON or OFF as required. These types of quantum decoders are commonly used as 'memory address decoders' in microprocessor memory applications.

It can be said that a quantum decoder is a quantum demultiplexer with an additional data line that is used to enable the quantum decoder. An alternative way of looking at the quantum decoder circuit is to regard inputs A, B, and C as address signals. Each combination of A, B, or C defines a unique memory address.

10.2 The design of a quantum decoder

In this subsection a quantum n-to-2^n decoder circuit is shown. The optimized quantum n-to-2^n decoder circuit is described using a 2-to-4 quantum decoder and quantum Fredkin gates.

10.2.1 The quantum decoder circuit

A 2-to-4 decoder generates four logical AND functions $\acute{A}\acute{B}$, $\acute{A}B$, $A\acute{B}$, AB. The quantum 2-to-4 decoder design generates all four necessary AND function using one quantum Peres gate and three CNOT quantum gates, as shown in figure 10.3. A 3-to-8 decoder is designed using one 2-to-4 quantum decoder circuit and four quantum Fredkin gates, as shown in figure 10.4. Using the same approach, an n-to-2^n decoder can be designed. The design of a 4-to-16 decoder is presented in figure 10.5. The generalized algorithm for constructing a quantum n-to-2^n decoder circuit is presented in algorithm 10.1.

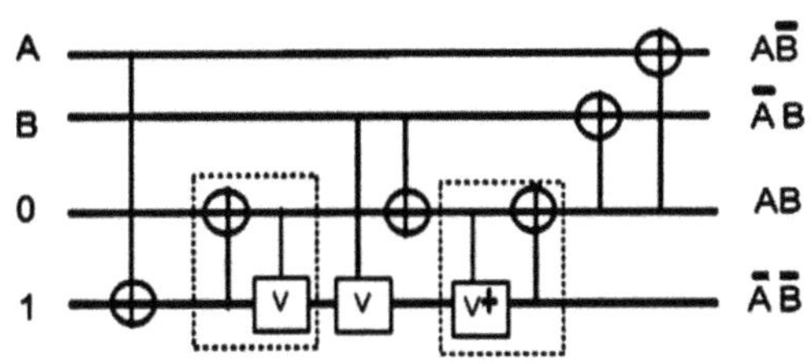

Figure 10.3. The quantum 2-to-4 decoder circuit. Reproduced with permission from [5]. Copyright 2013 IEEE.

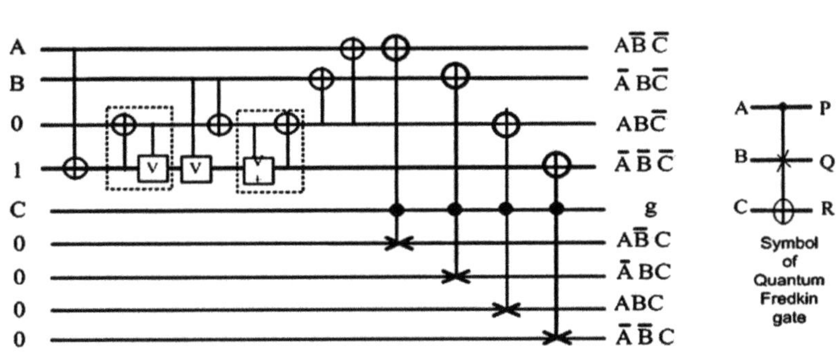

Figure 10.4. The quantum 3-to-8 decoder circuit. Reproduced with permission from [5]. Copyright 2013 IEEE.

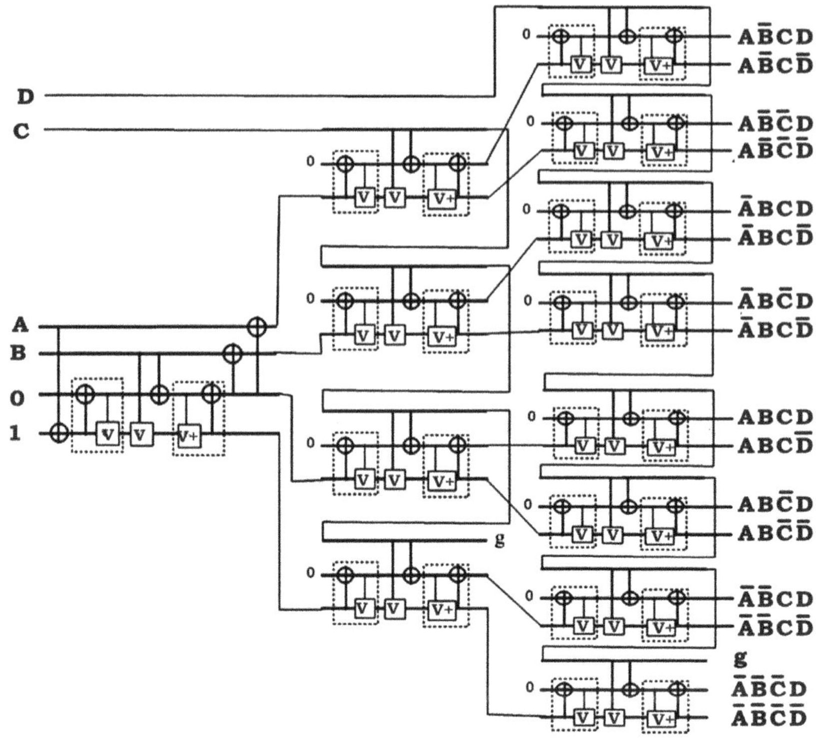

Figure 10.5. The quantum 4-to-16 decoder circuit.

INPUT: n-bit operand $A(A_n, A_{n-1}, \ldots, A_1)$ [here, n starts from 3]
OUTPUT: 2^n-bit operand $D(D_{2n}, D_{2n-1}, \ldots, D_1)$

1: START
2: **if** n/2=0 **then**
3: Design needs $3(\lceil 2^{n-5} \rceil)$ number of quantum *2-to-4*
4: decoder circuits and $(2^n/2)$ number of quantum Fredkin gates
5: **else**
6: Design needs $3(\lceil 2^{n-6} \rceil)$ number of quantum *2-to-4*
7: decoder circuits and $(2^n/2)$ number of quantum Fredkin gates
8: **end if**
9: END

Algorithm 10.1. Constructing an n-to-2^n quantum decoder circuit.

10.2.2 Analysis of the properties of the circuits

In this section the properties of a quantum n-to-2^n decoder circuit are presented.

Property 10.1. A quantum n-to-2^n decoder circuit can be realized with $[7 \times (3 \times \lceil 2^{n-5} \rceil) + 5 \times (2^n/2)]$ quantum gates, when $n \geqslant 4$ and $n/2 = 2$, where n is the number of input bits.

Proof. The above statement is proved by mathematical induction. A 2-to-4 decoder circuit requires one quantum Peres gate and three quantum CNOT gates. Thus the number of quantum gates of a 2-to-4 decoder circuit is seven, where the numbers of quantum Peres gates and quantum CNOT gates are 4 and 1, respectively. A 4-to-16 decoder circuit consists of three 2-to-4 decoder circuits and eight quantum Fredkin gates, where the number of quantum Fredkin gates is 5. Thus the total number of quantum gates required to construct 4-to-16 decoder circuit is $NOQG_{4\text{-to-}16} = 5 \times NOQG_{Fredkin} + 7 \times NOQG_{2\text{-to-}4} = 5 \times (2^4/2) + 7 \times (3 \times \lceil 2^{4-5} \rceil) = 61 = 5 \times (2^4/2) + 7 \times (3 \times \lceil 2^{4-5} \rceil)$. Thus the statement holds for base case $n = 4$. Assume that the statement holds for $n = k$. Thus a quantum k- to-2^k decoder circuit can be realized with quantum gates when $k \geqslant 4$ and $k/2 = 2$.

A $(k + 1)$-to-2^{k+1} decoder circuit requires $7 \times (3 \times (\lceil 2^{k+1-5} \rceil))$ quantum 2-to-4 decoders and quantum Fredkin gates. As a result, the total number of quantum gates required to construct a $(k + 1)$-to-2^{k+1} decoder circuit is $(7 \times (3 \times \lceil 2^{k+1-5} \rceil) + 5 \times (2^{k+1} + 2^n/2))$ when $k \geqslant 4$ and $k/2 = 2$. Thus the statement holds for $n = k + 1$. Therefore a quantum n-to-2^n decoder circuit can be realized with $[7 \times (3 \times \lceil 2^{n-5} \rceil) + 5 \times (2^n/2)]$ quantum gates when $n \geqslant 4$ and $n/2 = 2$.

Similarly, a quantum n-to-2^n decoder circuit can be realized with $[7 \times (3 \times \lceil 2^{n-6} \rceil) + 5 \times (2^n/2)]$ quantum gates when $n \geqslant 4$ and $n/2 \neq 2$.

Property 10.2. A quantum n-to-2^n decoder circuit produces $[7 \times (3 \times \lceil 2^{n-5} \rceil) + 5 \times (2^n/2)]\Delta$ quantum delay when $n \geqslant 4$ and $n/2 = 2$, where Δ is the unit delay.

Proof. From property 10.1 it is found that the number of quantum gates of an n-to-2^n decoder circuit are $[7 \times (3 \times \lceil 2^{n-5} \rceil) + 5 \times (2^n/2)]$ when $n \geqslant 4$ and $n/2 = 2$. Therefore the total delay of a quantum n-to-2^n decoder circuit is $[7 \times (3 \times \lceil 2^{n-5} \rceil) + 5 \times (2^n/2)]\Delta$ quantum delay when $n \geqslant 4$ and $n/2 = 2$.

Similarly, a quantum n-to-2^n decoder circuit produces $[7 \times (3 \times \lceil 2^{n-6} \rceil) +5 \times (2^n/2)]\Delta$ quantum delay when $n \geqslant 4$ and $n/2 \neq 2$.

Property 10.3. A quantum n-to-2^n decoder circuit requires $350 \times (3 \times \lceil 2^{n-5} \rceil) + 250 \times (2^n/2)$ Å area when $n \geqslant 4$ and $n/2 = 2$, where Å is the unit of measuring area.

Proof. From property 10.1 a quantum 2-to-4 decoder has seven quantum gates and a quantum Fredkin gate has five quantum gates. Thus a quantum n-to-2^n decoder circuit requires $[7 \times (3 \times \lceil 2^{n-5} \rceil) + 5 \times (2^n/2)]$ quantum gates when $n \geqslant 4$ and $n/2 = 2$. Therefore the total area for a quantum n-to-2^n decoder circuit is $[50 \times 7 \times (3 \times \lceil 2^{n-5} \rceil) + 5 \times (2^n/2)] = 350 \times (3 \times \lceil 2^{n-5} \rceil) + 250 \times (2^n/2)$ Å when $n \geqslant 4$ and $n/2 = 2$.

Similarly, a quantum n-to-2^n decoder circuit requires $350 \times (3 \times \lceil 2^{n-6} \rceil) + 250 \times (2^n/2)$ Å area when $n \geqslant 4$ and $n/2 \neq 2$, where Å is the unit of measuring area.

Property 10.4. A quantum n-to-2^n decoder circuit requires $[996.1 \times (3 \times \lceil 2^{n-5} \rceil) + 711.5 \times (2^n/2)]$ meV power when $n \geqslant 4$ and $n/2 = 2$, where meV is the unit of measuring power.

Proof. From property 10.1 a 2-to-4 quantum decoder has seven quantum gates and a quantum Fredkin gate has five quantum gates. Thus a quantum n-to-2^n decoder circuit requires $[7 \times (3 \times \lceil 2^{n-5} \rceil) + 5 \times (2^n/2)]$ quantum gates when $n \geqslant 4$ and $n/2 = 2$. Therefore the total power for a quantum n-to-2^n decoder circuit is $(142.3 \times [7(3 \times \lceil 2^{n-5} \rceil) + 5 \times (2^n/2)])$ meV power when $n \geqslant 4$ and $n/2 = 2$, where meV is the unit of measuring power.

Similarly, a quantum n-to-2^n decoder circuit requires $[996.1 \times (3 \times \lceil 2^{n-6} \rceil) +711.5 \times (2^n/2)]$ meV power when $n \geqslant 4$ and $n/2 \neq 2$, where meV is the unit of measuring power. □

Property 10.5. A quantum n-to-2^n decoder circuit can be realized with $(2(3 \times \lceil 2^{n-5} \rceil) + (2^n/2))$ constant inputs when $n \geqslant 4$ and $n/2 = 2$, where n is the number of input bits.

Proof. The above statement is proved by mathematical induction. The quantum 2-to-4 decoder circuit requires two constant inputs and a quantum Fredkin gate requires one constant input. A 4-to-16 decoder circuit requires six constant inputs from three 2-to-4 decoder circuits and eight constant inputs from eight quantum Fredkin gates. Thus the total number of constant inputs required to construct a 4-to-16 decoder circuit ($NOQG_{4\text{-to-}16}$) is $NOCI_{4\text{-to-}16} = NOCI_{Fredkin} + NOCI_{2\text{-to-}4} = (2^4/2) + 2 \times (3 \times \lceil 2^{4-5} \rceil) = 14 = (2^4/2) + 2 \times (3 \times \lceil 2^{4-5} \rceil)$. Thus the statement holds for base case $n = 4$. Assume that the statement holds for $n = k$. Thus a quantum k-to-2^k decoder circuit can be realized with $(2(3 \times \lceil 2^{k-5} \rceil) + (2^k/2))$ constant inputs when $k \geqslant 4$ and $k/2 = 2$.

A $(k+ 1)$-to-2^{k+1} decoder circuit requires $(2^{k+1}/2)$ constant inputs for quantum Fredkin gates and $(2(3 \times \lceil 2^{k+1-5} \rceil))$ constant inputs for quantum 2-to-4 decoders. As a result, the total number of constant inputs required to construct a $(k + 1)$-to-2^{k+1} decoder circuit is $(2(3 \times \lceil 2^{k+1-5} \rceil) + (2^{k+1}/2))$ when $n \geqslant 4$ and $n/2 = 2$. Thus the statement holds for $n = k + 1$. Therefore a quantum n-to-2^n decoder circuit can be realized with $(2(3 \times \lceil 2^{n-5} \rceil) + (2^n/2))$ constant inputs when $n \geqslant 4$ and $n/2 = 2$, where n is the number of input bits.

Similarly, a quantum n-to-2^n decoder circuit can be realized with $(2(3 \times \lceil 2^{n-6} \rceil) + (2^n/2))$ constant inputs when $n \geqslant 4$ and $n/2 \neq 2$, where n is the number of input bits. □

Property 10.6. A quantum n-to-2^n decoder circuit requires $(3\Omega + 4\sigma)[3 \times \lceil 2^{n-5} \rceil) + (2^n/2)]$ quantum gate calculation complexity when $n \geqslant 4$ and $n/2 = 2$, where σ is the CNOT gate calculation complexity and Ω is the controlled-V or controlled-V$^+$ gate calculation complexity.

Proof. From property 10.1, a quantum 2-to-4 decoder and a quantum Fredkin gate both have 4σ CNOT quantum gate calculation complexity and 3Ω controlled-V (controlled-V$^+$) quantum gate calculation complexity. Thus the total quantum gate calculation complexity for an n-to-2^n decoder circuit is $(3\Omega + 4\sigma)[3 \times \lceil 2^{n-5} \rceil) + (2^n/2)]$ quantum gate calculation complexity when $n \geqslant 4$ and $n/2 = 2$, where σ is the CNOT gate calculation complexity and Ω is the controlled-V or controlled- V$^+$ gate calculation complexity.

Similarly, a quantum n-to-2^n decoder circuit requires $(3\Omega + 4\sigma)[3 \times \lceil 2^{n-6} \rceil) + (2^n/2)]$ quantum gate calculation complexity, when $n \geqslant 4$ and $n/2 \neq 2$, where σ is the CNOT gate calculation complexity and Ω is the controlled-V or controlled-V$^+$ gate calculation complexity.

10.3 Summary

This chapter presents the design methodology of an n-to-2^n quantum decoder circuit, where n is the number of input bits. A technique to calculate the quantum gate complexity of quantum circuits has been shown. The efficiency of the design is proved by several properties. The decoder designed using quantum logic has obvious advantages of low power dissipation. Thus it can be applied very well in wireless sensors, network components, and quantum signal processing, etc.

Critical thinking questions

1. Explain the properties of a quantum decoder.
2. How many 3 × 8-line quantum decoders with an enable input line are needed to construct a 6 × 64-line quantum decoder without using any additional logic gates? Explain in detail.
3. How many logic gates are needed to represent the output of a 2-to-4 quantum decoder? Use a figure to explain.
4. What will happen if in a 2-to-4 quantum decoder, the enable pin is set to 1 and the inputs A and B are set to 0 and 0, respectively? Describe briefly.
5. Describe the applications of a quantum decoder.

References

[1] Aspencore Binary decoder *Electronics Tutorials* https://www.electronics-tutorials.ws/combination/comb_5.html (Accessed: 5 December 2018)

[2] Weisstein E W Square root *Wolfram MathWorld* http://mathworld.wolfram.com/SquareRoot.html (Accessed: 22 December 2018)

[3] Chrzanowska-Jeske M 1993 Architecture and synthesis issues in FPGAs *Proc. NORTHCON'93 Electrical and Electronics Convention* pp 102–5

[4] Haghparast M and Monfared A T 2017 Novel quaternary quantum decoder, multiplexer and demultiplexer circuits *Int. J. Theor. Phys.* **56** 1694–707

[5] Lisa N J and Babu H M H 2013 A compact realization of a reversible quantum *n*-to-2*n* decoder *4th Int. Conf. on Electronics Information and Emergency Communication* (Piscataway, NJ: IEEE) pp 90–3

[6] Sharmin F, Polash M M A, Shamsujjoha M, Jamal L and Babu H M H 2011 Design of a compact reversible random access memory *4th IEEE Int. Conf. on Computer Science and Information Technology* **10** 103–7

[7] Voyiatzis I, Gizopoulos D and Paschalis A M 2005 Accumulator-based test generation for robust sequential fault testing in DSP cores in near-optimal time *IEEE Trans. Very Large Scale Integr. VLSI Syst.* **13** 1079–86

[8] Yunru L, Yeqing Y, Shaojie C and Lijun C 2011 Reversible watermarking algorithm for wireless sensor network using method of multi-node negotiation *Int. Conf. on E-Business and E-Government* (Piscataway, NJ: IEEE) pp 1–4

IOP Publishing

Quantum Computing (Second Edition)

A pathway to quantum logic design

Hafiz Md Hasan Babu

Chapter 11

The quantum square root circuit

Learning objectives

- Explain the square root function.
- Develop an algorithm for an n-qubit quantum square root.
- Understand the characteristics of the square root function.
- Make a circuit analysis of the quantum square root.
- Learn how to create a quantum square root circuit design.
- Determine the complexity of the quantum gate easily.

A square root of x is a number r such that $r^2 = x$. It is also written in the form $x^{1/2}$ or, in particular, $\sqrt{x}$. This chapter is focused on identifying a useful quantum square root circuit. The quantum conditions and design of a quantum square root circuit are analyzed. These circuits consist of quantum adder/subtractor circuits and CNOT gates.

11.1 The properties of a quantum square root function

The principal of a quantum square root function $f(x) = \sqrt{x}$ (usually just referred to as the 'square root function') is a function that maps the set of nonnegative real numbers onto itself. In geometrical terms, the square root function maps the area of a square to its side length.

The quantum square root of x is rational if and only if x is a rational number that can be represented as a ratio of two perfect squares. The square root function maps rational numbers into algebraic numbers (a superset of the rational numbers).

For all real numbers

$$\sqrt{x^2} = |x| = x = \begin{matrix} x & \text{if } x \geqslant 0 \\ -x & \text{if } x < 0. \end{matrix}$$

For all nonnegative real numbers x and y, $\sqrt{xy} = \sqrt{x}\sqrt{y}$ and $\sqrt{x} = x^{\frac{1}{2}}$.

The quantum square root function is continuous for all nonnegative x and differentiable for all positive x. If f denotes the square root function, its derivative is given by

$$f'(x) = \frac{1}{2\sqrt{x}}.$$

The Taylor series of $\sqrt{1} + x$ about $x = 0$ converges for $|x| \leqslant 1$:

$$\sqrt{1 + x} = \sum_{n=0}^{\infty} \frac{(-1)^n(2n)!}{(1 - 2n)(n!)(4^n)} x^n = 1 + \frac{1}{2}x - \frac{1}{8}x^2 + \frac{1}{16}x^3 - \frac{5}{128}x^4 + \ldots$$

The quantum square root of a nonnegative number is used in the definition of the Euclidean norm (and distance), as well as in generalizations such as Hilbert spaces. It defines an important concept of standard deviation using the theory of probability and statistics. It has a major use in the formula for roots of a quadratic equation, such as quadratic fields and rings of quadratic integers which are based on square roots. These are important in algebra and have uses in geometry. Square roots frequently appear in mathematical formulas elsewhere, as well as in many physical laws.

11.2 The design of a quantum square root circuit

In this section an n-bit quantum square root circuit is shown. The optimized n-bit quantum square root circuit is designed using QAS circuits and CNOT gates.

11.2.1 The quantum adder/subtractor circuit

A QAS circuit is constructed using a Haghparast–Navi gate (HNG) quantum circuit and CNOT quantum gates. It is constructed directly at the quantum level. A one-bit QAS circuit is a combinational quantum circuit which adds three bits X, Y, Z (where Z is the carry-in for the circuit) and generates a sum $S(X \oplus Y \oplus Z)$ and a carry-out C_{out}, $((X \oplus Y)Z \oplus XY)$, or subtracts $X - Y - Z$ resulting in a difference D, $(X \oplus Y \oplus Z \oplus 1)$, and a borrow B_{out}, $((X \oplus Y)Z \oplus XY \oplus 1)$. Thus the inputs to this circuit are data signals, X, Y, and Z, a control signal A/S and a constant input bit 0. The outputs are sum, carry, A/Sg, which is an A/S control signal propagating to the next block, and two garbage outputs g_1 and g_2. A CNOT quantum gate with a control signal A/S and a target Y, as well as a full-adder implemented by a HNG quantum circuit, contribute to the QAS block. When A/S is set to 0, then CNOT passes the true copy of Y. When A/S is 1, then Y' is produced at the output of CNOT, and $X + Y' + Z$ is performed, where $'$ denotes the complement sign. The A/Sg output of each block is used as a control signal A/S to the consecutive block. To obtain a true copy of input Y for further operation, another CNOT gate is added between lines A/S and $A/S \oplus Y$ with control at A/Sg, which performs the operation $A/S \oplus Y \oplus A/S = Y$. The structure of a QAS circuit is shown in figure 11.1.

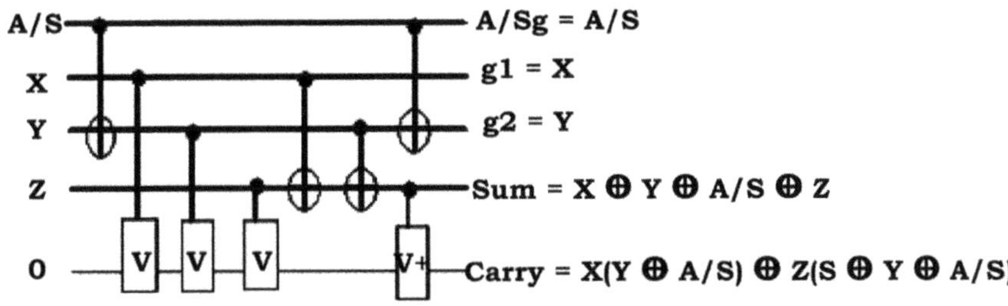

Figure 11.1. The QAS circuit.

11.2.2 The quantum square root circuit

An array structure of a non-restoring square root circuit uses classical controlled adder/subtractor blocks. The $(2 \times 4) = 8$-bit non-restoring circuit realizes a digit-by-digit scheme, where at each iteration computed in each row, only one digit of the square root is performed. Based on this structure the quantum square root circuit is created using the QAS circuit presented in figure 11.1. A few additional CNOT gates are added to provide fan-out signals. In the quantum square root circuit, the QAS circuit is incorporated from figure 11.1 and reuses the copy of input signal Y of the previous stage for the next stage square root operation. A $(2 \times 4) = 8$-bit quantum square root circuit with the input $x_1x_2x_3x_4x_5x_6x_7x_8$, the four-bit square root output $q_1q_2q_3q_4$, and the eight-bit remainder output $r_1r_2r_3r_4r_5r_6r_7r_8$ are presented in figure 11.2. The order of the QAS signal propagation is in the direction from right to left. Hence the right-most QAS2 circuit in the top row of figure 11.2 takes the control input A/S. In the classical design, first the A/S signal is set to '1' in order to calculate the first digit of the square root. A/S input to 1 is also set. The outputs A/Sg and C_{out} of the QAS2 circuit are connected to the left QAS1 circuit. To provide the inverted A/S as well as A/S itself, each inverter in a classical implementation is replaced by a CNOT. In addition, since each square root bit from each row (q_i) is the control signal for the next row as shown in figure 11.2, it is necessary to fan-out q_i. In quantum implementation a CNOT gate is used to generate a copy of a required signal. For example, to obtain a copy of q_1, a CNOT gate is also used with the I/O mapping: $q_1, 0 \rightarrow q_1, q_1 \oplus 0$.

A two-qubit quantum square root circuit with four-bit input $x_1x_2x_3x_4$, two-qubit square root output q_1q_2, and four-qubit remainder output $r_1r_2r_3r_4$ is constructed using six QAS circuits and six CNOT gates arranged in two rows. Therefore, an n-qubit quantum square root circuit with the $2n$-qubit input $x_1\ x_2, \ldots, x_{2n}$, the n-qubit square root output $q_1q_2, \ldots, q_n$ and the $2n$-qubit remainder output $r_1r_2, \ldots, r_{2n}$ can be constructed using $[\sum_{i=1}^{n}(i+2) - 1]$ QAS circuits and $4n - 2$ CNOT gates arranged in n rows, presented in figure 11.3, where n is the number of output qubits. In addition, $4n - 2$ CNOT gates are used to control the fan-out signals. The generalized algorithm for constructing a compact n-qubit quantum square root circuit is presented in algorithm 11.1.

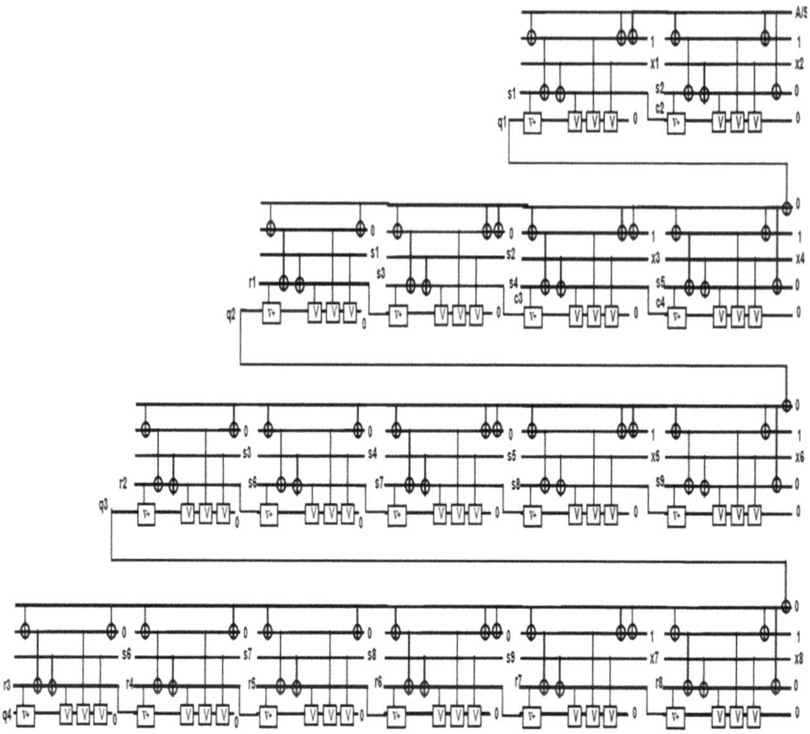

Figure 11.2. The (2 × 4) = 8-qubit quantum square root circuit.

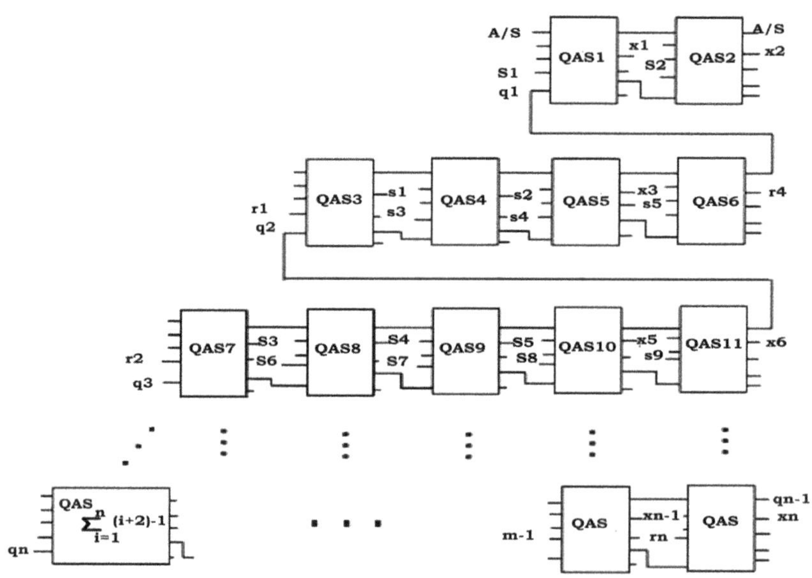

Figure 11.3. Block diagram of an *n*-qubit quantum square root circuit.

INPUT: $2n$-qubit operand X
OUTPUT: n-qubit quotient q

```
START
for i=1 to ∑_{k=1}^{n} (k+2) − 1 do
  if i = 1 then
    for j=2i to 1 do
      if j>1 then
        Input:={A/S,1,X_j,0,0} and Output:={A/S, g, g, S_j,C_j}
      else
        Input:={A/S,1,X_j,C_{j+1},0} and Output:={A/S, g, g, r_j,q_j}
      end if
    end for
  else if i>1 then
    for j=2i to 1 do
      if j=2i then
        Input:={A/S,1,X_j,0,0} and Output:={A/S, g, g, S_j,C_j}
      else if j>2i − 2 then
        Input:={A/S,1,X_j,C_{j+1},0} and Output:={A/S, g, g, S_j,C_j}
      else if j>1 then
        Input:={A/S,0,S_{j−2i},C_{j+1},0} and Output:={A/S, g, g, S_j,C_j}
      else
        Input:={A/S,0,S_{j−2i},C_{j+1},0} and Output:={A/S, g, g, r_j,q_j}
      end if
    end for
  end if
end for
END
```

Algorithm 11.1. Constructing an n-qubit quantum square root circuit.

11.2.3 Analysis of the properties of the quantum circuit

In this subsection the properties of an n-qubit quantum square root circuit are presented.

Property 11.1. An n-qubit quantum square root circuit can be realized with $[8[\sum_{i=1}^{n}(i + 2) - 1] + 4n - 2]$ quantum gates, where n is the number of output bits.

Proof. Property 11.1 is proved by mathematical induction.

A two-qubit square root requires six QAS circuits where the number of quantum gates of the QAS circuit is eight. Moreover, this circuit needs an extra six CNOT gates. Thus the total number of quantum gates required to construct a two-qubit quantum square root (NOG_2) circuit is

$$\begin{aligned} NOG_2 &= \sum_{i=1}^{2}(i + 2) - 1] + (4 \times 2 - 2) \\ &= 8 \times [(1 + 2) + (2 + 2) - 1] + 6 \\ &= 8 \times [3 + 4 - 1] + 6 \\ &= 54. \end{aligned}$$

Thus the statement holds for base case $n = 2$.

Assume that the statement holds for $n = k$. Thus a quantum k-qubit square root circuit can be realized with $[8[\sum_{i=1}^{k}(i+2) - 1] + 4k - 2]$ quantum gates.

A $(k + 1)$-qubit quantum square root circuit requires $[8[\sum_{i=1}^{k+1}(i+2) - 1]]$ QAS circuits and $[4(k+1) - 2]$ CNOT gates. As a result, the total number of quantum gates required to construct a $(k + 1)$-qubit quantum square root circuit is $[8[\sum_{i=1}^{k+1}(i+2) - 1] + 4(k+1) - 2]$.

Thus the statement holds for $n = k + 1$.

An n-qubit quantum square root circuit can be realized with $8[\sum_{i=1}^{n}(i+2) - 1] + 4n - 2$ quantum gates, where n is the number of output bits.

Property 11.2. An n-qubit quantum square root circuit produces the following quantum delay:

$(8[\sum_{i=1}^{n}(i+2) - 1] + 4n - 2 - [\sum_{i=1}^{n}(i+2) - 1])\Delta$, where n is the number of output qubits and Δ is the unit of delay.

Proof. A quantum square root circuit consists of QAS circuits and CNOT gates. In a one-qubit QAS circuit, 8Δ delay is required. Thus the total delay for an n-qubit QAS circuit is $(8[\sum_{i=1}^{n}(i+2) - 1])\Delta$ delay. A quantum square root circuit requires $(4n - 2)\Delta$ additional delay. As there are $[\sum_{i=1}^{n}(i+2) - 1]$ controlled-V $^+$ gates at the same level with a CNOT gate of an n-qubit quantum square root circuit, $[\sum_{i=1}^{n}(i+2) - 1]\Delta$ delay can be deducted from the total delay of the quantum square root circuit. Therefore, the total delay of a quantum square root circuit is

$(8[\sum_{i=1}^{n}(i+2) - 1] + 4n - 2 - [\sum_{i=1}^{n}(i+2) - 1])\Delta$.

Example 11.1. When $n = 4$ in figure 11.1, the total delay of a (2×4)-qubit = 8-bit quantum square root circuit is $(8[\sum_{i=1}^{4}(i+2) - 1] + 4 \times 4 - 2 - [\sum_{i=1}^{4}(i+2) - 1])$ $\Delta = 133\Delta$.

Property 11.3. An n-qubit quantum square root circuit requires $[(4 + 4)[\sum_{i=1}^{n}(i+2) - 1] + 4n - 2]$ quantum gate calculation complexity, where n is the number of output qubits, σ is CNOT gate calculation complexity, and Ω is the controlled-V or controlled-V$^+$ gate calculation complexity.

Proof. From property 11.1, it is found that a quantum square root circuit consists of a QAS circuit and CNOT gates. A one-bit QAS circuit has eight quantum gates. Thus an n-qubit quantum square root circuit requires $8[\sum_{i=1}^{n}(i+2) - 1]$ QAS circuits. Moreover, a quantum square root circuit needs an extra $4n - 2$ CNOT gates. Therefore, the total area for an n-bit quantum square root circuit is $((8[\sum_{i=1}^{n}(i+2) - 1] \times 50) + (4n - 2) \times 50))$Å $= 400[\sum_{i=1}^{n}(i+2) - 1] + 200n - 100$ Å.

Property 11.4. An n-qubit quantum square root circuit produces $1138.4[\sum_{i=1}^{n}(i+2)-1]+569.2n-284.6$ meV, where n is the number of output bits and meV is the unit of measuring power.

Proof. From property 11.1, it is found that a quantum square root circuit consists of a QAS circuit and CNOT gates. A one-bit QAS circuit has eight quantum gates. Thus an n-qubit quantum square root circuit has $8[\sum_{i=1}^{n}(i+2)-1]$ QAS circuits. Moreover, a quantum square root circuit needs an extra $4n-2$ CNOT gates. Therefore, the total power for an n-qubit quantum square root circuit is

$$\left(\left(8\left[\sum_{i=1}^{n}(i+2)-1\right]\times 142.3\right)+(4n-2)\times 142.3\right)\text{meV}$$

$$=1138.4\left[\sum_{i=1}^{n}(i+2)-1\right]+569.2n-284.6\text{ meV}.$$

11.3 Summary

This chapter presents a design methodology of an n-qubit quantum square root circuit, where n is the number of output qubits. An algorithm is shown for design the compact n-qubit quantum square root circuit using a quantum adder/subtractor circuit. A new technique to calculate the quantum gate complexity of quantum circuits is shown in this chapter. The circuit has been constructed with the optimum number of quantum gates, garbage outputs, delays, quantum gate calculation complexity, area, and power. The design is more scalable and performs much better than its counterparts, e.g. the n-qubit quantum square root circuit requires $(8[\sum_{i=1}^{n}(i+2)-1]+4n-2)$ quantum gates, $(8[\sum_{i=1}^{n}(i+2)-1]+4n-2-[\sum_{i=1}^{n}(i+2)-1])\Delta$ delay, $((4\Omega+4\sigma)[\sum_{i=1}^{n}(i+2)-1]+4n-2)$ quantum gate calculation complexity, $(400\left[\sum_{i=1}^{n}(i+2)-1\right]+200n-100\right)A^{o}$ area, and $(1138.4[\sum_{i=1}^{n}(i+2)-1]+569.2n-284.6)$ meV power, where n is the number of output qubits, σ is the CNOT gate calculation complexity, Ω is the controlled-V (controlled-V^{+}) gate calculation complexity, ρ is the NOT gate calculation complexity, Δ is the unit delay, Å is the unit of measuring area, and meV is the unit of measuring power. The efficiency of the design is also proved by several properties. The square root is the basic arithmetic unit of a computer system. For example, numerical analysis, complex number computations, statistical analysis, computer graphics, and radar signal processing are among the fields where quantum square root circuits can be used.

Critical thinking questions

1. Describe the characteristics of the quantum square root function.
2. Discuss some applications for the quantum adder/subtractor circuit.
3. Analyze some features of the n-qubit square root circuit.

4. From the circuit shown in figure below, calculate the total delay, when $n = 4$.

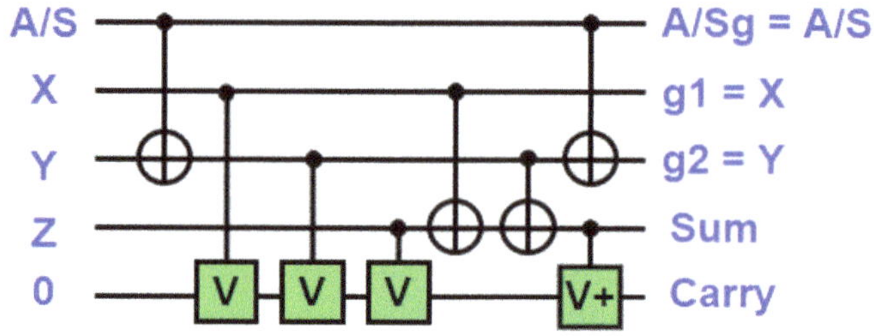

References

[1] Square root *Wikipedia* https://en.wikipedia.org/wiki/Square_root (Accessed: 6 December 2018)

[2] Semiconductor Industry Association 2006 International technology roadmap for semiconductors, 2005 edition ITRS http://www.itrs.net/

[3] Balandin A A and Wang K L 1999 Implementation of quantum controlled-NOT gates using asymmetric semiconductor quantum dots *Quantum Computing and Quantum Communications* (Berlin: Springer) pp 460–7

[4] Black P E, Kuhn D R and Williams C J 2002 Quantum computing and communication *Advances in Computers* vol 56 (Amsterdam: Elsevier) pp 189–244

[5] Gandhi S M, Devishree J and Mohan S S 2014 A new reversible SMG gate and its application for designing two's complement adder/subtractor with overflow detection logic for quantum computer-based systems *Computational Intelligence, Cyber Security and Computational Models* (Berlin: Springer) pp 259–66

[6] Hamacher V C, Vranesic Z G, Zaky G, Vransic Z and Zakay S 1996 *Computer Organization* pp 224–238 (New York: McGraw-Hill)

[7] Jain V K and Lin L 1998 Nonlinear DSP coprocessor cells-one and two cycle chips *Proc. 1998 IEEE Int. Symp. on Circuits and Systems* vol 2 (Piscataway, NJ: IEEE) pp 264–7

[8] Samavi S, Sadrabadi A and Fanian A 2008 Modular array structure for non-restoring square root circuit *J. Syst. Archit.* **54** 957–66

[9] Shor P W 1999 Polynomial-time algorithms for prime factorization and discrete logarithms on a quantum computer *SIAM Rev.* **41** 303–32

[10] Sultana S and Radecka K 2011 Reversible implementation of square-root circuit *18th IEEE Int. Conf. on Electronics, Circuits and Systems* (Piscataway, NJ: IEEE) pp 141–4

[11] Sultana S, Radecka K and Pang Y 2011 A study on relating redundancy removal in classical circuits to reversible mapping *29th IEEE Int. Conf. on Computer Design* (Piscataway, NJ: IEEE) pp 206–11

IOP Publishing

Quantum Computing (Second Edition)

A pathway to quantum logic design

Hafiz Md Hasan Babu

Chapter 12

Quantum latches and counter circuits

Learning objectives

- Define a latch with a brief description.
- Understand the uses of latches.
- Describe the properties of latches.
- Discuss the characteristics of quantum counter circuits.
- Find out ways to construct quantum SR latch, quantum D latch, quantum T latch, and quantum J–K latch circuits.
- Construct a synchronous and an asynchronous quantum counter circuit with descriptions.

A quantum latch is a circuit that has two stable states and can be used to store state information. It is the basic storage element in sequential logic. A quantum counter is a quantum register which produces a specified output sequence or counts the clock pulses arriving at the clock input. It has a wide variety of applications in almost all fields, such as industrial, domestic, security, surveillance, and communication, to name just a few. The properties along with the designs for the quantum latches and quantum counters are discussed in this chapter.

12.1 The properties of quantum latches

The current model of computing is based on the concept of a state machine. Such a machine exhibits different behaviors depending on the value of its current state. Thus some structure is required to keep track of this state. The basic building block used in memory of current computers is the quantum flip-flop, which themselves are built out of quantum latches. There are various types of quantum latches, but the basic type upon which other designs are generally based is the quantum set–reset (SR) latch. The primary inputs of such a structure are $|S\rangle$ (set) and $|R\rangle$ (reset). A diagram illustrating a quantum NOR gate implementation of a block diagram of a quantum SR latch is given in figure 12.1.

doi:10.1088/978-0-7503-5683-1ch12

Table 12.1. The next state values for the quantum SR latch.

Inputs		Next state	
S	R	Q^{+}	$\overline{Q^{+}}$
0	0	Q	$\overline{Q}$
0	1	0	1
1	0	0	1
1	1	Not permitted	

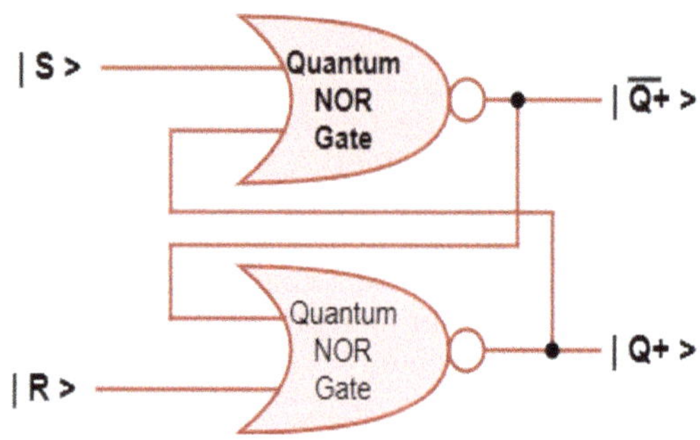

Figure 12.1. The block diagram of a quantum SR latch.

A variety of quantum latches has been introduced in several research works. This chapter examines the behavior of some of these quantum latches. Quantum realizations of latches are also included in this chapter. These quantum latches may be used to emulate a traditional SR latch. The behavior of the quantum SR latch is characterized by the truth table in table 12.1.

12.2 The design of quantum latches

In this section, some quantum latches are shown. Sections 12.2.1–12.2.4 describe the quantum SR latch, quantum D latch, quantum T latch, and quantum J–K latch, respectively.

12.2.1 The quantum SR latch

The quantum SR latch can be designed using two cross-coupled NOR gates or two cross-couple NAND gates. The S input sets the latch to 1, while the R input resets the quantum latch to 0. The cross-coupled NOR gate strategy was used to design the Fredkin gate based quantum SR latch and the cross-coupled NAND gate strategy was used to design the quantum Toffoli gate based quantum SR latch. The modified Toffoli gate (MTG) based quantum SR latch was also introduced. These quantum SR latch designs do not have an enable signal (clock).

Table 12.2. The modified truth table of the quantum SR latch.

S	R	Q	Q^+
0	0	0	0
0	0	1	1
0	1	0	0
0	1	1	0
1	0	0	1
1	0	1	1
1	1	0	0
1	1	1	1

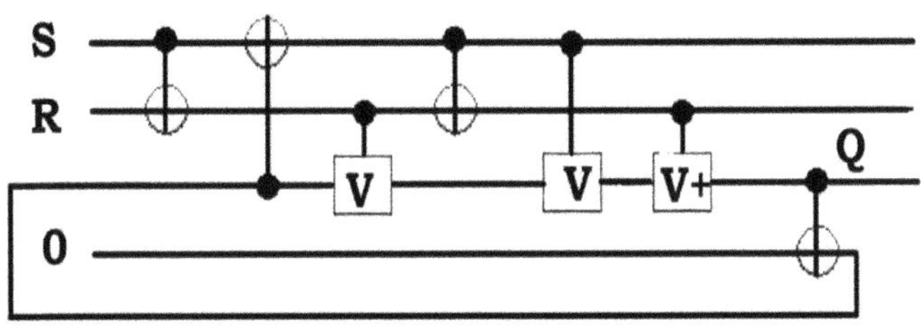

Figure 12.2. The quantum SR latch.

The characteristic equation of the quantum SR latch can be written as $Q^+ = \mathrm{S} + \overline{R} \cdot Q$. From the characteristic equation it is observed that for five-input combinations ($S = 0$, $R = 0$, $Q = 1$), ($S = 1$, $R = 0$, $Q = 0$), ($S = 1$, $R = 0$, $Q = 1$), ($S = 1$, $R = 1$, $Q = 0$), and ($S = 1$, $R = 1$, $Q = 1$), it will be $Q^+ = 1$. In order to design a quantum SR latch that can work for all input combinations, the output Q is modified for the selective input combinations as follows. When ($S = 1$, $R = 1$, $Q = 0$) the output Q^+ is assigned the value of 0 and when ($S = 1$, $R = 1$, $Q = 1$) the output Q^+ is assigned the value of 1. Thus when $S = 1$ and $R = 1$, we have $Q^+ = Q$. After the modifications in the output for these two selective input combinations, it is observed that now for only four input combinations ($S = 0$, $R = 0$, $Q = 1$), ($S = 1$, $R = 0$, $Q = 0$), ($S = 1$, $R = 0$, $Q = 1$), and ($S = 1$, $R = 1$, $Q = 1$), will it be $Q^+ = 1$.

The modified truth table of the quantum SR latch is shown in table 12.2. From table 12.1 a new characteristic equation $Q^+ = (S \oplus Q) \bullet (S \oplus R) \oplus Q$ is derived for the quantum SR latch. The quantum SR latch is shown in figure 12.2.

12.2.2 The quantum D latch

The characteristic equation of the quantum D latch can be written as $Q^+ = D \cdot E + \overline{E} \cdot Q$. When the enable signal (clock) is 1, the value of the input D is reflected at the output, that is $Q^+ = D$. When $E = 0$ the quantum latch maintains its previous state, that is $Q^+ = Q$.

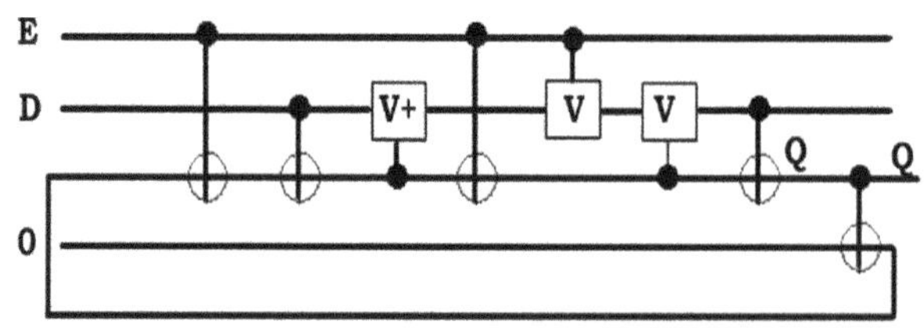

Figure 12.3. The quantum D latch.

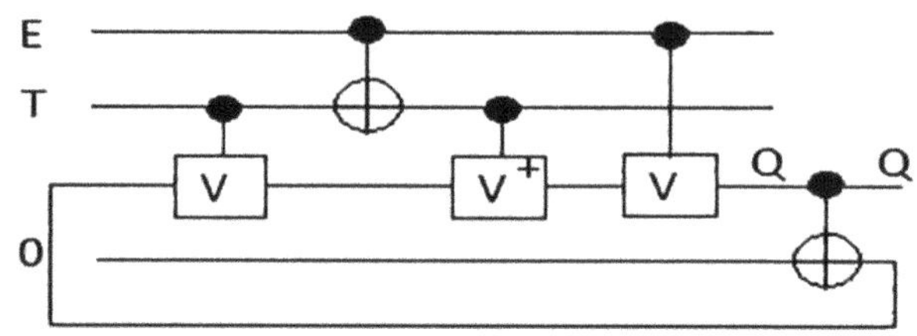

Figure 12.4. The quantum T latch. Reproduced with permission from [7]. Copyright 2010 IEEE.

This can be understood as follows. From the characteristic equation of the quantum D latch $Q^+ = D \cdot E + \overline{E} \cdot Q$, it can be seen that for four input combinations ($E = 0$, $D = 0$, $Q = 0$), ($E = 0$, $D = 1$, $Q = 0$), ($E = 1$, $D = 0$, $Q = 0$), and ($E = 1$, $D = 0$, $Q = 1$), the output Q^+ is 0. The addition of one garbage output can resolve only two output positions since one qubit can produce only two distinct output combinations. The quantum D latch is shown in figure 12.3.

12.2.3 The quantum T latch

The characteristic equation of the quantum T (toggle) latch can be written as $Q^+ = (T \cdot Q) \cdot E + \overline{E} \cdot Q$. However, the same result can also be obtained from $Q^+ = (T \cdot E) \oplus Q$. The T latch is a complementing quantum latch which complements its value when $T = 1$, that is when $T = 1$ and $E = 1$ then $Q^+ = \overline{Q}$. When $T = 0$, the quantum T latch maintains its state and there will be no change in the output. The quantum T latch characteristic equation can be directly mapped to the quantum Peres gate and the fan-out at output Q can be avoided by cascading the quantum Feynman gate (it is seen that $T \cdot E \oplus Q$ matches the template of the quantum Peres gate). The design is shown in figure 12.4. It is seen that for the four input combinations ($E = 0$, $T = 0$, $Q = 0$), ($E = 0$, $T = 1$, $Q = 0$), ($E = 1$, $D = 0$, $Q = 0$), and ($E = 1$, $T = 1$, $Q = 1$), the output Q^+ is 0.

The design of a quantum T latch has both Q output and the complementary $\overline{Q}$ output. In this section a quantum T latch is introduced based on quantum Peres and quantum Feynman gates with the outputs Q and $\overline{Q}$. The quantum T latch is shown in figure 12.5.

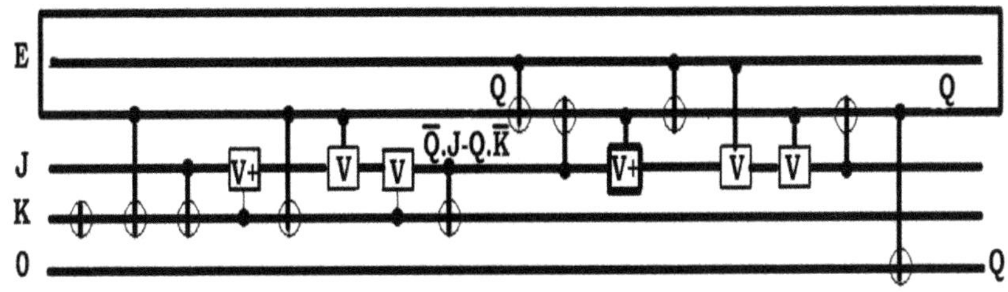

Figure 12.5. The quantum J–K latch.

12.2.4 The quantum J–K latch

From the characteristic equation of the quantum J–K latch for the eight input combinations ($E = 0$, $J = 0$, $K = 0$, $Q = 0$), ($E = 0$, $J = 0$, $K = 1$, $Q = 0$), ($E = 0$, $J = 1$, $K = 0$, $Q = 0$), ($E = 0$, $J = 1$, $K = 1$, $Q = 0$), ($E = 1$, $J = 0$, $K = 0$, $Q = 0$), ($E = 1$, $J = 0$, $K = 1$, $Q = 0$), ($E = 1$, $J = 0$, $K = 1$, $Q = 1$), and ($E = 1$, $J = 1$, $K = 1$, $Q = 1$), the output Q^+ is 0.

The quantum J–K latch has the capability of setting the output, resetting the output, or complementing the output depending on the value of J and K. When E (clock) is 1, the J input can set the output to 1 (when $J = 1$), the reset input can reset the output to 0 (when $K = 1$), and when both J and K are set to 1 the output Q is complemented.

The quantum J–K latch does not produce the complemented output $\overline{Q}$. Thus a quantum realization of the quantum J–K latch is illustrated in figure 12.5 which produces both the output Q and its complement $\overline{Q}$. In this design the Feynman gate is used to generate the complement of the output Q.

12.3 The properties of quantum counter circuits

The important feature of the quantum counter circuit is that different subsystems of a digital system can be controlled by the sequence generated; timing signals can be generated and clocks of different frequencies can be generated. A variety of counters are available. They can be classified as quantum synchronous and asynchronous counters based on clock input, positive edged or negative edged counters based on the clock trigger, binary or decade counters based on the counts, up, down or up/down counters based on the direction of the count, and J–K, T, or D based counters based on the quantum latches used in the quantum counter circuit.

The primitive logic blocks used in quantum counters are quantum latches and combinational gates. Quantum asynchronous or quantum ripple counters do not use a common clock and the output of one stage affects the clock input of the next stage. In a quantum synchronous counter the clock is given simultaneously to all the quantum latches. The total qubits of the counter do not change simultaneously but synchronously at the rising or falling edge of the clock, depending on the type of edge triggered latches that are used in the counter.

12.4 The design of quantum counters

In this section the design of synchronous and asynchronous quantum counter circuits is shown.

12.4.1 The quantum asynchronous counter

In a quantum asynchronous counter the quantum T latches are arranged in such a way that the output of one latch is connected to the clock input of the next higher order latch. The output of a latch triggers the next latch. The latch holding the least significant qubit receives the incoming count pulse. A four-qubit quantum asynchronous counter is shown in figure 12.6.

12.4.2 The quantum synchronous counter

In a quantum synchronous counter, the clock pulses are applied to the inputs of all the latches at a time. A latch is complemented depending on the input value T and the clock pulse. The latch in the least significant position is completed with every clock pulse. A latch in another position is complemented only when all the outputs of preceding latches produce 1, and same strategy is followed here to implement the quantum synchronous counter. Figure 12.7 shows the four-qubit quantum synchronous counter.

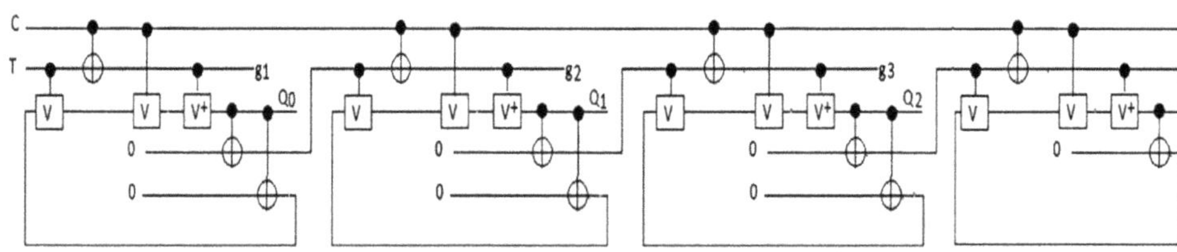

Figure 12.6. The quantum asynchronous counter.

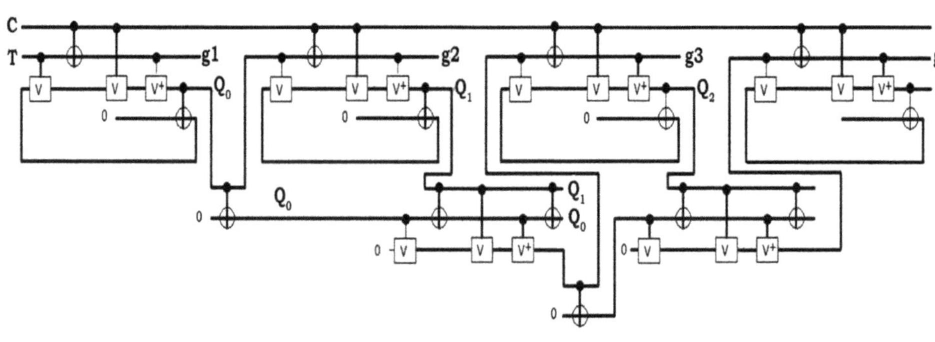

Figure 12.7. The four-qubit quantum synchronous counter.

12.5 Summary

This chapter presents the design methodologies of four quantum latches (quantum SR latch, quantum D latch, quantum T latch, and quantum J–K latch) and two quantum counter circuits (quantum asynchronous counter and quantum synchronous counter). A further application would be to use the latches towards the design of complex quantum sequential circuits such as flip-flops, storage registers, and shift registers, etc. In addition, the quantum counters can also be used to build quantum sequential circuits of higher order and complex design of quantum computers. These designs form the basis of a quantum arithmetic circuit design and thus contribute to reducing power dissipation.

Critical thinking questions

1. Describe the characteristics of quantum latches.
2. How many stable states can a quantum latch contain? Explain in detail.
3. What happens when both of the quantum SR latches' inputs are high? Explain with an example.
4. Give an explanation of the analysis process of the primary step of the quantum SR latch.
5. Describe several types of quantum latches and explain each one in detail.
6. Explain the properties of the quantum counter.
7. Describe the differences between a quantum asynchronous counter and a quantum synchronous counter.

References

[1] Chuang M-L and Wang C-Y 2008 Synthesis of reversible sequential elements ACM *J. Emerging Technol. Comput. Syst.* **3** 4

[2] Picton P 1996 Multi-valued sequential logic design using Fredkin gates *Mult.-Valued Log. J.* **1** 241–51

[3] Rice J E 2008 An introduction to reversible latches *Comput. J.* **51** 700–9

[4] Thapliyal H and Ranganathan N 2010 Design of reversible sequential circuits optimizing quantum cost, delay, and garbage outputs *ACM J. Emerging Technol. Comput. Syst.* **6** 14

[5] Thapliyal H and Vinod A P 2007 Design of reversible sequential elements with feasibility of transistor implementation *IEEE Int. Symp. on Circuits and Systems* (Piscataway, NJ: IEEE) pp 625–8

[6] Thapliyal H and Zwolinski M 2006 Reversible logic to cryptographic hardware: a new paradigm *49th IEEE Int. Midwest Symp. on Circuits and Systems* (Piscataway, NJ: IEEE) pp 342–6

[7] Thapliyal H and Ranganathan N 2010 Design of reversible latches optimized for quantum cost, delay and garbage outputs *23rd International Conference on VLSI Design* (Piscataway, NJ: IEEE) pp 235–240

IOP Publishing

Quantum Computing (Second Edition)
A pathway to quantum logic design
Hafiz Md Hasan Babu

Chapter 13

The quantum controlled ternary barrel shifter

Learning objectives

- Examine ternary quantum gates.
- Learn about the right rotation, left rotation, right shift, logical left shift, arithmetic right shift, and logical left shift of a quantum shifter circuit.
- Comprehend the quantum ternary Peres gate.
- Construct a quantum ternary barrel shifter.
- Become familiar with the quantum ternary modified Fredkin gate.
- Mention the properties of a quantum ternary barrel shifter.
- Talk about the characteristics of the quantum ternary circuit.

Most mathematical operations in quantum computing need shifting and rotation operations. Some of these applications are convolution, correlation, multiplication, etc. This requirement has made the use of shifters an important functional part of modern computers. To make these shifters faster, barrel shifters are being designed. Barrel shifters can shift data in one single clock cycle and they do not have an inbuilt memory, and this makes the response of the systems faster when compared to other flip-flop shifters.

A barrel shifter is an n input and n output quantum logic circuit in which k select lines control the shift operation. A barrel shifter can be unidirectional, allowing data to be shifted only to the left (or the right), or bidirectional, which allows data to be rotated or shifted in both directions. A barrel shifter with n inputs and k select lines is called an (n, k) barrel shifter. Among the different designs of barrel shifter, the logarithmic barrel shifter is the most widely used because of its simple design, smaller area, and the elimination of the decoder circuitry. The conventional irreversible design of a logarithmic barrel shifter is shown in figure 13.1. An n-bit logarithmic barrel shifter contains $\log_2(n)$ stages, where the ith stage either shifts over 2^i bits or leaves the data unchanged. Each stage of a logarithmic barrel shifter is controlled by a control bit. If the control bit is set to one, the input data will be shifted in the associated stage, otherwise it remains unchanged.

doi:10.1088/978-0-7503-5683-1ch13

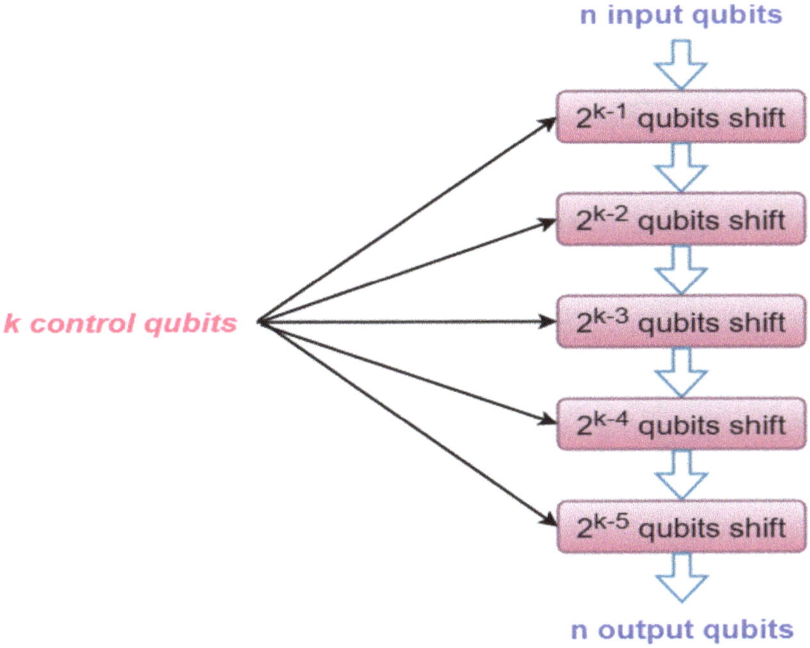

Figure 13.1. The block diagram of the (n, k) logarithmic barrel shifter.

13.1 Ternary quantum gates

The elementary logical operation of quantum computing can be performed using conditional two-qubit logic gates, which are called quantum Muthukrishnan–Stroud gates (quantum M–S gates). The generalized quantum ternary gate (GQTG) is built on top of quantum M–S gates. The ternary quantum logics are $|0\rangle$, $|1\rangle$, and $|2\rangle$. A quantum M–S gate for ternary logic is presented in figure 13.2, where the input A acts as the control input and input B works as the controlled input. The output of the circuit P is equal to the input A. If $A = 2$ the other output Q is the Z-transform, where $Z \in \{+1, +2, 12, 01, 21\}$ of the input B, otherwise $Q = B$. The Z-transforms are as follows: +1 and +2 indicate increasing by one and two, respectively, and 12, 01, and 02 stand for swap the last, first, and middle two values.

The diagram of a GQTG and its representation by a quantum M–S gate are illustrated in figure 13.3. The controlling input A controls an abstract ternary multiplexer (a conditional gate) which operates on the controlled input B. From the value of A, a ternary shift operation is employed on B.

13.1.1 The quantum ternary Peres gate

The quantum Peres gate is an important concept in universal computing and the ternary quantum version of this gate is also important for ternary quantum logic synthesis. The quantum ternary Peres gate (QTPG) is a 3×3 quantum gate, mapping (A, B, C) to $(P = A, Q = A + B, R = AB + C)$, where A, B, C are the inputs

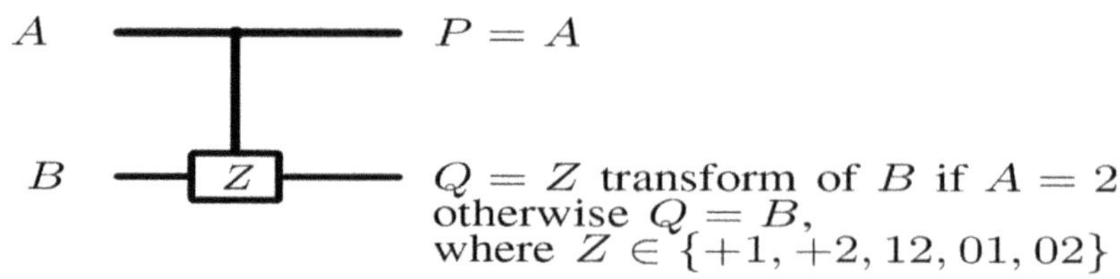

Figure 13.2. The quantum M–S gate.

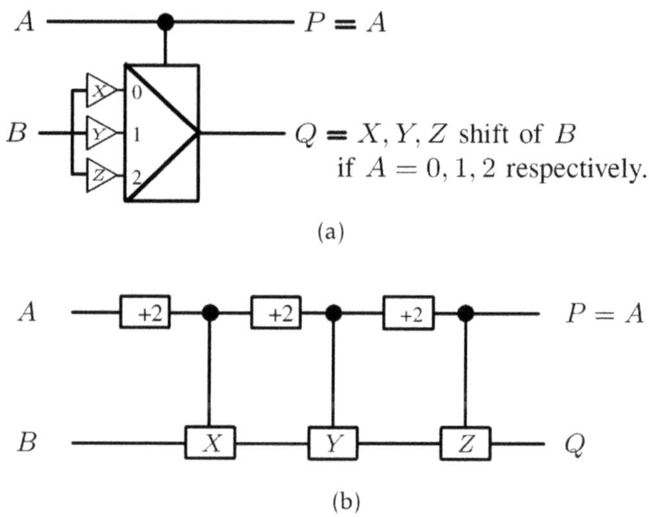

Figure 13.3. The GQTG using quantum M–S gates. (a) The GQTG. (b) The quantum realization of an M–S gate.

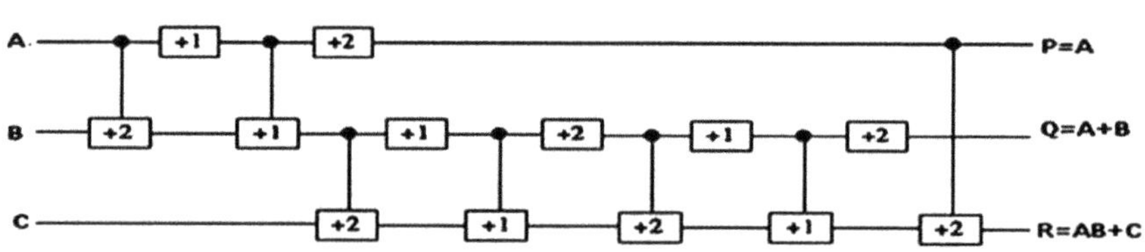

Figure 13.4. The realization of a QTPG using quantum M–S gates.

and P, Q, R are the respective outputs. The QTPG is implemented using seven quantum M–S gates and six quantum shift gates. Thus the number of quantum gates is 13. The quantum M–S and quantum shift gates based on QTPG are shown in figure 13.4.

13.1.2 The quantum ternary modified Fredkin gate

The quantum Fredkin gate (see figure 2.4 in section 2.2.3) is a binary gate, and the quantum ternary modified Fredkin gate (QTMFG) is a multi-valued sequential logic

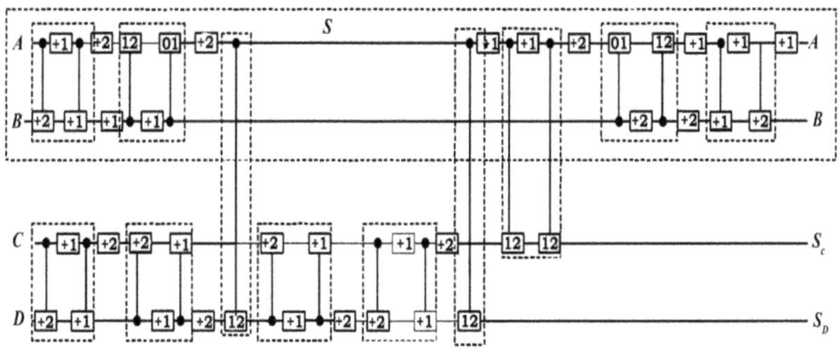

Figure 13.5. The realization of a QTMFG using quantum M–S gates.

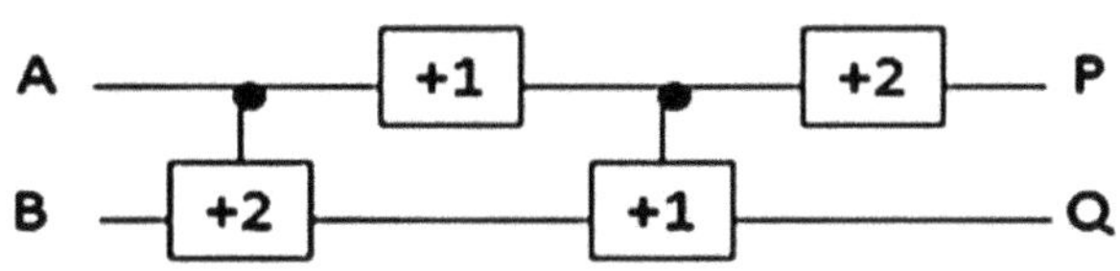

Figure 13.6. The realization of a QTFG using quantum M–S gates.

gate. The QTMFG is a universal gate for quantum multi-valued sequential logic. Figure 13.5 presents a realization of the QTMFG using quantum M–S gates.

13.2 The properties of ternary quantum circuits

The ternary quantum gate calculation complexity refers to the number of ternary quantum gates (quantum M–S gates, quantum shift gates) used to synthesize the given circuit. Let ϵ be the quantum M–S gate calculation complexity and γ be the quantum shift gates calculation complexity. For example, the quantum ternary Feynman gate (QTFG) (figure 2.2 in in section 2.2.1) has two quantum M–S gates and two quantum shift gates. Therefore, the quantum gate calculation complexity is $2\epsilon + 2\gamma$, which is depicted in figure 13.6.

13.3 The quantum barrel shifter

As shown in figure 13.1, a barrel shifter has n input and n output logic circuits in which k select lines control the bit shift operation. The design of a quantum barrel shifter presents design methodologies which can perform six operations: logical right shift, arithmetic right shift, right rotate, logical left shift, arithmetic left shift, and left rotate. For illustration, all of these operations are shown in table 13.1 for an eight-bit logarithmic barrel shifter where the eight-bit input data are denoted as $i_7, i_6, i_5, i_4, i_3, i_2, i_1, i_0$, where i_7 is the sign bit, and the shift or rotate operation is performed by three bits, and X denotes the shifted result.

All the operations that can be performed by a logarithmic barrel shifter, as shown in table 13.1, are as follows.

Table 13.1. Operations of an eight-bit barrel shifter with three-bit shift value.

Operation performed	X (final output[a])
Logical right shift	$000i_7i_6i_5i_4i_3$
Arithmetic right shift	$i_7i_7i_7i_7i_6i_5i_5i_3$
Right rotation	$i_2i_1i_0i_7i_6i_5i_4i_3$
Logical left shift	$i_4i_3i_2i_1i_0000$
Arithmetic left shift	$i_7i_3i_2i_1i_0000$
Left rotation	$i_4i_3i_2i_1i_0i_7i_6i_5$

[a] The input data are $i_7i_6i_5i_4i_3i_2i_1i_0$.

13.3.1 Logical right shift

As illustrated in table 13.1 a three-bit logical right shift operation right shifts the input data by three bits and sets the left-most three bits to zero. Thus the output will be $000i_7i_6i_5i_4i_3$.

13.3.2 Arithmetic right shift

A three-bit arithmetic right shift operation right shifts the input data bit by three bits and sets the left-most three bits to the sign bit (i_7). Thus the output will be $i_7i_7i_7i_7i_6i_5i_4i_3$
.

13.3.3 Right rotation

A three-bit right rotation operation performs a right shift operation on input data by three bits. Further, the left-most three bits are set to the right-most three bits of the original input data. As illustrated in table 13.1, the final output after the right rotation operation will be $i_2i_1i_0i_7i_6i_5i_4i_3$ as the input data $i_7i_6i_5i_4i_3i_2i_1i_0$ are shifted three times to the right and the three left-most bits ($i_7i_6i_5$) are set to the three right-most bits ($i_2i_1i_0$) of the original data.

13.3.4 Logical left shift

A three-bit logical left shift operation left shifts the input data by three bits and sets the right-most three bits to zero. Thus the final output will be $i_4i_3i_2i_1i_0000$.

13.3.5 Arithmetic left shift

In the arithmetic left shift operation the sign bit of the input data remains intact and the remaining bits are logically left shifted by three bits. As illustrated in table 13.1 the output will be $i_7i_3i_2i_1i_0000$ as the input data $i_7i_6i_5i_4i_3i_2i_1i_0$ are logically left shifted by three bits and the sign bit (i_7) remains intact.

13.3.6 Left rotation

A three-bit left rotation operation performs a left shift operation on the input data by three bits. Further, the right-most three bits are set to the left-most three bits of the original input data. As illustrated in table 13.1, the final output after the left rotation operation will be $i_4i_3i_2i_1i_0i_7i_6i_5$ as the input data $i_7i_6i_5i_4i_3i_2i_1i_0$ are shifted three times to the left and the three right-most bits ($i_2i_1i_0$) are set to the three left-most bits ($i_7i_6i_5$) of the original data.

13.4 The design of a quantum ternary barrel shifter

In this section two new designs for a quantum ternary barrel shifter are shown. The design of the quantum ternary barrel shifter and the properties of the designed circuits are shown in detail.

13.4.1 The optimized quantum ternary barrel shifter

Two new designs for a quantum ternary barrel shifter are shown in this subsection. Unidirectional and bidirectional quantum barrel shifters are designed. A unidirectional barrel shifter shifts data only to the left (or right). A four-bit quantum ternary unidirectional barrel shifter's block diagram, shown in figure 13.7, has two stages (k_i, $i = 0, 1$) and it is constructed using QTFGs, QTMFGs, and QTPGs. A bidirectional logarithmic logical shifter is a non-rotating barrel shifter which can shift input data to the left or right. It has a control signal (K) for determining the direction of the shift. If the K signal is set to zero, the logical shifter will work as a logical right shifter, otherwise it will work as a logical left shifter. For example, a four-bit quantum ternary bidirectional barrel shifter's block diagram, illustrated in figure 13.8, has two stages (k_i, $i = 0, 1$) and it is constructed using QTFGs, QTMFGs, and QTPGs. The QTFG is utilized to avoid fan-out. Moreover, the QTMFGs and QTPGs are applied as 2-to-1 multiplexers and AND gates, respectively. The (4, 2) quantum ternary unidirectional and bidirectional logical barrel shifters, which are depicted in figures 13.7 and 13.8, respectively, take n_3, n_2, n_1, n_0 as data inputs and k_1, k_0 as select inputs. The (4, 2) quantum ternary unidirectional and bidirectional logical barrel shifters can be generalized for quantum (n, k) ternary unidirectional and bidirectional logical barrel shifters,

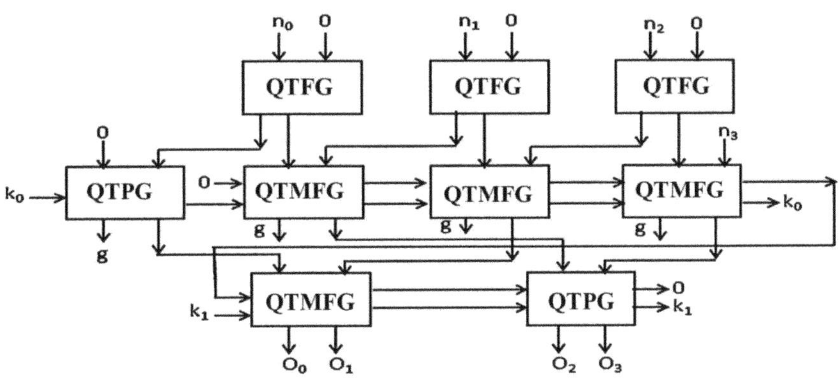

Figure 13.7. The block diagram of a quantum ternary unidirectional barrel shifter.

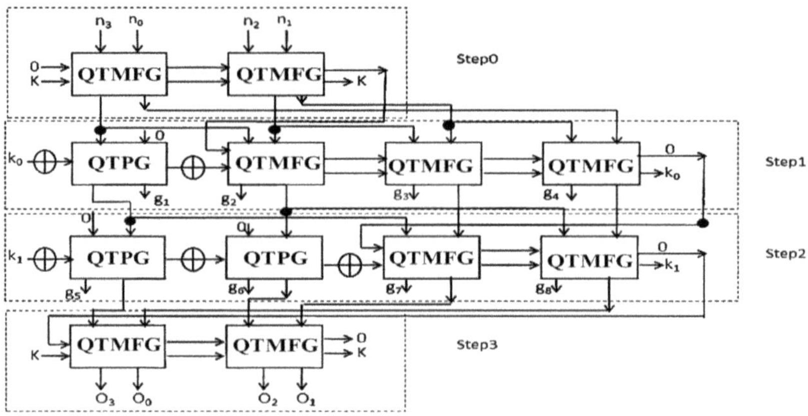

Figure 13.8. The block diagram of a (4, 2) quantum ternary bidirectional barrel shifter.

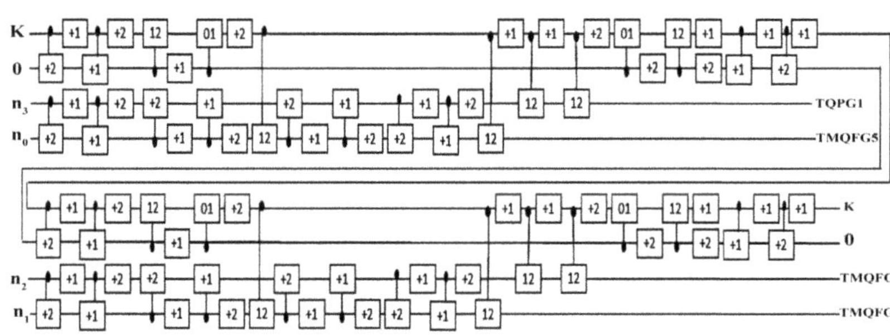

Figure 13.9. The realization of a (4, 2) quantum ternary bidirectional barrel shifter (step 0) using quantum M–S gates. Reproduced with permission from [3]. Copyright 2015 IEEE.

respectively. The ternary gate representation of the (4, 2) quantum ternary bidirectional barrel shifter is depicted from figures 13.9–13.12 for each step.

13.4.2 The properties of the designed circuit

In this subsection the properties of the (n, k) quantum ternary barrel shifter are presented.

Property 13.1. A QTPG is utilized for AND operation in the design. Every stage in the barrel shifter is denoted by i. Thus the number of required QTPGs in each stage is equal to $2^i - 1$, $i = 0, 1, \ldots, k - 1$. Thus in the design, k stages will have a total of $N_{\text{QTPG.TU}} = (\sum_{i=0}^{k-1} 2^i) - 1 = 2^k - 2$ QTPGs. The QTMFGs are responsible for shifting/rotating of data input bits. Each stage has n QTMFGs except for the last stage. The last stage requires only $n/2$ QTMFGs. Thus there are $k - 1$ stages which require n QTMFGs and the last stage requires $n/2$ QTMFGs, but there are $2^k - 2$ QTMFGs which are replaced by QTPGs. Hence in the design the total number of

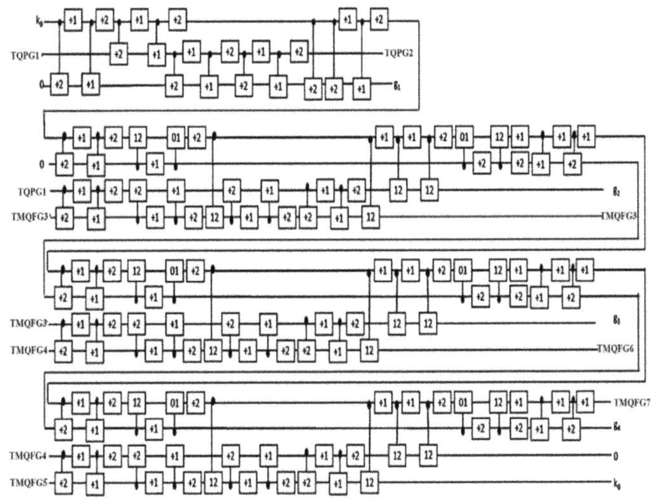

Figure 13.10. The realization of a (4, 2) quantum ternary bidirectional barrel shifter (step 1) using quantum M–S gates. Reproduced with permission from [3]. Copyright 2015 IEEE.

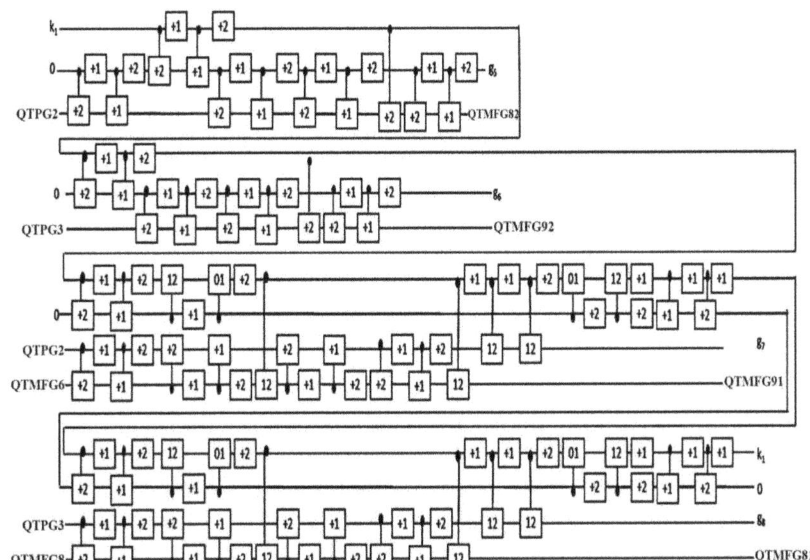

Figure 13.11. The realization of a (4, 2) quantum ternary bidirectional barrel shifter (step 2) using quantum M–S gates. Reproduced with permission from [3]. Copyright 2015 IEEE.

QTMFGs is $N_{\mathrm{QTMFG.TU}} = ((n(k-1) + n/2) - (2^k - 2))$. QTFGs are used for copying signals. The number of QTFGs in the input stage is $(n-1)$. Hence the design can be realized with $N_{\mathrm{QTFG.TU}} = (n - 1)$ QTFGs. Therefore the design requires $N_{\mathrm{QTPG.TU}} = 2^k - 2$, $N_{\mathrm{QTMFG.TU}} = ((n(k-1) + n/2) - (2^k - 2))$, $N_{\mathrm{QTFG.TU}} = n - 1$

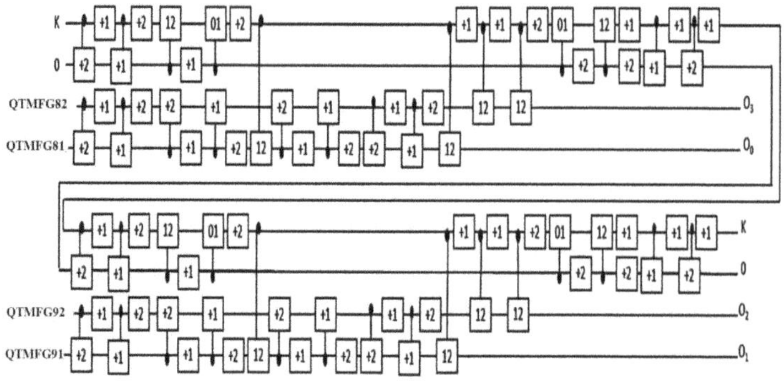

Figure 13.12. A realization of a (4, 2) quantum ternary bidirectional barrel shifter (step 3) using quantum M–S gates. Reproduced with permission from [3]. Copyright 2015 IEEE.

ternary quantum gates, where n is the number of bits, k is the total number of stages of the circuit, and $N_{\text{QTPG.TU}}$, $N_{\text{QTMFG.TU}}$, and $N_{\text{QTFG.TU}}$ are the total number of ternary quantum Peres, ternary modified quantum Fredkin, and quantum ternary Feynman gates, respectively.

Property 13.2. The QTPG is utilized as an AND gate in the design. Every stage in the barrel shifter is denoted by i. Thus the number of required QTPGs in each stage is equal to 2^i, $i = 0, 1, \ldots, k - 1$. Thus in the design k stages will have a total of $N_{\text{QTPG.TB}} = \sum_{i=0}^{k-1} 2^i = 2^k - 1$ QTPGs. The input stage and the output stage of the circuit need $n/2$ QTMFGs and the barrel shifter requires $(n - 2^i)$ QTMFGs for the rest of the stages. Thus an n-bit bidirectional barrel shifter with k stages will have a total of $N_{\text{QTMFG.TB}} = \sum_{i=0}^{k-1}(\frac{n}{2} + \frac{n}{2} + n - 2^i) = \sum_{i=0}^{k-1}(n + n - 2^i) = (k + 1)n - \sum_{i=0}^{k-1}(2^i) = (k + 1)n - (2^k - 1)$ MQFGs. Fan-out gates must be used for copying the signal in the quantum circuit. One of the fan-out gates is a QTFG. The number of QTFGs in each stage is $(n - 2^i)$. Hence the circuit can be realized with $N_{\text{QTFG.TB}} = \sum_{i=0}^{k-1}(n - 2^i)$ QTFGs. Therefore, the design requires $N_{\text{QTPG.TB}} = 2^k - 1$, $N_{\text{QTMFG.TB}} = (k + 1)n - (2^k - 1)$, and $N_{\text{QTFG.TB}} = \sum_{i=0}^{k-1}(n - 2^i)$ ternary quantum gates, where n is the number of bits, k is the total number of stages of the circuit, and $N_{\text{QTPG.TB}}$, $N_{\text{QTMFG.TB}}$, and $N_{\text{QTFG.TB}}$ are the total number of ternary quantum Peres, ternary modified quantum Fredkin, and quantum ternary Feynman gates, respectively.

Property 13.3. The ternary quantum unidirectional and bidirectional barrel shifters are realized by QTPGs, QTFGs, and QTMFGs with the number of quantum gates being 13, 4, and 41, respectively. Thus the optimized (n, k) quantum ternary unidirectional and bidirectional barrel shifters require $(13N_{\text{QTPG.TU}} + 41N_{\text{QTMFG.TU}} + 4N_{\text{QTFG.TU}})$ and $(13N_{\text{QTPG.TB}} + 41N_{\text{QTMFG.TB}} + 4N_{\text{QTFG.TB}})$ quantum gates, respectively, where n is the number of bits and k is the total number of stages of the circuit.

Property 13.4. The ternary quantum unidirectional and bidirectional barrel shifters are realized using QTPGs, QTFGs, and QTMFGs. A QTPG has 7ϵ quantum M–S gate calculation complexity and 6γ quantum shift gate calculation complexity, a QTFG has 2ϵ quantum M–S gate calculation complexity and 2γ quantum Shift gate calculation complexity; and a QTMFG has 20ϵ quantum M–S gate calculation complexity and 21γ quantum shift gate calculation complexity. Thus the total quantum gate calculation complexities are $(7N_{\text{QTPG.TU}\epsilon}+6N_{\text{QTPG.TU}\gamma} + 20N_{\text{QTMFG.TU}\epsilon}+21N_{\text{QTMFG.TU}\gamma}+2N_{\text{QTFG.TU}\epsilon}+2N_{\text{QTFG.TU}\gamma})$ and$(7N_{\text{QTPG.TB}\epsilon}+6N_{\text{QTPG.TB}\gamma}+20N_{\text{QTMFG.TB}\epsilon}+21N_{\text{QTMFG.TB}\gamma}+2N_{\text{QTFG.TB}\epsilon}+2N_{\text{QTFG.TB}\gamma})$ for the ternary quantum unidirectional and bidirectional barrel shifters, respectively, where ϵ is the M–S quantum gate calculation complexity and γ is the shift quantum gate calculation complexity.

Property 13.5. The optimized (n, k) ternary unidirectional and bidirectional barrel shifters produce at least $n(k - 1)$ and nk garbage outputs, respectively, where n is the number of bits and k is the total number of stages of the circuit.

Proof. The above statement has been proved by the method of contradiction. Suppose (n, k) ternary quantum unidirectional and bidirectional barrel shifters do not produce at least $n(k - 1)$ and nk garbage outputs, respectively, where n is the number of bits and k is the total number of stages of the circuit.

In every stage of the ternary quantum unidirectional and bidirectional barrel shifters, QTPGs and QTMFGs produce one garbage output, except for the first and last stages. In other words, in each row the quantum barrel shifter produces at least n garbage outputs. There are $k - 1$ and k numbers of stages in the unidirectional and bidirectional quantum barrel shifters, respectively. Thus $(k - 1)$ stages of a unidirectional quantum barrel shifter will have a total of $n(k - 1)$ garbage outputs, whereas k stages of a bidirectional quantum barrel shifter produce nk garbage outputs. Thus the optimized (n, k) unidirectional and bidirectional quantum ternary barrel shifters produce at least $n(k - 1)$ and nk garbage outputs, respectively.

This contradicts the supposition that (n, k) unidirectional and bidirectional quantum ternary barrel shifters do not produce at least $n(k - 1)$ and nk garbage outputs, respectively. Hence the supposition is false and the property 13.5 is true and this completes the proof.

13.5 Summary

In this chapter two new design methodologies for a quantum ternary barrel shifter are presented. The optimized shifters are designed using quantum ternary Feynman gates, quantum ternary modified Fredkin gates, and quantum ternary Peres gates. The quantum ternary barrel shifters will be useful in a wide variety of applications in signal processing systems. A general use of barrel shifters is in the hardware employment of floating-point arithmetic. For a floating-point add or subtract operation, the significants of the two numbers must be associated, which requires shifting the smaller number to the right and increasing its exponent until it matches the exponent of the larger number. This is performed by subtracting the exponents and using the barrel shifter to shift the smaller number to the right by the difference in one cycle. If a shifter were used, shifting by n-bit positions would need n clock cycles.

Critical thinking questions

1. Is the ternary quantum gate universal? If so, give an example with an explanation.
2. Describe the characteristics of ternary quantum gates.
3. Describe the differences between quantum ternary gates and quantum two-valued gates.
4. Write a brief description of each type of quantum barrel shifter and describe their characteristics.
5. Describe the procedure to optimize a quantum ternary barrel shifter.

References

[1] Khan A I, Nusrat N, Khan S M, Hasan M and Khan M H A 2007 Quantum realization of some ternary circuits using Muthukrishnan–Stroud gates *37th Int. Symp. on Multiple-Valued Logic* (Piscataway, NJ: IEEE) p 20

[2] Kotiyal S, Thapliyal H and Ranganathan N 2010 Design of a ternary barrel shifter using multiple-valued reversible logic *10th IEEE Conf. on Nanotechnology* (Piscataway, NJ: IEEE) pp 1104–8

[3] Lisa N J and Babu H M H 2015 A compact representation of a quantum controlled ternary barrel shifter *2015 IEEE Int. Symp. on Circuits and Systems* (Piscataway, NJ: IEEE) pp 2145–8

[4] Muthukrishnan A and Stroud C R Jr 2000 Multivalued logic gates for quantum computation *Phys. Rev.* A **62** 052309

[5] Nia N H 2012 Design of an optimized reversible ternary and binary bidirectional and normalization barrel shifters for floating point arithmetic *Life Sci.* J **9** 1904–15

[6] Peres A 1985 Reversible logic and quantum computers *Phys. Rev.* A **32** 3266

IOP Publishing

Quantum Computing (Second Edition)

A pathway to quantum logic design

Hafiz Md Hasan Babu

Chapter 14

Quantum RAM, quantum ROM, and quantum cache memory

Learning objectives

- Define quantum random access memory (RAM).
- Discuss quantum memory units.
- Construct a RAM with a quantum n-to-2n decoder.
- Describe the procedure of a quantum RAM.
- Explain the algorithm of a quantum decoder.

Quantum random access memory (RAM) is a highly versatile device for storing and accessing information. It consists of an array of memory cells (write-enabled master–slave flip-flop) where information is stored in the form of qubits. Each cell is associated with a unique address, i.e. a number which identifies the location of the cell. The main characteristic of quantum RAM is that each memory cell can be addressed separately as well and hence the designation is 'random access'. To access a cell, its address must be provided by a decoder. The device will then output the content of the memory cell.

If the quantum decoder has n selection qubits, the quantum RAM is capable of addressing 2^n different memory locations: when given an n-qubit address k, quantum RAM returns the qubit string f_k which is stored in the memory slot of the database labeled k. Figure 14.1 presents the general architecture of a $2^n \times m$ quantum RAM, where n is the selection bits of the quantum decoder and m is the number of qubits of input data.

This chapter focuses on the construction procedure and the complexities of quantum RAM, and it describes each of the underlying components in the quantum circuitry to build quantum RAM.

14.1 The quantum n-to-2^n decoder

A quantum decoder is a collection of quantum logic gates set up in such a way that, for an input combination, all output terms are low except one. These terms are the

doi:10.1088/978-0-7503-5683-1ch14

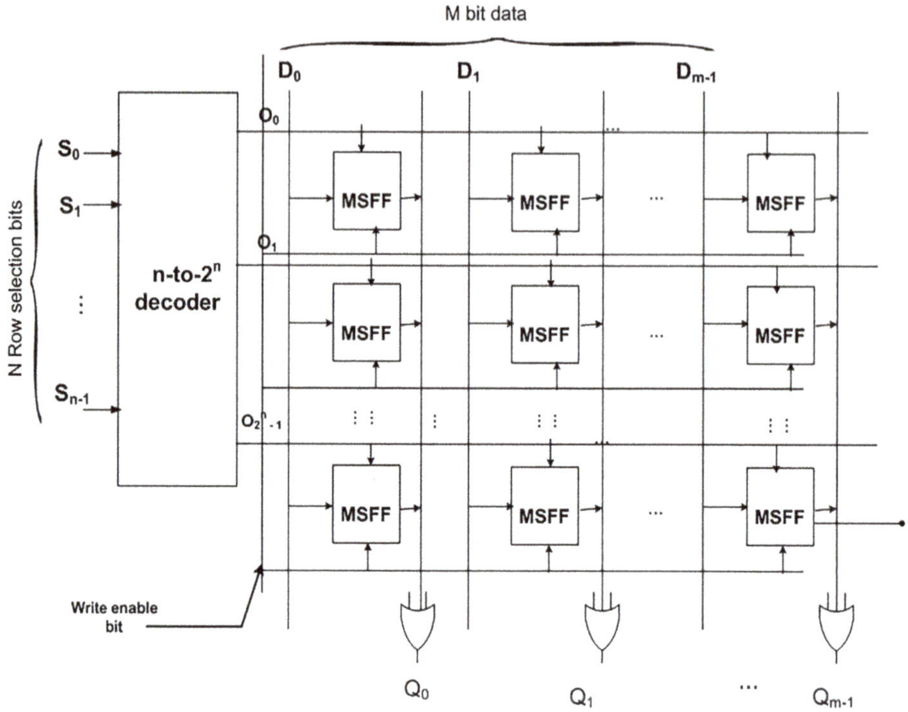

Figure 14.1. General architecture of RAM.

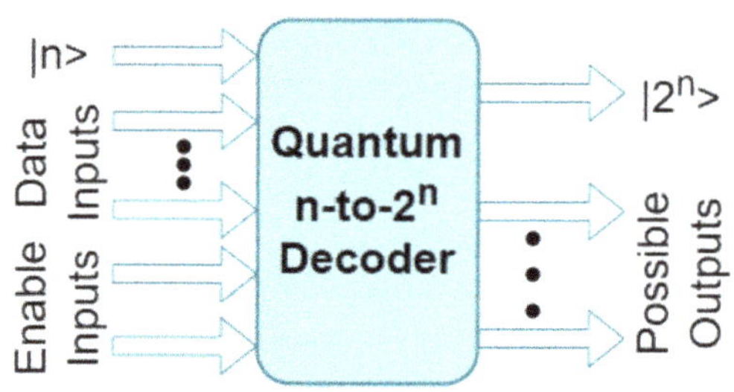

Figure 14.2. Block diagram of a quantum n-to-2^n decoder.

miniterms which use a variable only once. Thus, when an input combination changes, two outputs will change. Let there be n inputs, then the number of outputs will be 2^n. In other words, there will be one line at the output for each possible input. Figure 14.2 presents the block diagram of a quantum n-to-2^n decoder. Table 14.1 presents the truth table of a quantum 2-to-4 decoder circuit.

A quantum decoder can be realized using the quantum Feynman double gate (QF2G) and quantum Fredkin gate (QFRG). Figure 14.3 illustrates the QRFG and QF2G. Figure 14.3(a) presents the quantum Fredkin gate and figure 14.3(b) presents the symbol for a quantum Fredkin gate. The quantum cost of the 2-to-4 decoder circuit is 12.

Table 14.1. Truth table of the quantum 2-to-4 decoder.

Input			Output			
EN	A	B	Y_3	Y_2	Y_1	Y_0
0	X	X	0	0	0	0
1	0	0	0	1	0	1
1	0	1	0	0	1	0
1	1	0	0	1	0	0
1	1	1	1	0	0	0

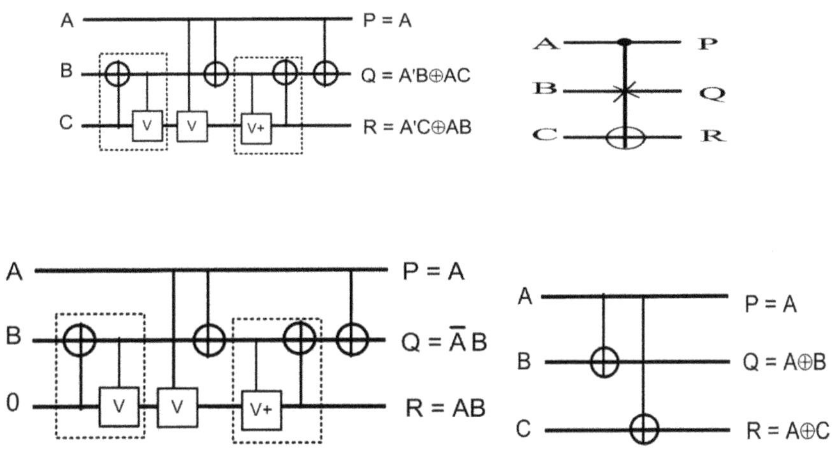

Figure 14.3. (a) The quantum Fredkin gate. (b) The symbol for a quantum Fredkin gate. (c) The simultaneous generation of two terms. (d) The quantum Feynman double gate.

The quantum 1-to-2 decoder (a) and the quantum 2-to-4 decoder (b) are shown in figure 14.4, A higher order quantum decoder, such as the 3-to-8 decoder can be designed with a 2-to-4 quantum decoder and an additional four QFRGs, as shown in figure 14.5. In the same way an n-to-2^n quantum decoder can be designed with $(n - 1)$-to-2^{n-1} quantum decoders and an additional 2^{n-1} QFRGs. Algorithm 14.1 presents the construction of an n-to-2^n quantum decoder.

Line 5 of the algorithm assigns the input to the QF2G for the first control bit (S_0), whereas line 7 assigns the first two inputs to the QFRG for all the remaining control bits. Lines 8–9 assign the third input to the FRG for $n = 2$, while lines 10–12 assign the third input to the QFRG through a recursive call to the previous quantum decoder for $n = 2$. Line 15 returns the outputs. The complexity of this algorithm is $O(n)$, where n is the number of data qubits. The quantum decoder is used to address the row of memory cells of the QRAM.

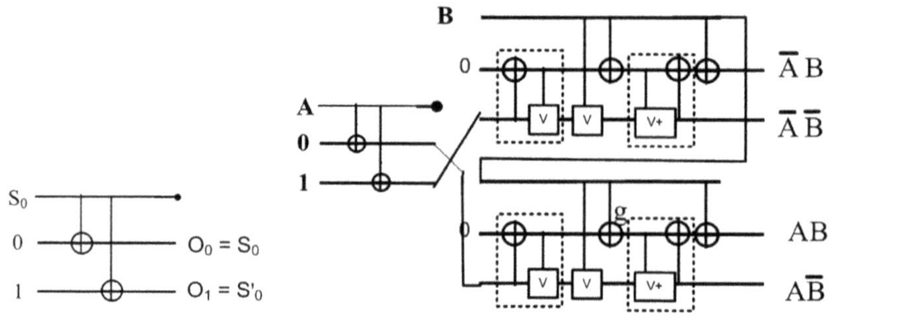

Figure 14.4. Quantum decoders for $n = 1$ and $n = 2$. (a) The quantum 1-to-2 decoder. (b) The quantum 2-to-4 decoder.

Input: $S_0S_1, \cdots, S_{n-1}, QF2G, QFRG$
Output: $n - to - 2^n$ quantum decoder circuit

START
i = input, o = output
for j = 0 to n-1 **do**
  if j=0 **then**
    $S_j \rightarrow first.i.QF2G,\ 1 \rightarrow second.i.QF2G,\ 0 \rightarrow third.i.QF2G$
  else
    $S_j \rightarrow first.i.QFRG,\ 0 \rightarrow second.i.QFRG$
    if j=2 **then**
      $third.o.QF2G \rightarrow third.i.QFRG$
    else
      call *Quantum Decoder(j-1)*, *Quantum Decoder.o.j* $\rightarrow$ $third.i.QFRG_j$
    end if
  end if
end for
return *QFRG.o.3* and *QFRG.o.2* $\rightarrow$ *desired output* and remaining *QF2G.o* and *QFRG.o* $\rightarrow$ *garbage output.*
END

Algorithm 14.1. The quantum n-to-2^n decoder (S, QF2G, QFRG).

14.2 The quantum memory unit

In quantum RAM, there are write-enabled master–slave flip-flops (MSFFs). To design a write-enabled MSFF, first of all a D latch is needed. Figure 14.6 presents the quantum D latch, and it can be seen that a QFRG and a QF2G are used in this circuit. The QFRG is needed to produce Q, whereas QF2G produces Q^+ and Q. The equations for the latch are $Q_n = D.\ CLK + CLK'.\ Q_{n-1}$ and $Q^+ = Q_0$ (n is time varying), which are used to capture the logic level when the data are present and the

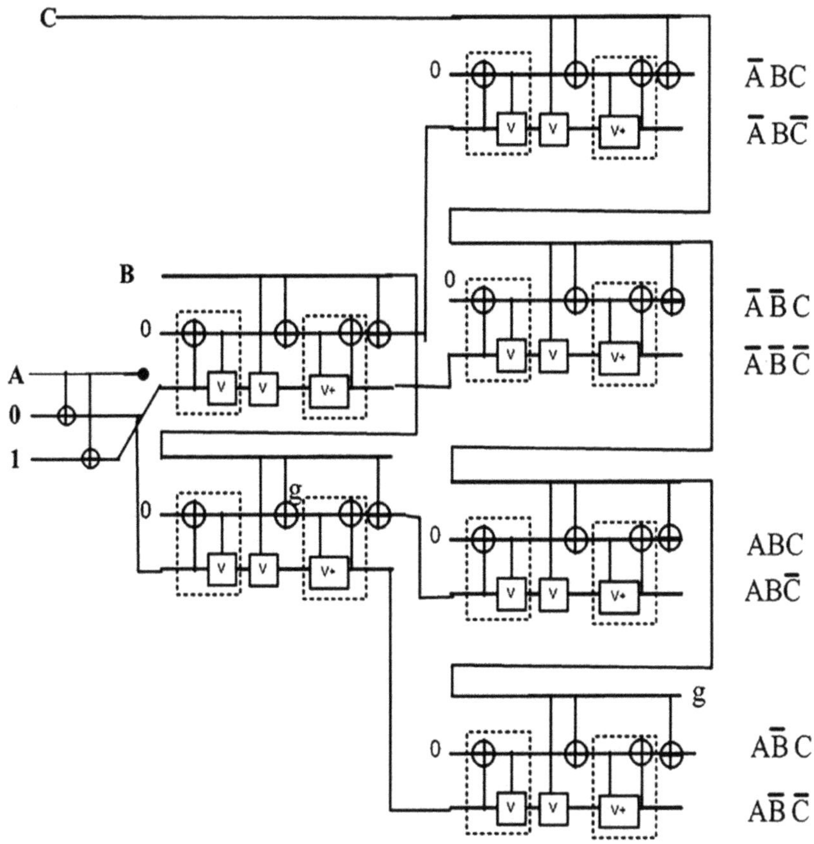

Figure 14.5. The quantum 3-to-8 decoder.

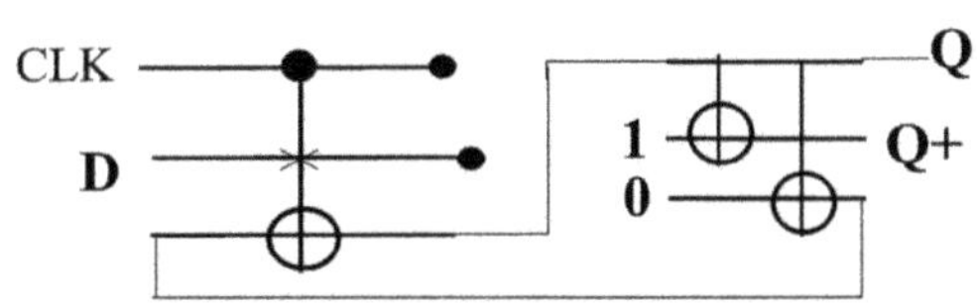

Figure 14.6. The quantum D latch.

clock is SET. A quantum D latch can be realized with seven quantum gates, two garbage outputs, and a quantum cost of 7.

After designing the quantum D latch, the quantum write-enabled MSFF (QMSFF) can be designed with two D latches and additional QFRGs and QF2Gs. Figure 14.7 illustrates the architecture of a QMSFF.

The write-enabled QMSFF works on two modes (read and write), since the data are both read from and written into the QRAM. The CLK signal provides the clock

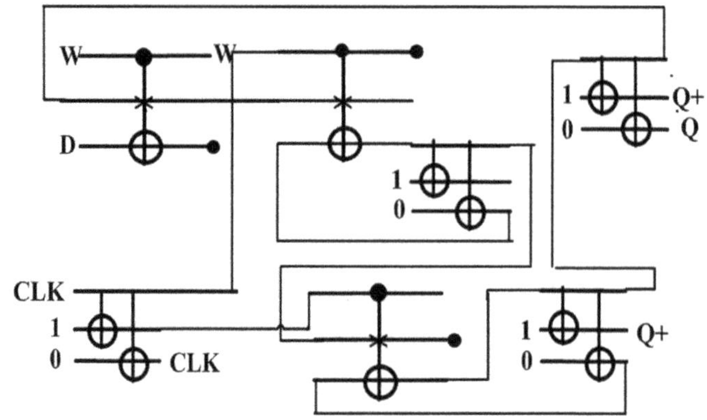

Figure 14.7. The write-enabled QMSFF.

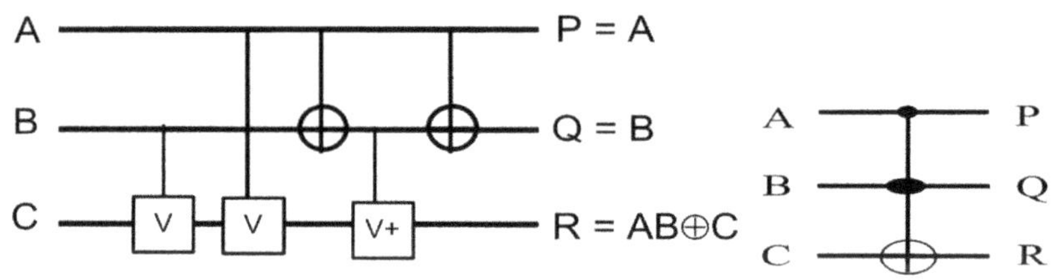

Figure 14.8. (a) The QTG. (b) The symbol for the QTG.

pulse, if CLK = 1, the D input is stored in the first latch, i.e. master flip-flop, but the second latch cannot change state. When CLK = 0 the first latch's output is stored in the second latch, i.e. the slave flip-flop, but the first latch cannot change state. The CLK signal is maintained by the first QF2G. In figure 14.7 W works as the write enable bit. It determines whether a read or a write operation is performed. When $W = 1$ a write operation is performed. The W and CLK outputs are propagated to the next MSFF if required and Q indicates the output of the MSFF.

The circuit consists of two D flip-flops connected together. When the clock is high, it can be found from the quantum circuit that the QMSFF consists of a QFRG, two QF2Gs, and two quantum D latches (the D latch is presented in figure 12.3 of section 12.2.2).

Considering the unused clock pulses as garbage outputs, this design has six garbage outputs and has a quantum cost of 23.

14.3 The construction procedure of QRAM

To design QRAM, the addressable components, i.e. the decoder and memory component MSFF, are replaced by the quantum decoder and QMSFF. Then the OR gates are replaced by Ex-OR gates. Additionally, quantum Toffoli gates (QTGs), as shown in figure 14.8, are used to propagate both the CLK and

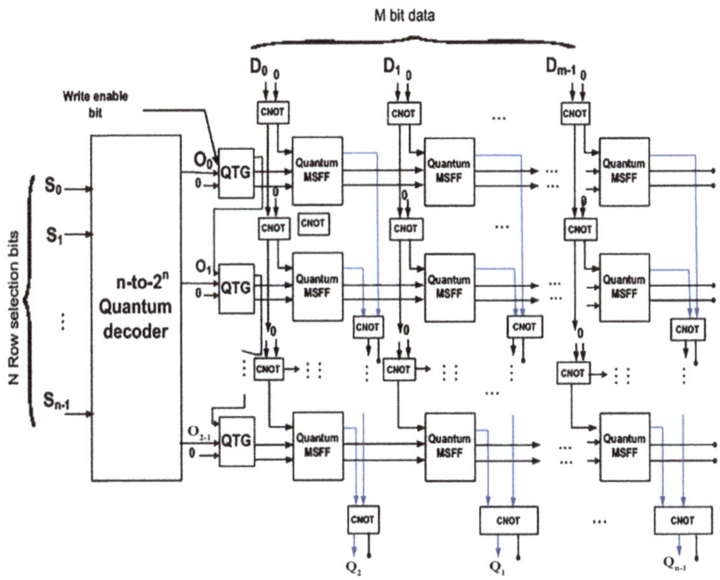

Figure 14.9. The block diagram of the $2^n \times m$ quantum RAM.

write-enabled bit W while diminishing the fan-out problem. Figure 14.9 presents the block diagram of $2^n \times m$ QRAM. The working principles of QRAM are as follows.

In the $2^n \times m$ RAM, n is the number of control bits of the quantum decoder and 2^n decoded outputs are the row selection bits. For a particular combination of n input variables of the configured quantum decoder, only one out of 2^n outputs is active at a time. This active output is propagated through the second output of the QTG, which acts as the CLK and hence selects only one row of the array of memory units as the CLK propagate through the row of the QMSFF.

The write bit W specifies whether there is a read or a write operation. When $W = 1$ or high, then the m data-inputs D_0 to D_{m-1} are written in the m flip-flops of the selected row as W is propagated by the third output of the QTG throughout the selected row.

When $W = 0$ or low, then the m outputs Q_0 to Q_{m-1} are the previously stored bits in the flip-flops, and they are read from the m flip-flops of the selected row. The use of a quantum CNOT gate (Ex-OR gate) eliminates the fan-out problem. These gates also perform the Ex-OR operations of the outputs of the flip-flops in a column.

In addition, for simplicity the quantum circuit of 4×2 quantum RAM is given in figure 14.10. This circuit is developed by combining the quantum 2-to-4 decoder (figure 14.5(a)), eight QMSFFs (figure 14.7), four QTGs (figure 14.8), and twelve quantum CNOT gates. In figure 14.10 the red lines indicate the output and the blue lines indicate the clock pulse.

14.4 Quantum ROM

Quantum read-only memory (ROM) is a non-volatile memory. This means it receives data and permanently writes it on a chip, and it lasts even after turning

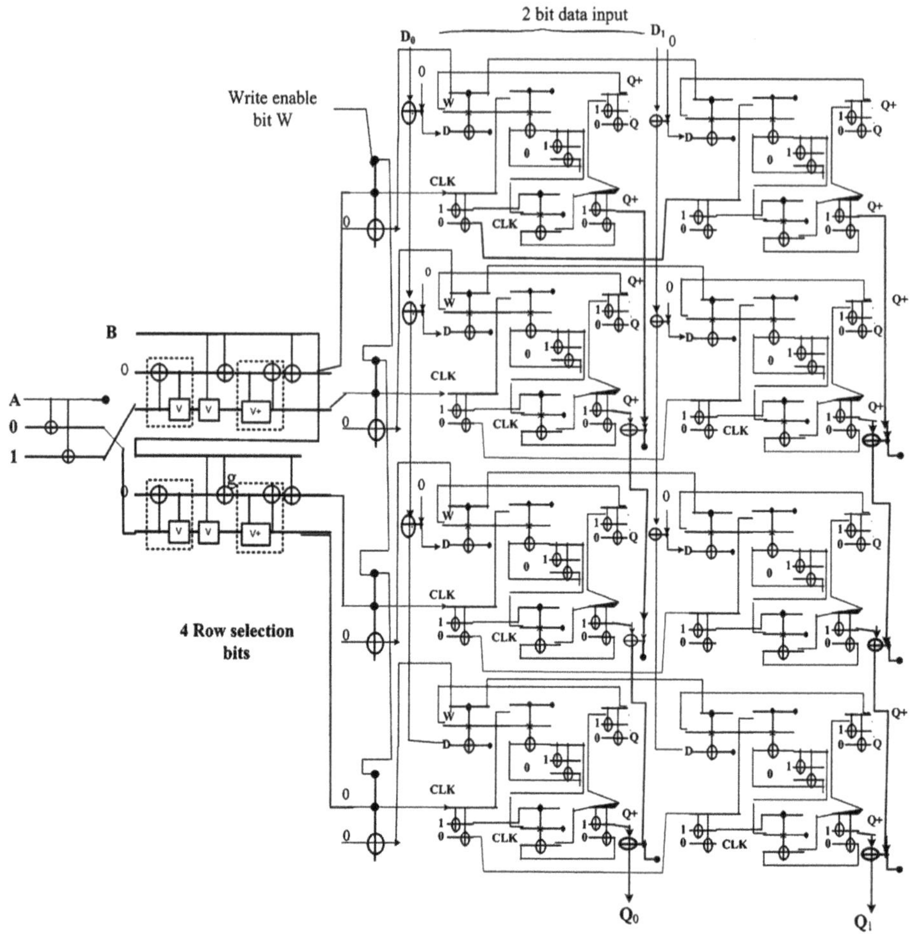

Figure 14.10. The 4 × 2 quantum RAM.

off the computer. The data are coded to not be overwritten. Thus ROM is used for things such as printer software or any start-up program. A quantum ROM stores the instructions that are required to start a computer. This operation is referred to as 'bootstrapping'. Quantum ROM chips are not only used in computers but can also be used in other electronic items.

In the architecture of quantum 4-to-2 ROM, as shown in figure 14.11, there are three quantum operations (NOT, AND, and OR) with three quantum gates (NOT, V, and V^+). The output of the decoder is operated with four quantum OR operations together and produces the desired output qubits $|F_1\rangle$ and $|F_2\rangle$ of a quantum 4-to-2 ROM. The working principle of the quantum ROM is as follows: (i) When input qubits $|A\rangle$, $|B\rangle = |0\rangle, |0\rangle$, the $|D_0\rangle$ line will open. Thus the value of $|D_0\rangle$ will be $|1\rangle$ and $|D_1\rangle$ to $|D_3\rangle$ will be $|0\rangle$. The output qubits of $|F_1\rangle$ and $|F_2\rangle$ perform OR operations with $|D_0\rangle = |1\rangle$, $|D_1\rangle = |0\rangle$, $|D_2\rangle = |0\rangle$, and $|D_3\rangle = |0\rangle$ to generate $|1\rangle$. (ii) When the input qubits $|A\rangle$, $|B\rangle = |1\rangle, |0\rangle$, the $|D_1\rangle$ line will open. Thus the value of

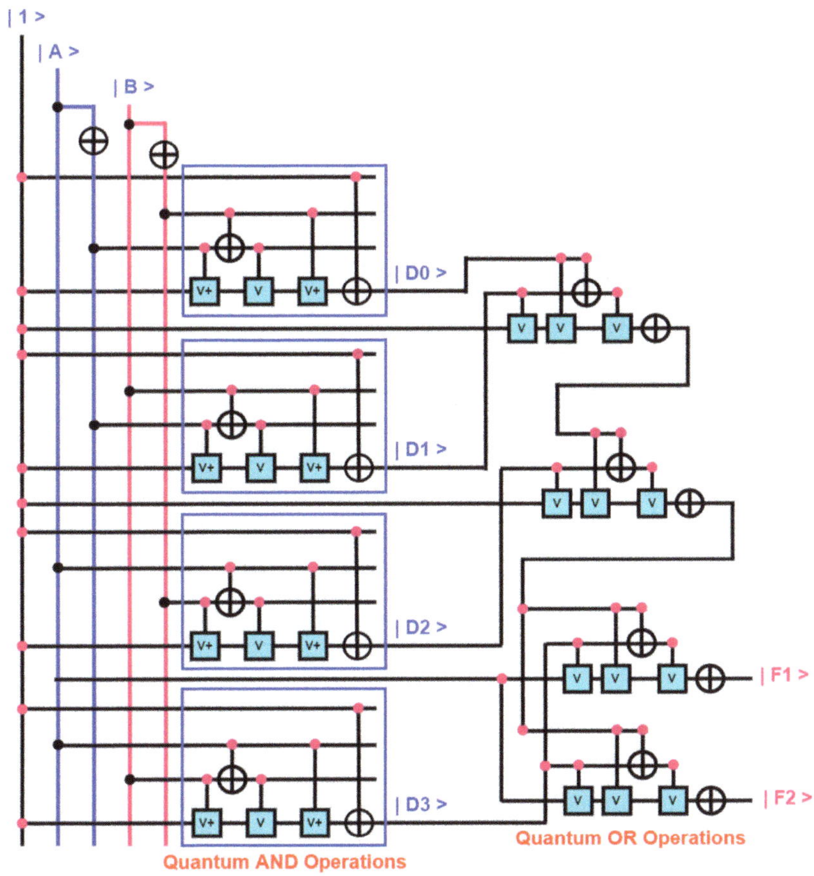

Figure 14.11. The 4-to-2 quantum ROM.

$|D_1\rangle$ will be $|1\rangle$ and $|D_0\rangle$, $|D_2\rangle$, and $|D_3\rangle$ will be $|0\rangle$. The output qubits of $|F_1\rangle$ and $|F_2\rangle$ perform OR operations with $|D_0\rangle = |0\rangle$, $|D_1\rangle = |1\rangle$, $|D_2\rangle = |0\rangle$, and $|D_3\rangle = |0\rangle$ to generate $|1\rangle$. (iii) When the input qubits $|A\rangle$, $|B\rangle = |0\rangle$, $|1\rangle$, the $|D_2\rangle$ line will open. Thus the value of $|D_2\rangle$ will be $|1\rangle$ and $|D_0\rangle$, $|D_1\rangle$, and $|D_3\rangle$ will be $|0\rangle$. The output qubits of $|F_1\rangle$ and $|F_2\rangle$ perform OR operations with $|D_0\rangle = |0\rangle$, $|D_1\rangle = |0\rangle$, $|D_2\rangle = |1\rangle$, and $|D_3\rangle = |0\rangle$ to generate $|1\rangle$. (iv) When the input qubits $|A\rangle$, $|B\rangle = |1\rangle$, $|1\rangle$, the $|D_3\rangle$ line will open. Thus the value of $|D_3\rangle$ will be $|1\rangle$ and $|D_0\rangle$ to $|D_2\rangle$ will be $|0\rangle$. The output qubits of $|F_1\rangle$ and $|F_2\rangle$ perform OR operations with $|D_0\rangle = |0\rangle$, $|D_1\rangle = |0\rangle$, $|D_2\rangle = |0\rangle$, and $|D_3\rangle = |1\rangle$ to generate $|1\rangle$.

14.5 Quantum cache memory

Quantum cache memory can be used in a quantum system due to its speed and reliability. It can pass and obtain data very frequently. The mapping and swapping techniques in the quantum cache memory are very optimized. The circuit diagram of a one-qubit quantum cache memory is depicted in figure 14.12. The circuit diagram

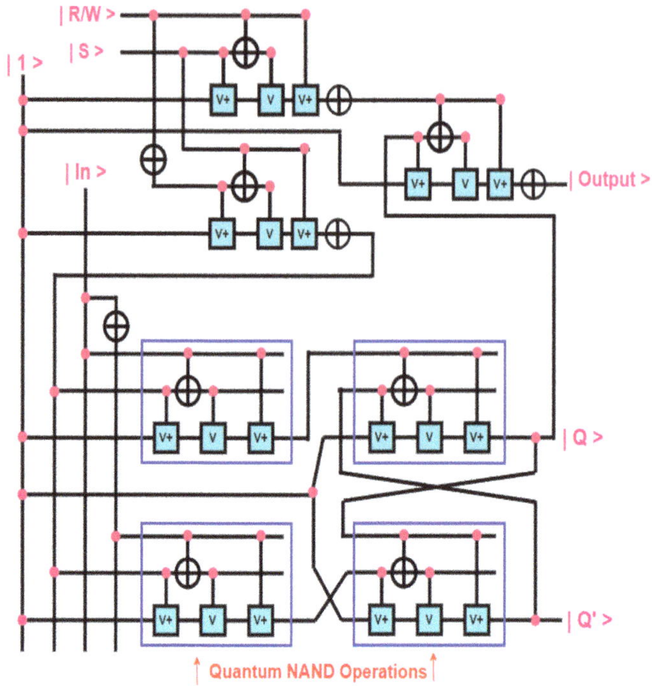

Figure 14.12. The circuit diagram of a one-qubit quantum cache memory.

of the one-qubit quantum cache memory consists of four quantum NAND and three quantum AND operations. Again, $|R/W\rangle = |1\rangle$ indicates the READ operation and $|R/W\rangle = |0\rangle$ indicates the WRITE operation. And here $|S\rangle$ indicates the selection of the qubit, where $|S\rangle = |1\rangle$ means the selection of memory.

Using quantum technology, the arrangement of a 4-to-2 qubit quantum cache memory is shown in figure 14.13. The quantum 4-to-2 cache memory consists of a quantum 2-to-4 decoder which decodes the input qubits into qubits with outputs. Here, R indicates the quantum cache memory cell. The circuit diagram of the quantum 4-to-2 quantum cache memory includes a 2-to-4 quantum decoder, eight quantum cache memory cells, and six quantum OR operations, where the inputs consist of two qubits and the $|R/W\rangle$ signal. In addition, DI inputs are used for writing the data into the quantum memory.

The four locations (00, 01, 10, 11) in the quantum cache memory are addressed by two qubits $|A\rangle$ and $|B\rangle$. In order to read from location 00, the address $|A\rangle|B\rangle = 00$ and $|R/W\rangle = |1\rangle$. The quantum decoder selects $|0\rangle$ as high. $|R/W\rangle = |1\rangle$ will apply $|0\rangle$ at the clock inputs of the two quantum cache memory cells of the top row and it will also apply $|1\rangle$ at the inputs of the output quantum AND operations, thus transferring the outputs of the two quantum D flip-flops to the inputs of the two quantum OR operations. The other inputs of the quantum OR operations will be $|0\rangle$. Thus, the outputs of the two quantum cache memory cells of the top row will be transferred to DO_1 and DO_0, which perform a READ operation. In the case of a

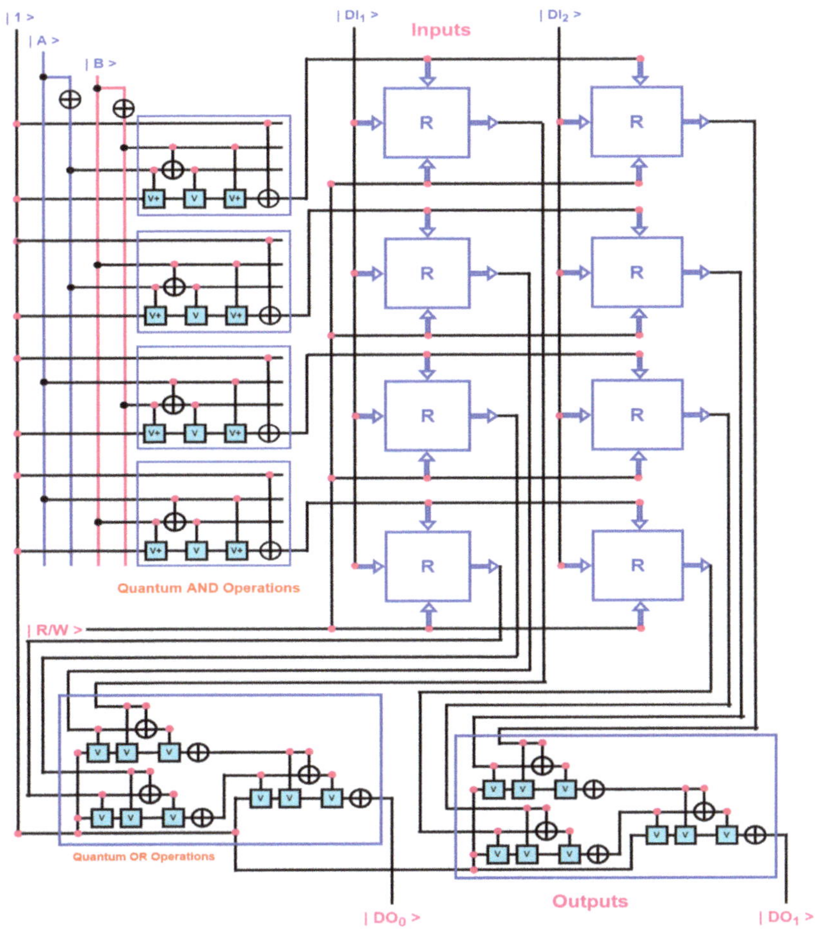

Figure 14.13. The circuit diagram of quantum 4-to-2 cache memory.

WRITE operation, the two-qubit data are presented at $|\mathrm{DI}_1\rangle$ $|\mathrm{DI}_0\rangle$. Suppose $|A\rangle$ $|B\rangle = |0\rangle$ $|0\rangle$. The top row is selected as $0_0 = 1$. Input bits at $|\mathrm{DI}_1\rangle$ and $|\mathrm{DI}_0\rangle$ will, respectively, be applied at the inputs of the quantum D flip-flops of the top row. Because $|R/W\rangle = |1\rangle$, the clock inputs of both the quantum D flip-flops of the top row are $|1\rangle$; thus, the D inputs are transferred to the outputs of the quantum flip-flops. Therefore, data at $|\mathrm{DI}_1\rangle$ $|\mathrm{DI}_0\rangle$ will be written into the quantum cache memory.

14.6 Summary

Quantum RAM is an essential component in quantum memory devices. This chapter focuses on the design procedure for QRAM. This chapter covers the design procedures of the quantum decoder step by step for addressing the RAM and the QMSFF (the unit memory component of RAM). A composite quantum realization of 4 × 2 quantum RAM is also given at the end of the chapter which will help the reader to understand quantum RAM better.

Critical thinking questions

1. Describe the properties of quantum RAM.
2. Briefly discuss some applications of quantum RAM.
3. What is the access time for quantum RAM?
4. How many memory locations can a quantum RAM chip access if it has n input address lines?
5. For read and write operations in a quantum RAM, which control signals are selected? Explain in detail.

References

[1] Asfestani M N and Heikalabad S R 2017 A novel multiplexer-based structure for random access memory cell in quantum-dot cellular automata *Physica* B **521** 162–7

[2] Fredkin E and Toffoli T 2001 Conservative logic *Collision-Based Computing* (Berlin: Springer) p 47–81

[3] Giovannetti V, Lloyd S and Maccone L 2008 Architectures for a quantum random access memory *Phys. Rev.* A **78** 52310

[4] Mahammad S N and Veezhinathan K 2010 Constructing online testable circuits using reversible logic *IEEE Trans. Instrum. Meas.* **59** 101–9

[5] Sayem A S M and Mitra S K 2011 Efficient approach to design low power reversible logic blocks for field programmable gate arrays *IEEE Int. Conf. on Computer Science and Automation Engineering* **4** 251–5

[6] Shamsujjoha M, Babu H M H and Jamal L 2013 Design of a compact reversible fault tolerant field programmable gate array: a novel approach in reversible logic synthesis *Microelectron. J.* **44** 519–37

[7] Sharmin F, Polash M M A, Shamsujjoha M, Jamal L and Babu H M H 2011 Design of a compact reversible random access memory *4th IEEE Int. Conf. on Computer Science and Information Technology* vol 10 pp 103–7

[8] Tanaka M, Sato R, Hatanaka Y and Fujimaki A 2016 High-density shift-register-based rapid single-flux-quantum memory system for bit-serial microprocessors *IEEE Trans. Appl. Supercond.* **26** 1–5

[9] Tayari M and Eshghi M 2011 Design of 3-input reversible programmable logic array *J. Circuits Syst. Comput.* **20** 283–97

[10] Thapliyal H and Ranganathan N 2010 Design of reversible sequential circuits optimizing quantum cost, delay, and garbage outputs ACM *J. Emerging Technol. Comput. Syst.* **6** 14

Chapter 15

The quantum arithmetic logic unit

Learning objectives

- Discuss the quantum arithmetic logic unit (ALU).
- Explain the design architecture of an ALU.
- Understand the purposes of ALUs.
- Learn three approaches for constructing ALUs.
- Examine the basic operations of ALUs.

The arithmetic logic unit (ALU) is a major component of the central processing unit of a computer system. It performs all processes related to arithmetic and logic operations that need to be performed on instruction words. An ALU executes one of the different micro-operations for performing the instructions on two operands A and B depending on the control signal inputs, and is suitable for quantum technology and embedded processors. The ALU can carry out Boolean functions, namely Ex-OR, AND, OR, NAND, NOR, and multiplexer (MUX), as well as basic arithmetic operations, such as addition (ADD) and subtraction (SUB). In this chapter three different designs of quantum ALUs are presented. The main basic operations that are required to realize an ALU are as follows:

- ADD (addition): $(A, B) \rightarrow (A, A + B)$.
- SUB (subtraction): $(A, B) \rightarrow (A, A - B)$.
- EX-OR (bitwise exclusive-OR): $(A, B) \rightarrow (A, A \oplus B)$.
- AND (bitwise-AND): $(A, B) \rightarrow (A, A \wedge B)$.
- OR (bitwise-OR): $(A, B) \rightarrow (A, A \vee B)$.
- NOP (no operation): $(A) \rightarrow (A)$.

15.1 The design of a quantum ALU

The component of the computer system that executes the majority of data-processing operations is called the central processing unit (CPU). The CPU is constructed using three major components, consisting of the control unit, ALU, and

doi:10.1088/978-0-7503-5683-1ch15

processor unit. The ALU plays an important role and performs the required micro-operations for executing the instructions. In order to design an effective quantum ALU, the number of logical and arithmetic calculations produced on the outputs must be considered in addition to the quantum cost and other evaluation parameters. The quantum ALU circuit produces a large number of calculations at the least quantum cost.

15.1.1 The first approach

The quantum ALU has eight inputs and eight outputs. The simple quantum gates employed in the implementation of all ALUs are covered in this chapter. In the first approach, the six logical calculations of the result output are AND, NOR, NAND, OR, ADD, and SUB. The inputs consist of three data inputs (A, B, and C_{in}), three select lines (S_0, S_1, S_2), and two constant inputs. The circuit can generate all selector inputs on the outputs (S_0, S_1, S_2). The eight outputs are S_0, S_1, and S_2 propagated to the output, C_{out}/borrow, result, and three garbage outputs. It produces two arithmetic and four logical operations.

The first approach for designing a one-bit quantum ALU is illustrated in figure 15.1 which shows five dotted rectangles, four V, two V+, and eight CNOT gates. A dotted rectangle is equivalent to a CNOT gate with a cost of one. The total quantum cost of the circuit is 19. The functions produced by the first approach are presented in table 15.1. From table 15.1, depending on the values in the selected inputs, the circuit produces six functions, namely ADD and SUB, which are arithmetic functions, and AND, NAND, OR, and NOR, which are logical functions.

15.1.2 The second approach

This quantum ALU has nine inputs and nine outputs and is illustrated in figure 15.2. The quantum ALU logic gates use two dotted rectangles, eight V, four V^{+}, and six CNOT gates. From the ALU three input values (A, B, and C_{in}), three select lines (S_0, S_1, S_2, S_3, S_4), and one constant input are selected. The principal outputs are F (as the product output) and C_{out}.

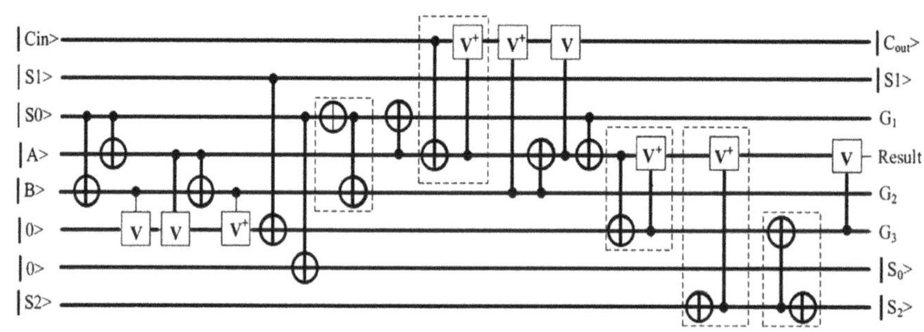

Figure 15.1. The quantum ALU (first approach). Reproduced with permission from [4]. Copyright 2016 Springer Nature.

Table 15.1. Different functions of the quantum ALU (first approach).

S_2	S_1	S_0	Result	Functions
0	0	0	(AB)	AND
0	0	1	$(A + B)$	NOR
0	1	0	(AB)	NAND
0	1	1	$(A + B)$	OR
1	0	0	$A \oplus B \oplus C_{in}$	ADD
1	0	1	$A \oplus B \oplus C_{in}$	SUB

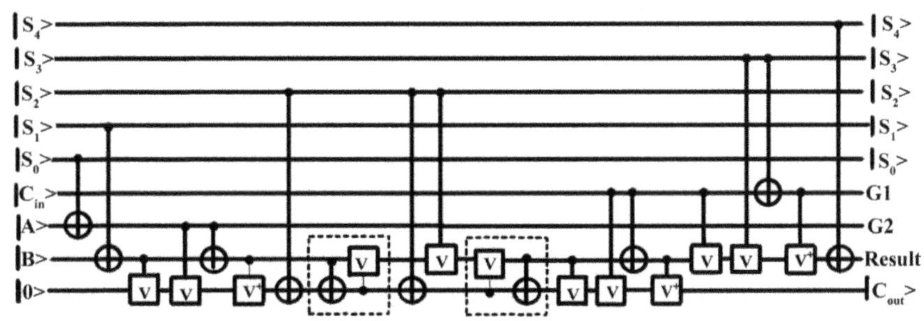

Figure 15.2. The quantum ALU (second approach). Reproduced with permission from [4]. Copyright 2016 Springer Nature.

In the second approach the total quantum cost is 24. There are two garbage values and one constant input. It has nine circuit lines and five selection lines. The presented design can be implemented with seven gates. Here it is clear that the inputs A, B, and C_{in} can be controlled depending on the values of the selection and carry input lines. The product output is F (as a final output) which will be obtained from 12 different operations, either arithmetic or logical, depending upon the values of C_{in} (as the carry input) and the selection lines. The other primacy output is carry out (C_{out}). The 12 operations generated in the second approach are given in table 15.2. A special function is selected through (S_0, S_1, S_2, S_3, S_4). The presented circuit has two main outputs (result and C_{out}). This circuit has five select control signals with a provision to realize 12 logical and arithmetic operations. The total elementary operations of the second ALU design are shown in table 15.2. By transforming the expressions based on the second approach, an extension of the ALU can perform other operations (such as NAND and NOR), which are the universal functions. Therefore, the second approach for the ALU circuit can implement several universal and arithmetic operations, namely ADD, SUB, AND, NAND, OR, NOR, EX-OR, and EX-NOR. Moreover, an n-bit ALU can be constructed by cascading n of these one-bit ALU circuits.

Table 15.2. Simple operations of the quantum ALU (second approach).

Result	C_{in}	S_0	S_1	S_2	S_3	S_4	Transformations	Description
$A + B$	0	0	0	0	0	0	$F = A + B + 0$	ADD without carry
$A + B + 1$	1	0	0	0	0	0	$F = A + B + 1$	ADD with carry
$A - B$	1	0	1	0	0	0	$F = A + B' + 1$	SUB
$A \oplus B$	0	0	0	0	1	0	$F = A \oplus B$	Exclusive-OR
$A \wedge B$	0	0	0	1	1	0	$F = A \wedge B$	Bitwise-AND
$A \vee B$	0	1	1	1	1	1	$F = (A' \wedge B')'$	Bitwise-OR
$(A \oplus B)'$	0	0	0	0	1	1	$F = (A \oplus B)'$	Exclusive-NOR
$A + 1$	1	0	0	0	0	0	$F = A + 0 + 1$	Self-increment of A
$A- 1$	1	0	1	0	0	0	$F = A + 1' + 1$	Self-decrement of A
A'	0	1	0	0	0	0	$F = A' + 0 + 0$	Bitwise negation of A
A'	1	1	0	0	0	0	$F = A' + 0 + 1$	Complement of A
A	0	0	0	0	0	0	$F = A + 0 + 0$	NOP (no operation)

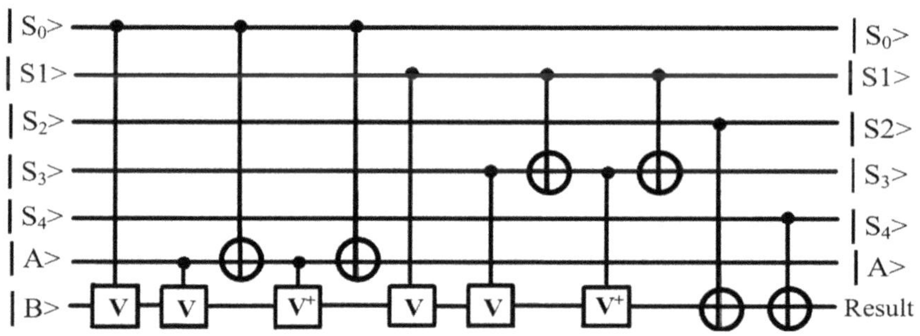

Figure 15.3. The quantum ALU (third approach). Reproduced with permission from [4]. Copyright 2016 Springer Nature.

15.1.3 The third approach

The quantum ALU has seven inputs and seven outputs. The third approach for a one-bit quantum ALU is illustrated in figure 15.3. Four V, two V$^+$, and six CNOT gates are employed to construct the ALU. From the ALU, two input values (A, B) and five selectors (S_0, S_1, S_2, S_3, S_4) are used for the input and the primary outputs.

In the third approach the total quantum cost is 12. There are no garbage values or constant inputs. The product output (result) will be obtained from the ten different operations presented in table 15.3. After transforming the expressions which are based on the third approach, an extension of the ALU can perform basic arithmetic–logical operations (ADD, SUB, NSUB, Ex-OR, NOP). The third approach is a garbage-free quantum ALU, which has no constant inputs.

Table 15.3. Basic operations of the quantum ALU (third approach).

S_0	S_1	S_2	S_3	S_4	ALU operations	Description
1	1	0	0	0	$F = A + B$	ADD
1	1	1	0	1	$F = B - A$	SUB
1	1	0	1	1	$F = A - B$	NSUB
1	0	0	0	0	$F = A \oplus B$	EX-OR
0	0	0	0	0	$F = B$	NOP
1	1	0	1	0	$F = B + A + 1$	ADD with carry
1	1	1	1	1	$F = B - A - 1$	SUB with borrow
1	1	0	0	1	$F = A - B - 1$	NSUB with borrow
1	0	1	0	0	$F = A \oplus B$	EX-NOR
0	0	1	0	0	$F = B$	Negation of B

15.2 Summary

In this chapter three approaches are explained for one-bit quantum arithmetic logic unit design, constructed using a combination of simple quantum gates. All the ALUs can handle many simple logical and arithmetic operations, namely addition, subtraction, bitwise-AND, bitwise-OR, NAND, NOR, exclusive-OR, exclusive-NOR, negation, and complement. In order to design a cost-effective quantum ALU, the number of logical and arithmetic calculations produced on the outputs must be considered. The ALU designs are evaluated in terms of the number of circuit lines, garbage outputs, constant inputs, and quantum cost. Moreover, the ALU designs are versatile approaches for the realization of quantum computing, having both remarkably low power consumption and nano-scaling. Therefore, the designs are implemented with simple quantum gates which have the scope to be realized in nanotechnology based systems, such as quantum cellular automata and other nano-scale architectures. The described circuits can be used in a wide variety of large quantum systems which enhance the performance of quantum computers. Furthermore, an optimized ALU is a requirement in computer systems and digital signal processors for the design of the instruction sets for more complex systems and low power VLSI design in nanotechnology, quantum computers, and programmable computing devices.

Critical thinking questions

1. Describe the characteristics of a quantum ALU.
2. Discuss some applications of a quantum ALU.
3. What is a quantum arithmetic logic unit and how does it work?
4. What kinds of operations can a quantum ALU perform?
5. Describe the simplest approach for designing a quantum ALU.

References

[1] Bennett C H 1973 Logical reversibility of computation *IBM J. Res. Dev.* **17** 525–32

[2] Dirac P A M 1981 *The Principles of Quantum Mechanics* (Oxford: Oxford University Press) Number 27

[3] Große D, Wille R, Dueck G W and Drechsler R 2008 Exact synthesis of elementary quantum gate circuits for reversible functions with don't cares *38th Int. Symp. on Multiple Valued Logic* pp 214–9

[4] Haghparast M and Bolhassani A 2016 Optimization approaches for designing quantum reversible arithmetic logic unit *Int. J. Theor. Phys.* **55** 1423–37

[5] Kaye P *et al* 2007 *An Introduction to Quantum Computing* (Oxford: Oxford University Press)

[6] Landauer R 1961 Irreversibility and heat generation in the computing process *IBM J. Res. Dev.* **5** 183–91

[7] Lim J, Kim D-G and Chae S-I 1999 Reversible energy recovery logic circuits and its 8-phase clocked power generator for ultra-low-power applications *IEICE Trans. Electron.* **82** 646–53

[8] Morrison M and Ranganathan N 2011 Design of a reversible ALU based on novel programmable reversible logic gate structures *IEEE Computer Society Annual Symp. on VLSI* (Piscataway, NJ: IEEE) pp 126–31

[9] Morrison M and Ranganathan N 2013 A novel optimization method for reversible logic circuit minimization *IEEE Computer Society Annual Symp. on VLSI* (Piscataway, NJ: IEEE) pp 182–7

[10] Morrison M A 2012 Design of a reversible ALU based on novel reversible logic structures *Thesis* University of South Florida, Tampa, FL

[11] Nielsen M A and Chuang I 2002 *Quantum Computation and Quantum Information* 10th edn (Leiden: Cambridge University Press)

[12] Moore G E 2000 Cramming more components onto integrated circuits *Readings in Computer Architecture* (San Francisco, CA: Morgan Kaufmann) p 56

[13] Thomsen M K, Glück R and Axelsen H B 2010 Reversible arithmetic logic unit for quantum arithmetic *J. Phys. A: Math. Theor.* **43** 382002

[14] Toffoli T 1980 Reversible computing *Int. Coll. on Automata, Languages, and Programming* (Berlin: Springer) pp 632–44

[15] Weste N H E and Harris D 2015 CMOS *CMOS VLSI Design: A Circuits and Systems Perspective* (India: Pearson Education)

IOP Publishing

Quantum Computing (Second Edition)
A pathway to quantum logic design
Hafiz Md Hasan Babu

Chapter 16

Quantum programmable logic devices

Learning objectives

- Define the quantum programmable array logic (PAL).
- Acquire knowledge about the quantum complex programmable logic device (CPLD) and explain how it works.
- Construct a quantum PAL with appropriate circuits and explain how it works.
- Discuss the importance and uses of quantum CPLDs.
- Learn about the uses of quantum PAL.
- Describe quantum field-programmable gate arrays (FPGAs) with their properties and working principles.
- Become familiar with the quantum programmable logic array (PLA) and discuss its working principles.
- Become aware of the significance and uses of quantum FPGAs.

Quantum programmable logic devices (PLDs) are the most crucial part of a quantum computer. One of the hot topics in the research field of quantum hardware design is how to construct a complex quantum system with low power consumption, low cost, and good performance. Quantum PLDs come in four primary varieties, each of which has a number of benefits, including lower development costs, minimal space, low power requirements, shorter design cycles, high switching speeds, high design security, simple design modifications, high reliability, and simple circuit testing.

It is important to note that the time requirements for logic design and design checking are very important factors for quality quantum circuits, even though the design is simple. For quantum PLDs, implementing new technology is easier and faster. Thus, the quantum computer is more dependable and faster due to all of the above benefits.

16.1 The quantum programmable array logic

A form of quantum PLD known as quantum programmable array logic (PAL) is used to implement a certain quantum logical function and has programmable quantum AND

doi:10.1088/978-0-7503-5683-1ch16

array operations and fixed quantum OR array operations. Only the quantum AND array may be programmed, making it simpler to operate. The quantum programmable read-only memory (PROM) and additional output quantum logic are the main components of quantum PAL, which are utilized to achieve a certain desired quantum function with fewer components.

16.1.1 The design procedure and working principles of quantum PAL

A quantum PAL with m inputs and n outputs is depicted in figure 16.1 Every quantum AND operation available in the quantum PAL is coupled to one of these inputs. The user can choose how the input lines and quantum AND operations are connected since the connection matrix is customizable. This implies that, depending on the logic, each and every input line must be connected to either a single or a multiple quantum AND operation. The logical 'and' functionality between the input lines becomes apparent as a result. With the aid of quantum PAL, the following expression is derived:

$|X\rangle = |AB\rangle + |AC'\rangle$

$|Y\rangle = |AB'\rangle + |BC'\rangle$.

The sum of products is represented by the two functions just mentioned. The above two equations require two quantum OR operations and four quantum AND operations, respectively. The second quantum OR operation, which creates $|Y\rangle$, will use the output of the first two quantum AND operations as its input, and the other two quantum AND operations' output will be the second quantum OR operation's output.

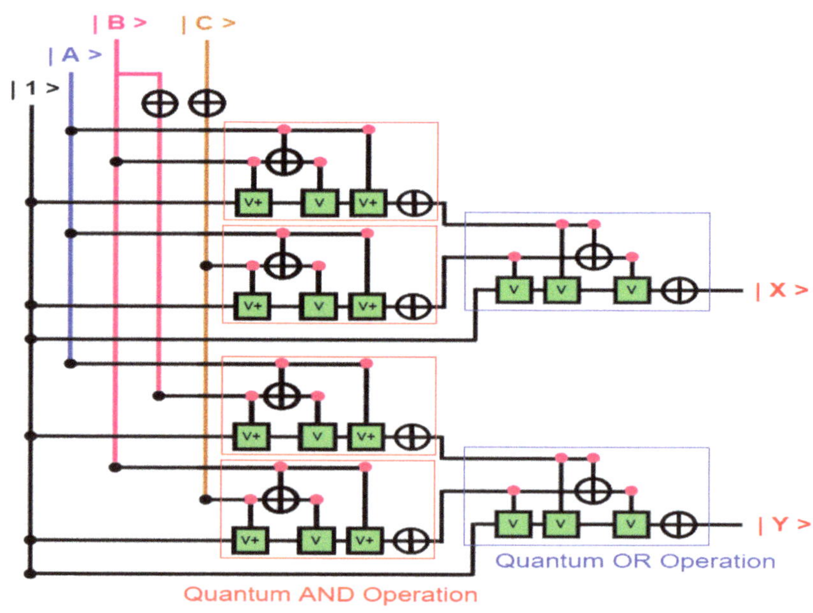

Figure 16.1. The quantum PAL.

Although it is programmed, the quantum AND array operation must operate with a fixed quantum OR array operation to meet the requirements of PAL.

16.1.2 The importance and applications of quantum PAL

In comparison to a silicon-based circuit, quantum PAL is much more cost-effective and extremely efficient. It provides both high security and reliability. Low power is needed for its functioning and its design is more versatile. A quantum PAL works more quickly than quantum programmable logic arrays (PLAs). The most popular version of a quantum programmable logic device (PLD) is called the quantum PAL, and it is utilized in different quantum circuits such as quantum counters, quantum state machines, quantum decoders, quantum synchronization circuits, quantum bus interfaces, quantum parallel-to-serial converters, quantum serial-to-parallel converters, quantum glue logic converters, and quantum combinational logic circuits.

16.2 The quantum programmable logic array

The quantum programmable logic device called the quantum programmable logic array (PLA) has the ability to operate in two different quantum array operations: programmable quantum AND and programmable quantum OR operation. In light of this, it is the most versatile quantum PLD. The inputs of quantum AND operation in this case are programmable. It follows that each quantum AND operation accepts inputs from quantum variables that are both normal and complemented. Therefore, it is possible to program any of the quantum inputs depending on the demand. By using these quantum AND procedures, it is possible to generate only the necessary quantum product terms.

The inputs of the quantum OR operation, in this case, can likewise be programmed. Due to the fact that all outputs of quantum AND operation are used as inputs for each quantum OR operation, it is easy to write any number of necessary product terms. As a result, the outputs of quantum PAL take the form of a sum-of-products.

16.2.1 The design procedure and working principles of quantum PLAs

A quantum PLA is constructed with m outputs and n inputs, as shown in figure 16.2. Fuse inputs enable the quantum AND array and quantum OR array operations by making them both programmable. Quantum AND matrices or arrays generate product terms by the quantum AND operation, which are then ORed to obtain the desired quantum function.

Let us use a quantum PLA to implement the following quantum function:

$|X\rangle = |AB\rangle + |AC'\rangle$; and

$|Y\rangle = |AB'\rangle + |BC\rangle + |AC'\rangle$.

The above two quantum functions are presented in sum-of-products form. The final quantum functions $|X\rangle$ and $|Y\rangle$ consist of two and three product terms, respectively. In each quantum function, the same product term, $|A\rangle|C'\rangle$, is utilized. For generating these two quantum functions, two programmable quantum OR operations and four programmable quantum AND operations are required.

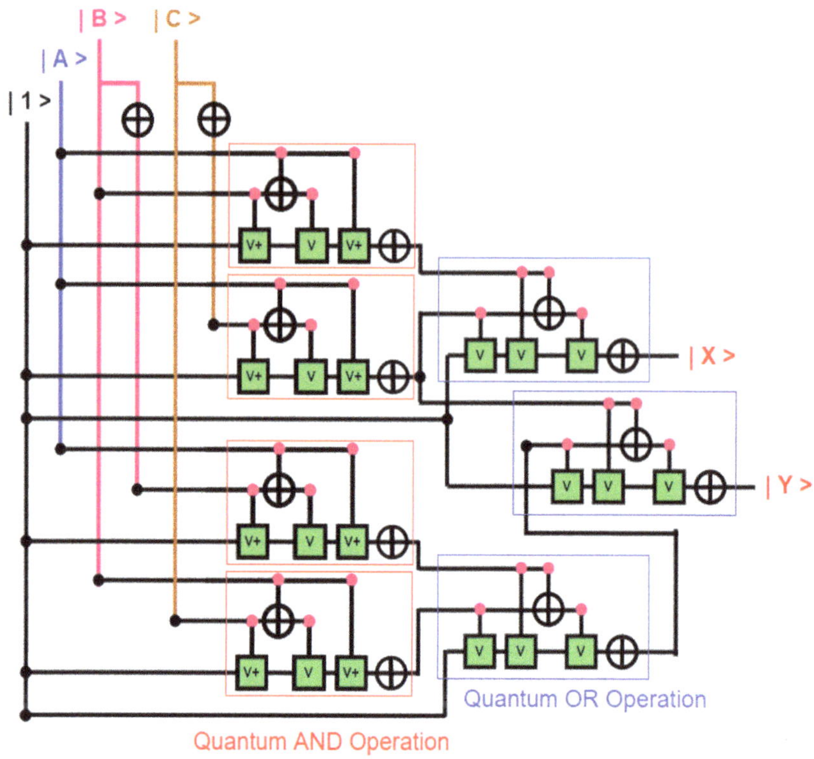

Figure 16.2. The quantum programmable logic array.

Both conventional and supplemented inputs of variables are accessible to the programmable AND operations. The inputs of each quantum AND operation in the preceding illustration include $|A\rangle$, $|B\rangle$, $|B'\rangle$, $|C\rangle$, and $|C'\rangle$. Therefore, it is necessary to program only the necessary product so that each quantum AND operation can produce one product term. Each programmed quantum OR operation has access to all of these product terms at its inputs. However, it is only required to program the necessary product terms to generate the corresponding quantum functions by each quantum OR operation.

16.2.2 The importance and applications of quantum PLAs

More versatility is offered by quantum PLAs since they offer programmable quantum AND operation arrays and programmable quantum OR operation arrays. Quantum PLAs have received widespread recognition as compact and space-saving alternatives to many complex circuits, most notably in feedback and control systems, where a lot of factor variables are required for the system to function effectively. The quantum PLA has a higher rate of speed and is more cost-effective. Its purpose is to provide datapath control. The quantum PLA is also used in quantum counter circuits. In addition, the quantum PLA serves as a quantum decoder. A programmed input/output (I/O) employs the quantum PLA as a quantum bus interface.

16.3 The quantum complex programmable logic device

A logic device with fully programmable quantum AND/OR array operations and macrocells is referred to as a complex programmable logic device (CPLD). The primary building elements of a CPLD are macrocells, which have complicated quantum logic operations, and quantum logic for implementing disjunctive normal form expressions. The quantum AND/OR array operations can perform a variety of quantum logic operations and are completely reprogrammable. Another way to think of macrocells is as functional building blocks that carry out quantum sequential or quantum combinatorial logic.

16.3.1 The design procedure and working principles of quantum CPLDs

Quantum CPLDs have a large number of logic blocks, each of which has 8–16 macrocells. Each logic block performs a distinct quantum function. Hence each logic block has a complete network of connected macrocells. These blocks could be linked together or not, depending on their intended usage. The inputs and outputs of these functional blocks are coupled by a global interconnection matrix (GIM). It is possible to change the contacts between the functional blocks by rearranging the connectivity matrix. The CPLD is connected to the outside by a few input and output blocks. Figure 16.3 presents the block diagram of the architecture of a CPLD.

The quantum CPLD architecture depicted in figure 16.3 has four quantum PLD function blocks. The links between the function blocks can be programmed. Interconnections between function blocks are made using a switch matrix. Additionally, the switch matrix of a quantum CPLD may or may not be entirely connected. It indicates that the quantum CPLD does not allow all connections

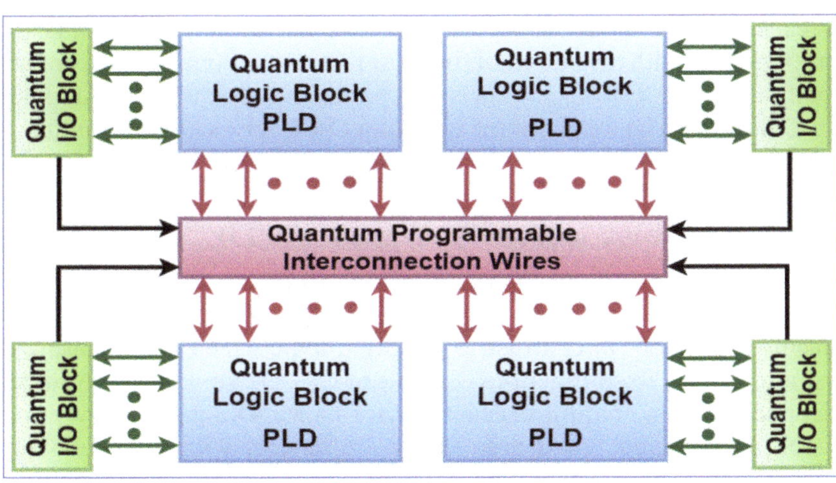

Figure 16.3. The block diagram of a quantum CPLD.

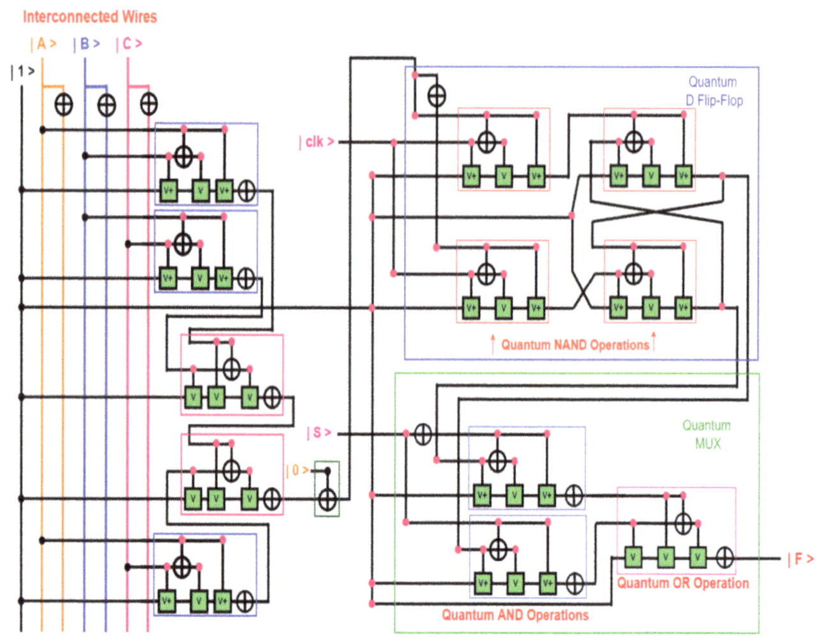

Figure 16.4. Programmable logic devices in a CPLD.

between outputs and inputs of the function block. A typical quantum PAL device has a complexity of only a few hundred logic operations, whereas CPLDs have thousands of quantum logic operations. The quantum realization of each PLD inside a quantum CPLD is shown in figure 16.4.

This quantum PLD consists of three quantum AND operations, two quantum OR operations, one quantum D flip-flop (D-FF), and one quantum multiplexer. The first quantum OR operation (the first pink box) receives inputs from the outputs of the first two blue boxes, which are quantum AND operations. The second quantum OR operation, which is related to quantum XOR operations, is then fed the output of the first quantum OR operation along with the output of the most recent quantum AND operations. The output of the quantum XOR operation is then passed into the quantum multiplexer and quantum D flip-flop. The output of the quantum D-FF serves as the quantum MUX's additional input. The final result is thus received via the quantum MUX.

16.3.2 The importance and applications of quantum CPLD

An electrical quantum component which is used to create reconfigurable quantum circuits is called a quantum PLD. A quantum PLD has an undetermined function at the moment of creation, in contrast to quantum logic which is built using discrete logic operations with fixed functions. Quantum CPLDs are perfect for demanding control applications that require excellent performance. Design-wise, they are simple and capable of reducing board area. They are devices with high reliability, low running costs, and low development cost. In addition, they create revenue more quickly.

16.4 The quantum field-programmable gate array

A two-dimensional array of cells known as a quantum field-programmable gate array (quantum FPGA) is an integrated quantum circuit which is built around a matrix of customizable logic blocks (CLBs) connected by programmable interconnects. It is made up of input/output pads, logic blocks with programmable interconnects, and reconfigurable interconnects. The logic blocks of a quantum FPGA could be memory components such as quantum flip-flops or quantum memory blocks. The logic blocks can carry out both straightforward and intricate computational quantum operations.

16.4.1 The design procedure and working principles of quantum FPGAs

The fundamental building block of a quantum FPGA is a quantum CLB, which is shown in figure 16.5. It is a logic cell that may be set up or programmed to carry out particular tasks. The connection block is joined to these building blocks. Quantum D flip-flops, quantum look-up tables (LUTs), and quantum 2-to-1 multiplexers are the three components of each quantum CLB. Quantum D flip-flops are employed as storage components. The right output is obtained by the multiplexer.

The brain of the quantum FPGA is a quantum LUT. It includes every logical result that could be produced by the design. The quantum LUT must be able to store all conceivable expression combinations. Quantum flip-flops, quantum LUTs, and quantum multiplexers are connected to create a quantum FPGA logic block.

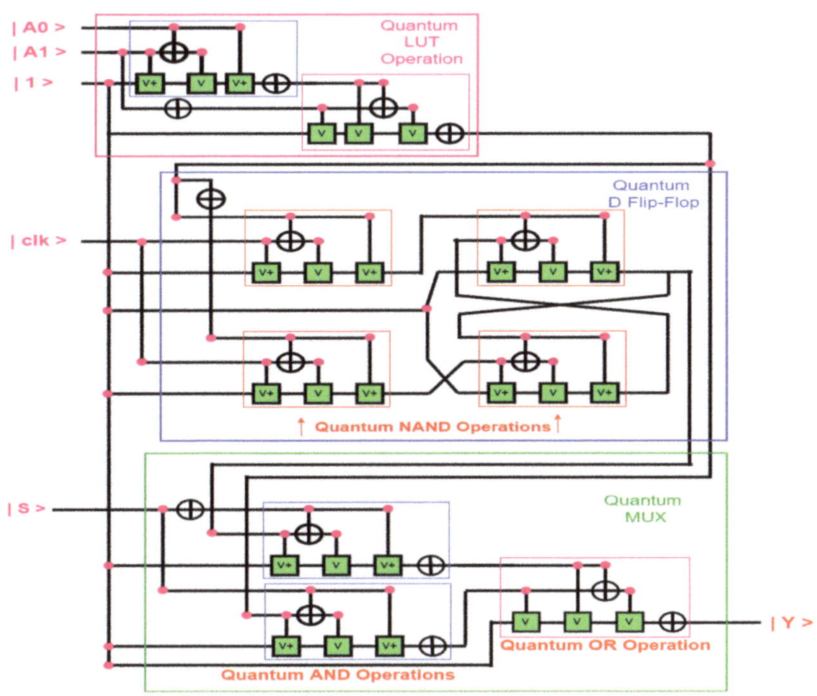

Figure 16.5. The configurable logic block.

A quantum D flip-flop, a quantum LUT, and a quantum multiplexer were used to construct the straightforward quantum FPGA logic block in figure 16.5. A straightforward two-input quantum LUT consists of one quantum AND operation, one quantum CNOT operation, and one quantum OR operation. The input of the quantum D flip-flop and multiplexer will be the output of the quantum LUT. The quantum D flip-flop is a sequential circuit made up of one CNOT and four quantum NAND operations. The output of the quantum D flip-flop is used as the input in the quantum multiplexer. Two quantum AND operations, one quantum CNOT operation, and one quantum OR operation are used to create a 2-to-1 quantum MUX. For the quantum FPGA logic block, the quantum multiplexer produces the desired output.

16.4.2 The importance and applications of FPGAs

One of the most widely used parts of quantum circuits is the quantum FPGA. The use of quantum FPGAs has many benefits. These benefits include reconfigurable computing, affordability, high density, fast registers, and resource routing. Its applications are distinctive. Additionally, it is versatile and reusable. Quantum FPGAs could be a perfect fit for various markets, including aerospace and military, due to their programmable nature. Quantum FPGAs are being used for next-generation full-feature consumer applications, such as high-performance convergent handsets, information appliances, home networking, and residential set-top boxes. It can also be very useful in applications for diagnosis, monitoring, and therapy.

16.5 Summary

Quantum PLDs are integrated circuits with quantum programmable logic. Both arrays of quantum AND operations and quantum OR operations are used in PLDs.

Four different PLDs have been presented in this chapter. These quantum PLDs are quantum CPLDs, quantum FPGAs, quantum PLAs, and quantum PALs. The design procedures and the working principles, along with the importance and applications of these PLDs have been discussed here.

Critical thinking questions

1. For quantum programmable logic functions, which type of quantum PLD should be used? Explain in detail.
2. Explain the differences between a quantum PLA and quantum PAL.
3. What are the advantages of quantum PLA over quantum ROM?
4. Which is more flexible, quantum PAL or a quantum PLA? Explain.
5. Describe the drawbacks of quantum programmable logic arrays.
6. What are the primary benefits of designing a circuit using quantum PLDs?
7. Explain the applications of quantum PLDs.
8. What happens once a quantum PAL has been programmed? Explain.
9. What is the main programming unit in a quantum FPGA? Discuss it in detail.
10. In a quantum LUT operation, how many combinations are supported?
11. Describe the three fundamental components of a quantum FPGA.

References

[1] Bocko M F, Herr A M and Feldman M J 1997 Prospects for quantum coherent computation using superconducting electronics *IEEE Trans. Appl. Supercond.* **7** 3638–41

[2] Cong W, Bičák J, Kubizňák D and Mann R B 2021 Quantum detection of inertial frame dragging *Phys. Rev.* D **103** 024027

[3] Karafyllidis I G 2005 Quantum computer simulator based on the circuit model of quantum computation *IEEE Trans. Circuits Syst. I: Regul. Pap.* **52** 1590–6

[4] Mia M S and Babu H M H 2014 An efficient approach to design a reversible fault tolerant programmable array logic *J. Bangladesh Electron. Soc.* **14** 71–81

[5] Ömer B 2005 Classical concepts in quantum programming *Int. J. Theor. Phys.* **44** 943–55

[6] Singla P and Malik N K 2012 A cost-effective design of reversible programmable logic array arXiv:1204.5525

[7] Vartiainen J J, Möttönen M and Salomaa M M 2004 Efficient decomposition of quantum gates *Phys. Rev. Lett.* **92** 177902

[8] Viamontes G F, Markov I L and Hayes J P 2004 High-performance QuIDD-based simulation of quantum circuits *Proc. Design, Automation and Test in Europe Conf. and Exhibition* vol 2 (Piscataway, NJ: IEEE) pp 1354–5

[9] Viamontes G F, Rajagopalan M, Markov I L and Hayes J P 2003 Gate-level simulation of quantum circuits *Proc. Asia and South Pacific Design Automation Conf.* pp 295–301

Chapter 17

The quantum processor circuit

Learning objectives

- Discuss the basic definition of a quantum processor.
- Learn about the fundamental concepts of and suitable circuits for the quantum instruction register (IR), quantum program counter (PC), quantum decoder, quantum multiplexer (MUX), and quantum accumulator.
- Define the fundamental elements of a quantum processor.
- Explain the quantum IR with its logical circuit.
- Acquire knowledge from the complete block diagram of a quantum processor.
- Define the functions of a quantum processor.
- Describe a quantum random access memory (RAM) with a suitable circuit.

The quantum logic circuitry that responds to and executes the fundamental instructions which power a computer is known as a quantum processor, also referred to as the quantum central processing unit (CPU). It interprets the majority of computer commands, and the quantum CPU is regarded as the primary and most significant integrated circuitry chip in a quantum computer. With this processor, a difficult mathematical operation, such as multiplication, division, or addition, is completed. The quantum computer's instructions and processing power come from the quantum processor. The faster a quantum computer can finish a task, the more potent and modern the quantum processor is. This chapter only focuses on the most powerful quantum processor components.

17.1 Introduction

In the 1970s, Richard Feynman asserted that it is possible to imitate physics with a computer. He made the initial discovery of nanotechnology, which uses individual atoms and molecules to create complex atomic structures. The method of technology employed to control individual states is known as 'quantum nanotechnology'. In the spring of 1988, Drexler taught the first formal course in nanotechnology at Stanford University. He raised the prospect of nanosized systems, and Drexler and Merkle developed a machine for computing energies called a molecular machine, which is a

doi:10.1088/978-0-7503-5683-1ch17

composition of static systems without knowledge of their dynamics. For many years, several research teams have attempted to model various parts of nanomachines. In addition, Peter Shor developed a method in 1994 that can factor a number N in $O((\log N))$ time and $O(\log N))$ space. This algorithm is known as Shor's algorithm. At an IBM demonstration in 2001, a team presented an algorithm that could factor 15 into 3 and 5. To do this, they employed a seven-qubit quantum computer. Shor's algorithm was used to construct a seven-qubit register. A physical mechanism for employing individual atomic ions, whose electrical states store the quantum bits of information, was then put forth by Cirac and Zoller. The altered quantum information can be carried by a single trapped ion. At the National Institute of Standards and Technology, David Wineland's team presented an ion-trapped quantum computer.

A two-level system can be used to implement quantum information or a qubit; the first level is the spin of an electron in a magnetic field, and the second level is to employ two levels of an atom. With the fidelity required for fault-tolerant quantum computing (QC) employing high threshold quantum error correction codes, qubits are initialized and handled using single-qubit gates, two-qubit gates, qubit state preparation, and readout.

The development of a practical quantum computer still faces numerous obstacles despite the great promise of trapped ions. To make a functional quantum processor, DiVincenzo demonstrated five essential requirements in 2000. He stated that a physical system must have the following characteristics. (i) Well-defined two-level quantum systems, or qubits, that can be isolated from their surroundings. (ii) The ability to initialize the system into a well-defined and determinate initial state. (iii) Qubit decoherence times that are significantly longer than the gate times. (iv) A collection of universal quantum gates that can be applied to any qubit (or pair of qubits, in the case of two qubits). (v) A qubit-specific measurement capability. Only a few qubit technologies, such as trapped ions, have so far met all of DiVincenzo's original requirements for high fidelity. 'Well-characterized' refers to a qubit in a number of distinct ways. Its physical characteristics should be known precisely, including the qubit's internal Hamiltonian (which determines its energy eigenstates, which are frequently, but not always, assumed to be the $|0_i$ and $|1_i$ states), the presence of and couplings to other qubit states, interactions with other qubits, and couplings to external fields that might be used to manipulate the qubit's state. The quantum computer's control system should be built in such a way that there is little chance that the system would ever enter states such as the third, fourth, etc, levels of the qubit.

17.2 Basic definitions

The term 'central processor' or 'main processor' refers to a CPU. A quantum computer's input/output operations can be carried out via a similar quantum electronic circuit to a digital computer's CPU. The computer program's instructions are carried out via basic mathematical and logical operations. The CPU is in charge of managing all types of data flow and instructions. The quantum CPU is made up of five main parts:

1. Quantum control unit (CU).
2. Quantum register.

3. Quantum arithmetic logic unit (ALU).
4. Quantum random access memory (RAM).
5. Quantum buses.

1. *Quantum control unit.* A quantum circuit that issues commands to a computer processor is called a control unit. Numerous selection circuits, including multiplexers, decoders, and other devices, make up the control unit.

2. *Quantum register.* The smallest collection of qubits that can store the locations that make up a quantum processor is a quantum register. Any type of data, including a qubit sequence, instructions, and a storage address, can be stored in a quantum register.

3. *Quantum ALU.* The quantum ALU was initially described by Oskin *et al* as a component of a quantum computer architecture. The program's arithmetic operations are managed by the arithmetic unit. The quantum ALU unit performs all types of mathematical operations, including addition, subtraction, division, and multiplication.

4. *Quantum RAM.* Quantum RAM, often known as volatile memory, is part of the quantum CPU that aids in boosting system performance. The primary function of quantum RAM is to temporarily store and access data.

5. *Quantum buses.* Whether computers are supercomputers, quantum computers, or conventional computers, buses are necessary for data transfer between processors and other parts. For this reason, the quantum processor that powers quantum computers will be discussed in this chapter, where buses will be employed similarly to other types of computers. Address bus, control bus, and data bus are the three types of buses that are covered briefly in the architecture of fundamental components.

17.3 The block diagram of a quantum processor

The CPU, which manipulates data and carries out commands, is the brain of a computer. Many entire circuits, including the instruction register (IR), program counter (PC), multiplexer, ALU, and RAM, etc, are included in it. Since it is a quantum processor, all circuits are constructed using quantum logic gates. The comprehensive block diagram of a quantum processor is shown in figure 17.1.

Only the fundamental CPU components are visible in this entire two-qubit processor. To conduct useful work, CPUs require two inputs: instructions and data. The IR's job is to instruct the CPU on what operations should be carried out on the data. Here qubit-like information is used to represent instructions. The memory stores the inputs to the CPU. As shown in figure 17.1, the instruction register receives RAM data from memory, and the CPU cycles through a cycle of fetching instructions. It is first fetched, then decoded, and then executed. The 'fetch–decode–execute' cycle can be used to describe this procedure. Data are transmitted from memory to the instruction register, which initiates the cycle. It should be noted that data sent from memory will always use the data bus to convey data. Selecting the machine language from the IR allows the unique qubit patterns to be removed

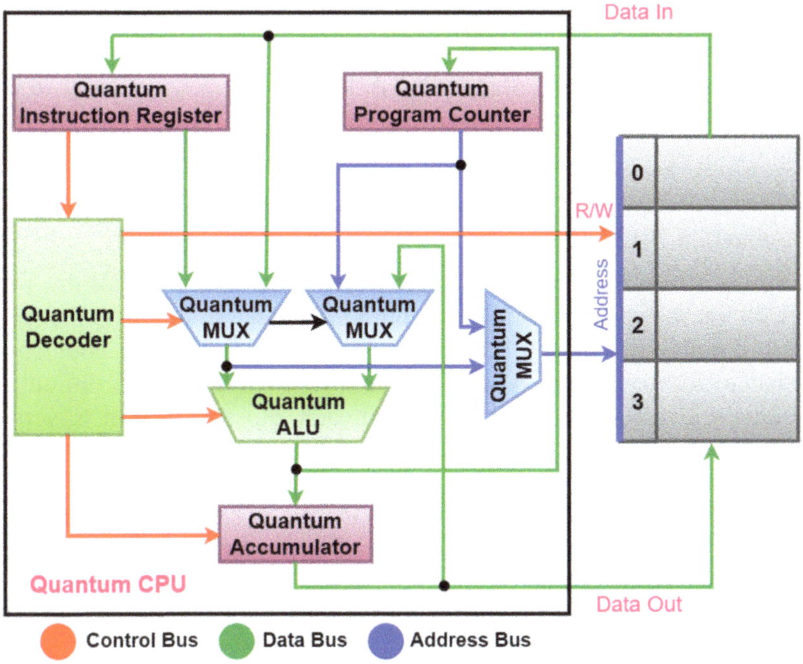

Figure 17.1. The quantum processor.

before being delivered to the quantum decoder. Decoding encoded data from one format to another is the decoder's main function. Because of the quantum decoder, the cycle's second stage can now function. The decoder indicates which qubit pattern will be used and activates the circuit required to carry out the specific operation. The following instruction will start the circuit working if the procedure was completed successfully. The instruction register is a special-purpose register that holds the current instruction being executed. When the CPU fetches an instruction from memory, it stores it in the instruction register. The program counter is increased by one memory address upon completion of the instruction. This is how this quantum processor operates overall.

17.4 The basic components of a quantum processor

The following components have been combined to create a whole quantum processor, such as a quantum RAM, quantum IR, quantum program counter, quantum decoder, quantum multiplexer, quantum ALU, and quantum accumulator. In the quantum CPU, data are also transferred from one component to another using buses. The data bus, address bus, and control bus are the three different types of buses.

The data bus, which transports data back and forth between the CPU and RAM, is bidirectional. The address bus, which connects other components such as primary storage and input/output devices to the processor, is unidirectional and is used to transfer memory addresses.

The final bus is a control bus, which is used to connect processors to other parts that check if everything is moving from one place to another smoothly or not. These are additional crucial CPU elements that must function properly in order to perform useful work.

Now, the other components of a quantum CPU are discussed below.

17.4.1 The quantum RAM

For the purpose of mimicking 4-to-2 qubit RAM, two address lines are required, each of which must have one ancillary bit and must be in CNOT form. This combination of address lines will be used as the input for 2-to-4 decoders, each of which consists of four quantum AND gates and has a single enable input.

This decoder provides four select lines, and each chosen line will traverse every RAM cell. Keep in mind that the RAM will calculate words as $2k$, where k is the address line, 2^k is the total number of n-bit words, and $k \times 2^k$ is the decoder combination. This two-qubit RAM consists of four distinct RAM cells, each of which has three inputs such as $|\text{In}_0\rangle$ or $|\text{In}_1\rangle$, read/write inputs, and a line selector. A quantum OR gate, which generates the final output, will take the output from four quantum RAM cells as its input. This is how a 4-to-2 qubit RAM is designed as a whole, as shown in figure 17.2.

The primary and volatile memory is the most crucial part of a CPU and stores data for a brief period of time. Figure 17.2 shows a 4-to-2 qubit RAM implementation. Each of the four independent 'words' of memory in this quantum RAM is two qubits wide. There are three inputs and one output for the quantum RAM cell. Figure 17.3 accurately and fully describes the circuit of a quantum RAM cell. Two quantum RAM cells that are set up so that both qubits may be accessed simultaneously make up a word. Two address lines are needed for four words of memory. The two-qubit address lines $|A_0\rangle$ and $|A_1\rangle$ are used as input to a 2-to-4 decoder, which chooses one of the four words. The decoder is activated by the memory-enabled input. None of the memory addresses will be chosen if the memory enable is set to $|0\rangle$, which is the case for all of the decoder's output. One of the four words is chosen, however, when the memory enables are $|1\rangle$. The value in the two address lines determines which word is chosen. The read/write input determines the operation once a word has been chosen.

The four qubits of the chosen word pass through the quantum OR gates to the output $|Z_0\rangle$ and $|Z_1\rangle$ terminals during the read operation. However, the data from the input lines is transferred into the four quantum cells of the chosen word during the write process. The non-selected quantum RAM cells become disabled and retain their original qubit. However, none of the words are chosen when the memory enable input that enters the decoder is identical to $|0\rangle$, and all quantum cells remain intact regardless of the read/write input's value. The RAM operates in the manner described below. The quantum RAM cell is described below.

The RS flip-flop has been used in the creation of the quantum RAM cell. Each word will include $m \times n$ total quantum cells, where m stands for words with n bits. The quantum cell contains one output line with the label 'Output', three input lines

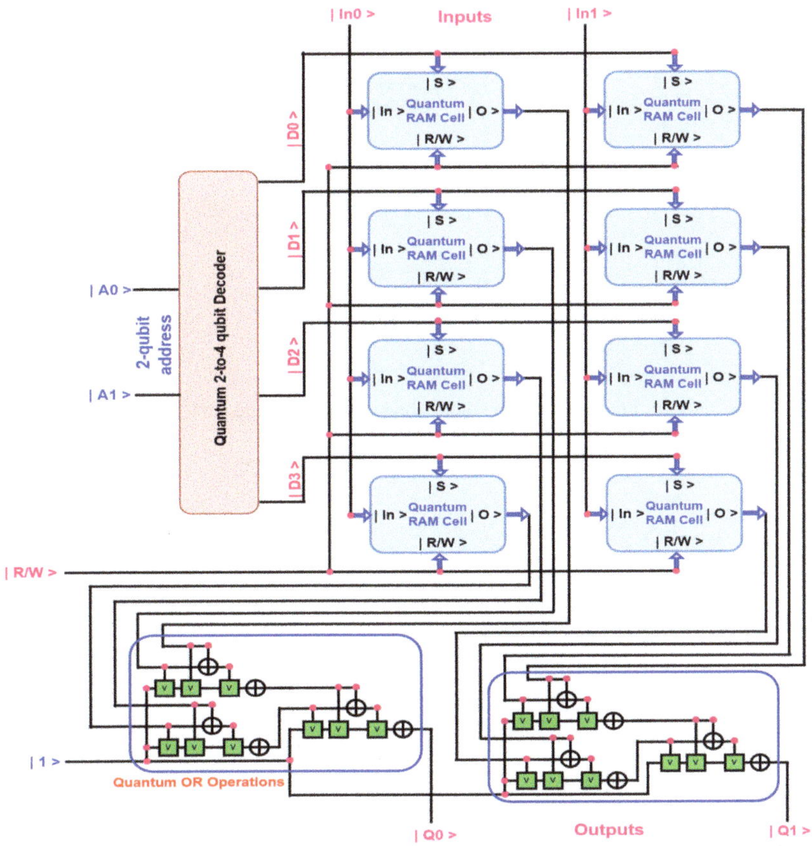

Figure 17.2. The 4-to-2 qubit quantum RAM.

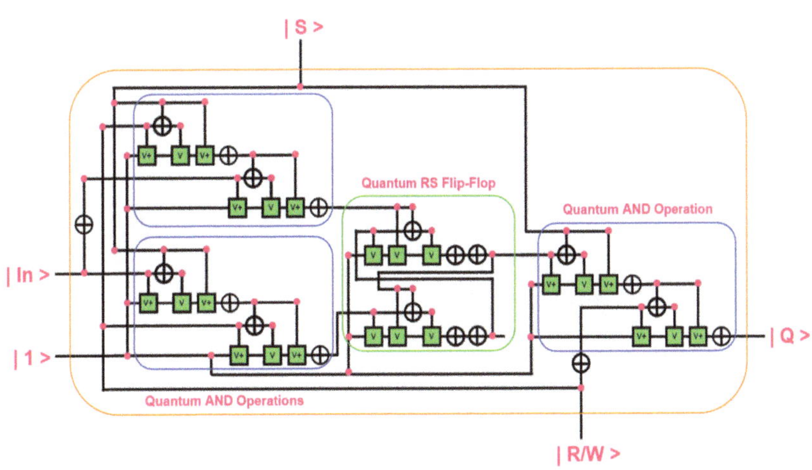

Figure 17.3. The quantum RAM cell.

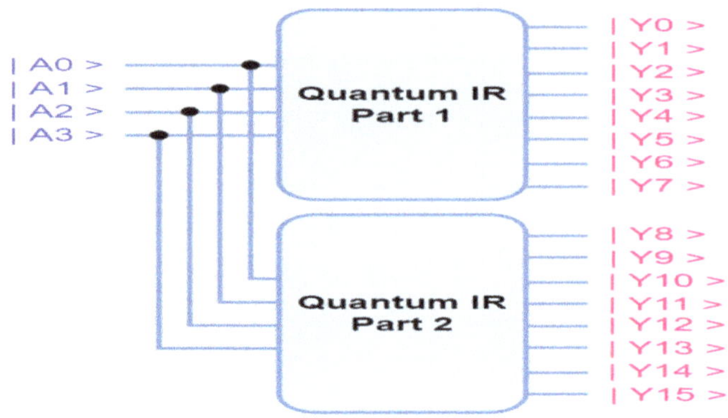

Figure 17.4. The block diagram of a quantum IR.

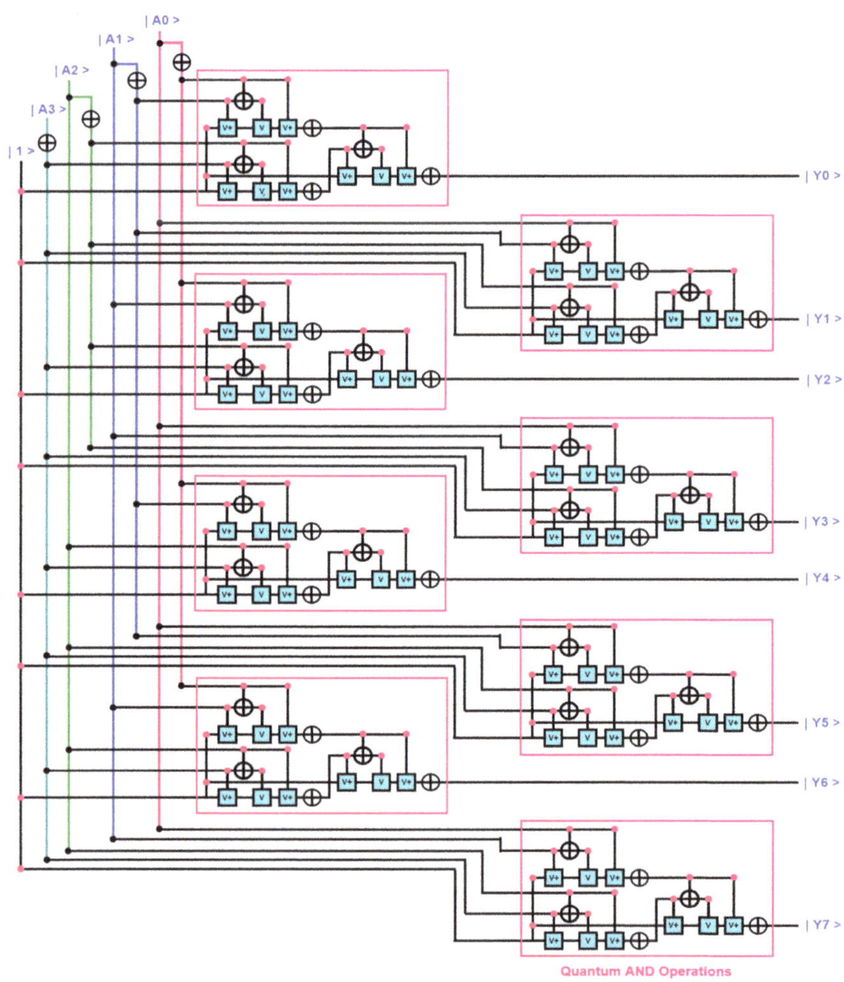

Figure 17.5. The four-qubit instruction register (first part).

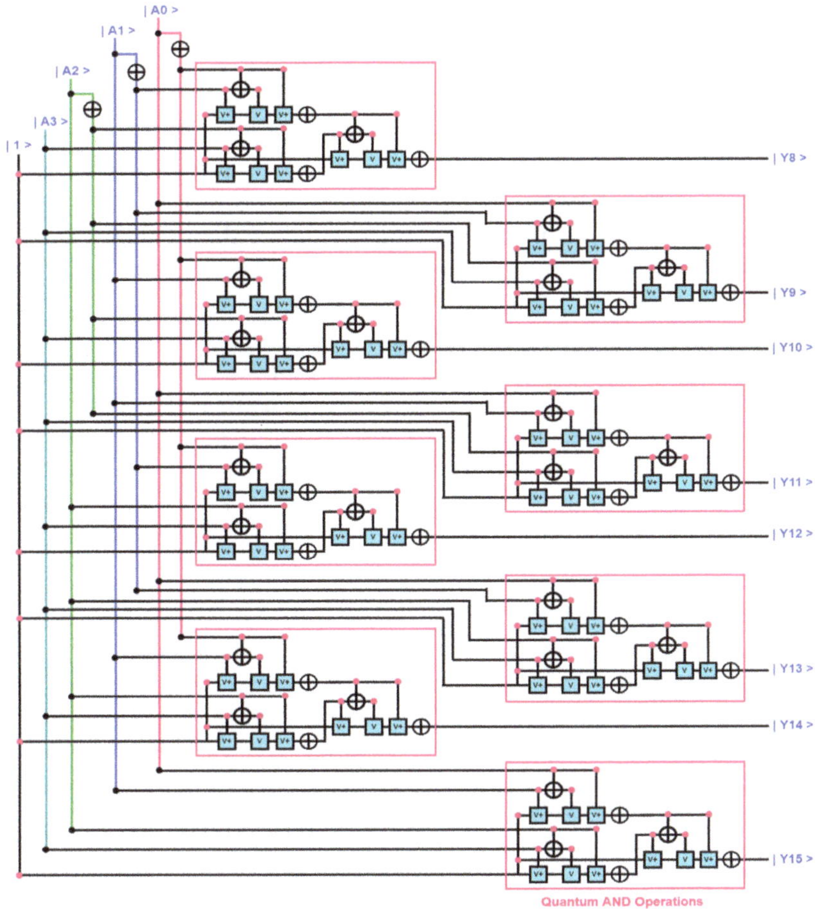

Figure 17.6. The four-qubit instruction register (second part).

labeled 'Select', 'Read/Write', and 'Input', and three input lines. Either reading or writing can be accessed using the 'choose' input. The cell executes the memory operation when the select line is high or $|1\rangle$. However, if the quantum cell's select line is low or set to $|0\rangle$, the cell is not motivated to execute a read from or write to. The following input is 'Read/Write', which will be handled by a system clock. The 'read' phase will be carried out if the clock value on the read/write line is $|0\rangle$, and the 'write' phase will be carried out if it is $|1\rangle$. Think about the selected cell for a moment. In such a scenario, if the clock value is $|0\rangle$, the contents of the cell must be read, and this time the output value will only be dependent on the flip-flop's Q value. However, if Q is low, the output of the cell will be $|0\rangle$, and if Q is high, the output of the cell will be $|1\rangle$. This occurs because the cell's output has a quantum AND gate with three inputs such as negated read/write, select, and Q, and both 'negated read/write' and 'select' are currently high.

17.4.2 The quantum instruction register

There are sixteen quantum AND operations in the quantum instruction register. It has been broken into two segments because it is a big circuit. Eight quantum AND operations are contained in the first part as shown in figure 17.4, and eight more are contained in the second part as shown in figure 17.5. The block diagram of this register is shown in figure 17.6.

A two-qubit CPU uses this instruction register; hence the instructions will be 4. A minimum set of four instructions, namely LOAD A, LOAD B, ADD A B, and OUT, can be defined.

The primary utility of the instruction registers, a type of special-purpose register, is to hold the instructions that the quantum computer is now executing. That instruction is now being performed and is stored in an instruction register. The instruction word is stored in the IR. Any instruction that the CPU retrieves from memory is first temporarily put in the instruction register.

A qubit word or code that specifies a particular operation to be carried out can be the instruction. The instruction is subsequently decoded and executed by the CPU.

17.4.3 The quantum program counter

The first of the two quantum D flip-flips that make up the quantum program counter (PC) is constructed using four quantum NAND gates, which are depicted using green boxes. According to figure 17.7, this quantum D flip-flop provides two outputs, one of which has been skipped because it is not necessary to display it. Four quantum NAND gates, which are restricted to a violet color box, make up another quantum D flip-flop. Additionally, this quantum D flip-flop produces two

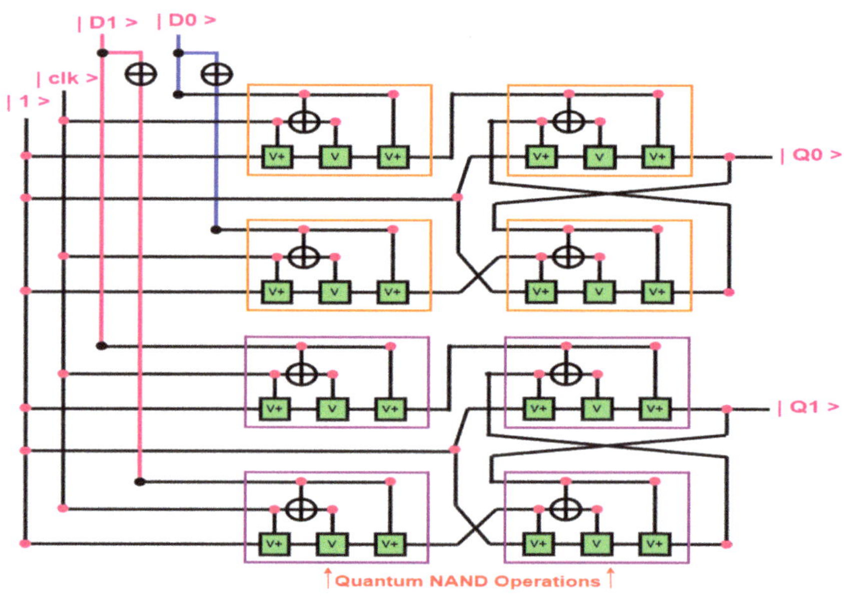

Figure 17.7. The two-qubit program counter register.

outputs. There are only two outputs that serve as inputs as shown in figure 17.7. Therefore, the remaining two outputs have been omitted. This two-qubit program counter is appropriate for processors with two qubits. The circuit structure of this quantum program counter is shown below.

This program counter is used to store the subsequent instruction that will be carried out. The program counter is increased by one after the current instruction is finished. In memory, each instruction and piece of data has a unique address. For instance, the program counter will initially be loaded with address 2 if a program starts with an instruction placed in memory location 2. The PC is increased by one to the following address, which is 3, when this command is carried out. A program's instructions always store themselves in the same order in the memory. Here, $|D_0\rangle$ and $|D_1\rangle$ stand for inputs from which the quantum D flip-flop processes and outputs the desired result.

17.4.4 The quantum decoder

The 2-to-4 quantum decoder, which is utilized in two-qubit CPUs, consists of four quantum AND operations and a single enable input. With and without CNOT form, $|A_0\rangle$ and $|A_1\rangle$ are inputs, and $|D_0\rangle$, $|D_1\rangle$, $|D_2\rangle$, and $|D_3\rangle$ are outputs. The circuit structure of a quantum 2-to-4 decoder is given below.

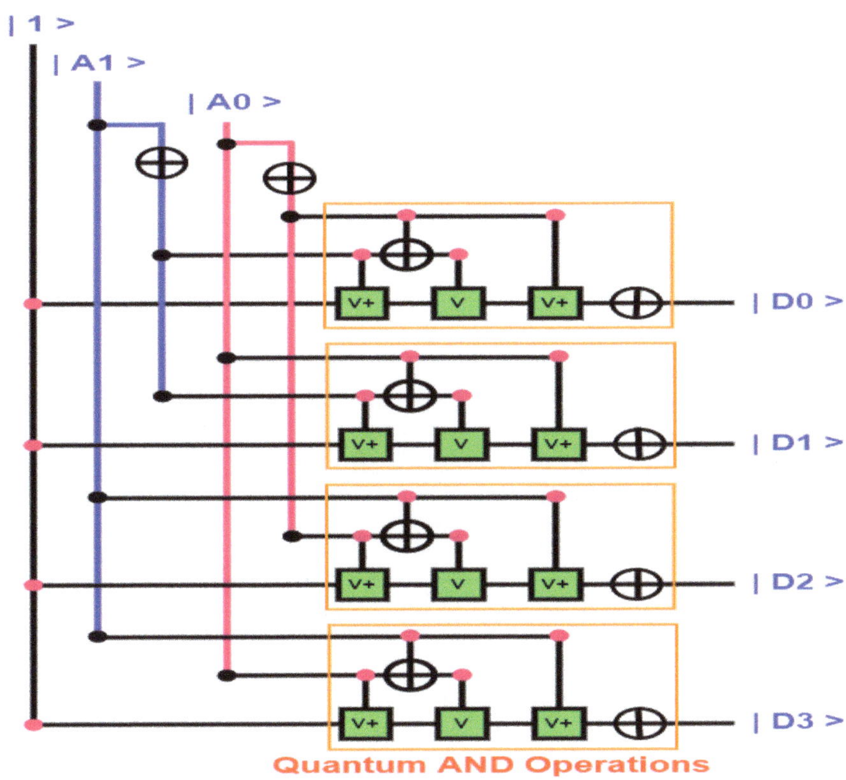

Figure 17.8. The quantum 2-to-4 decoder circuit.

It is a combinational quantum circuit with n input lines and a potential for 2^n output lines. Each output has a single product, and these quantum AND operations are used to produce this product. When the enable input is $|1\rangle$, the minimum of two input qubits are $|A_0\rangle$ and $|A_1\rangle$. However, if the enable input is set to $|0\rangle$, all of the decoder's outputs will be equal to $|0\rangle$, and if it is set to $|1\rangle$, just one of these four outputs will be active, i.e. $|1\rangle$.

17.4.5 The quantum multiplexer

Four data input lines, from $|D_0\rangle$ to $|D_3\rangle$, two select lines, $|S_0\rangle$ and $|S_1\rangle$, and one output line make up a 4-to-1 quantum multiplexer (MUX), as shown in figure 17.8.

There are other ways to present a 4-to-1 quantum MUX, however, in this case, three 2-to-1 quantum MUXs are used to design the 4-to-1 quantum MUX. The quantum MUX can be formulated as 2^n-to-1. The 2-to-1 quantum MUX is treated as one MUX. Any one of the four input lines may be chosen via the select lines $|S_0\rangle$ and $|S_1\rangle$ to connect the output lines.

Multiple inputs and a single output are provided by the multiplexer. Two quantum AND operations construct a quantum 2-to-1 MUX. A quantum OR gate will accept its input from the result of two quantum AND gates. From a quantum 2-to-1 MUX two least significant qubit outputs are produced, which are propagated. In this way, the desired result is obtained.

17.4.6 The quantum arithmetic logic unit

This is a two-qubit ALU that uses two to four quantum decoders to choose among sixteen quantum AND operations. Among the addition, multiplication, subtraction, and division quantum operations, only one of the quantum operations is carried out by the 2-to-4 quantum decoder (figure 17.9). The circuit structure of a quantum decoder is shown in figure 17.10.

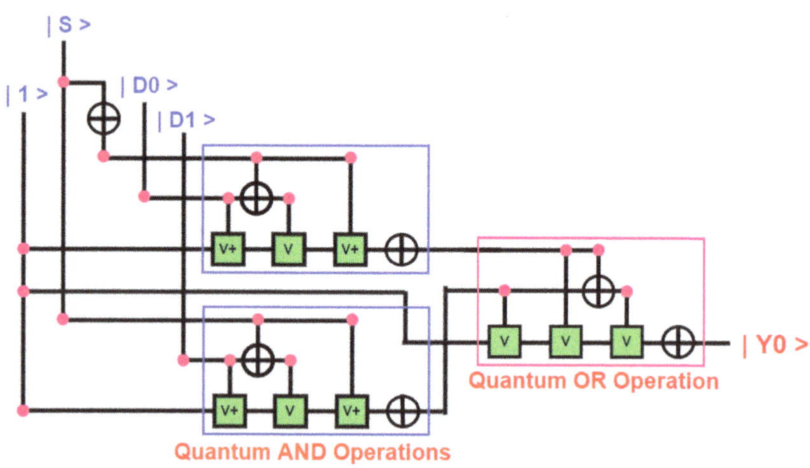

Figure 17.9. The 4-to-1 quantum MUX.

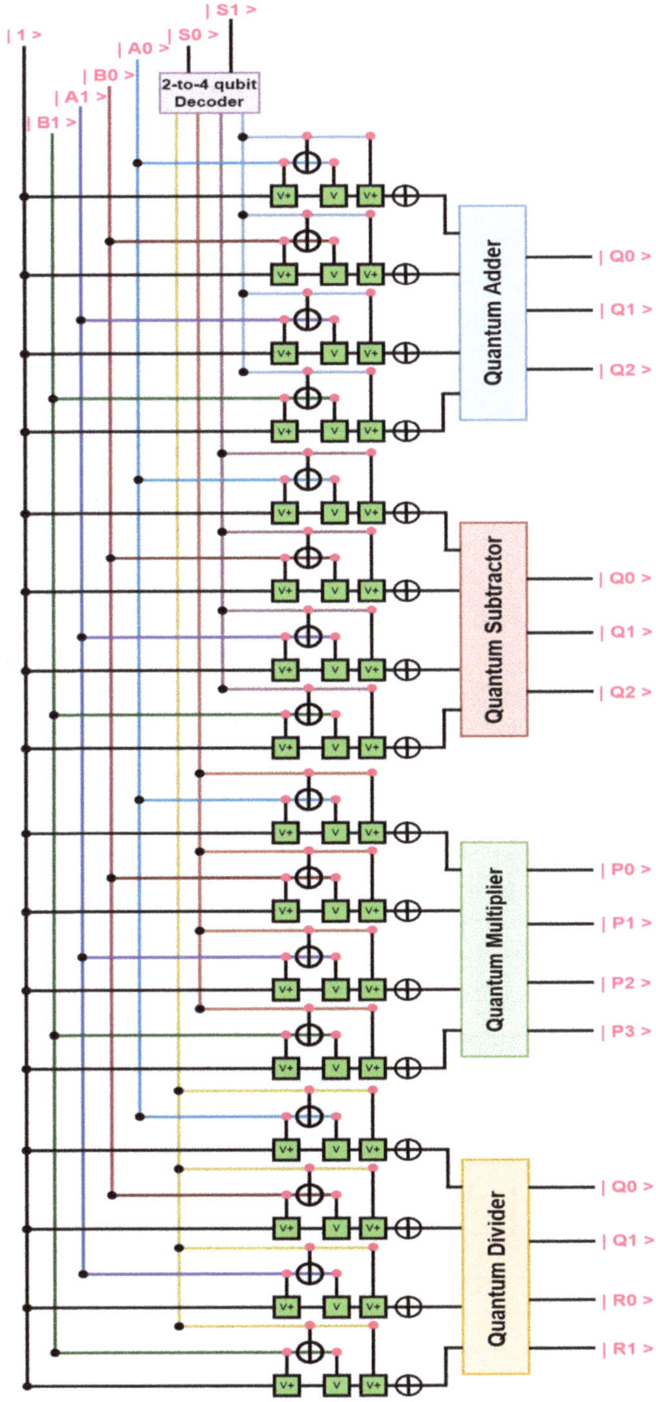

Figure 17.10. The two-qubit quantum ALU.

A quantum decoder is used to perform all quantum operations such as addition and subtraction, etc. A brief description of each quantum operation is given below.

1. *The quantum adder*. This quantum adder performs addition using quantum X-OR, AND, and OR operations. An illustration of this quantum operation is presented in figure 17.11.

Figure 17.11 shows that the first adder generates a sum and C_{out}, where C_{out} is the C_{in} of the following adder. Additionally, the entire adder has a C_{out} and the final output.

2. *The quantum subtractor*. A quantum subtractor works by subtracting two quantum numbers and is constructed using a quantum complete subtractor. The circuit is as follows. A combinational quantum circuit known as a complete subtractor may execute the subtraction of two qubits, where $|A\rangle$ and $|B\rangle$ are control bits, $|B_{\text{in}}\rangle$ is the borrow input, $|D\rangle$ is the difference output, and B_{out} is the borrow out. Here, the symbols $|A\rangle$ and $|B\rangle$ stand for minuend and subtrahend, respectively, while $|0\rangle$ is the constant target qubit. Figure 17.12 shows that the first full subtractor results in a difference, and B_{out} is the B_{in} of the following full subtractor. Finally, this entire subtractor has a B_{out} and a second output.

3. *The quantum multiplier*. Two quantum digits are required for a two-qubit quantum multiplier. For the main circuit of a quantum function representing a

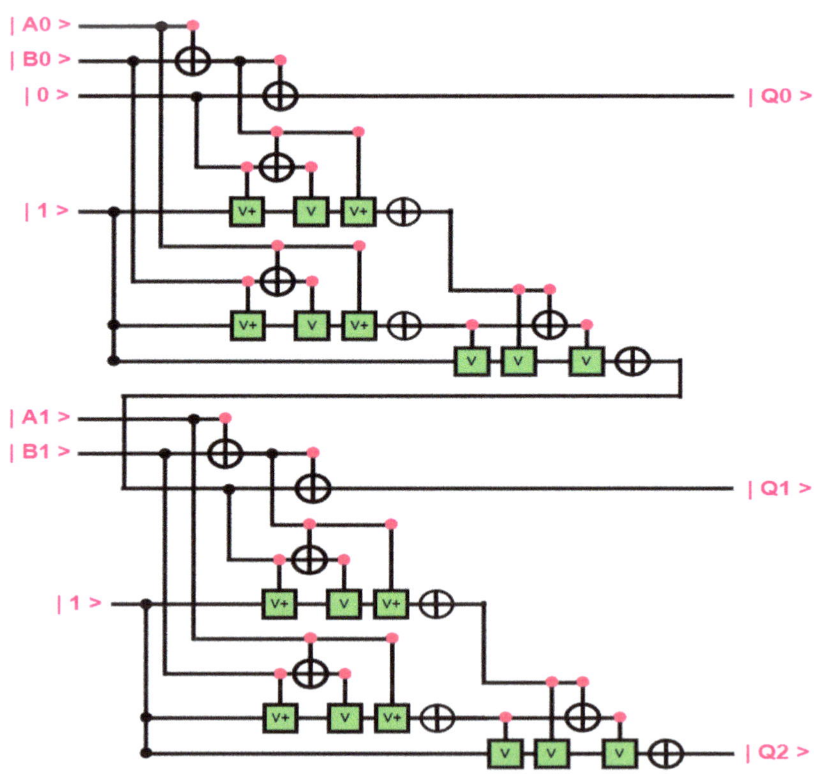

Figure 17.11. The quantum adder operation.

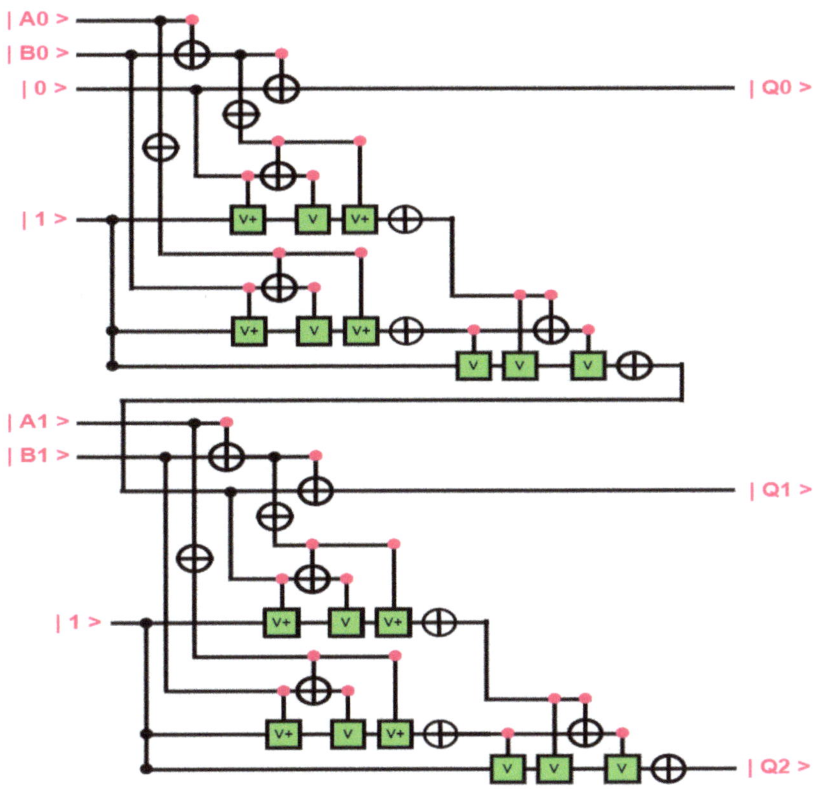

Figure 17.12. The quantum subtractor operation.

two-qubit multiplier, four quantum AND gates and two quantum half adders are required. In this case, the multiplication will be done by the quantum AND gates, and the partial product terms or carry will be added by quantum half adders (figure 17.13). Thus, the final quantum multiplier circuit is as follows.

It is first necessary to carry out quantum AND operations. Let the quantum digits $|A_1\rangle$, $|A_0\rangle$ and $|B_1\rangle$, $|B_0\rangle$ be the multiplicand and multiplier, respectively. Given that in a quantum circuit the target qubit, $|1\rangle$, is taken to be a constant qubit, in this circuit, the partial product terms will be added by the quantum half adders, while the quantum AND gate will execute the multiplication. These quantum operations result in a straightforward four-qubit output.

4. *The quantum divider*. First, the quantum AND logical operation is required for a two-qubit quantum divider. Three inputs have been utilized in this quantum AND operation. The third input of the first quantum AND gate has been connected to the second quantum AND gate operations once more to prevent dilution, allowing to obtain the output of quantum AND operations with just three inputs. The block diagram in figure 17.14 is a two-qubit divider.

First, there are four quantum AND gates in this circuit by taking three inputs. The third input of the first quantum AND gate is connected to the second quantum AND gate once again to prevent circuit dilution. This connection makes it possible

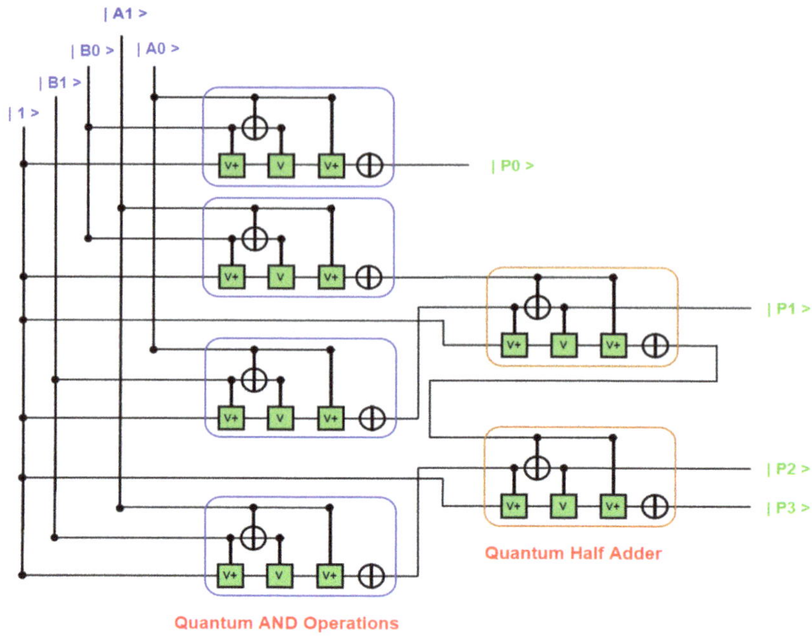

Figure 17.13. The quantum multiplier operation.

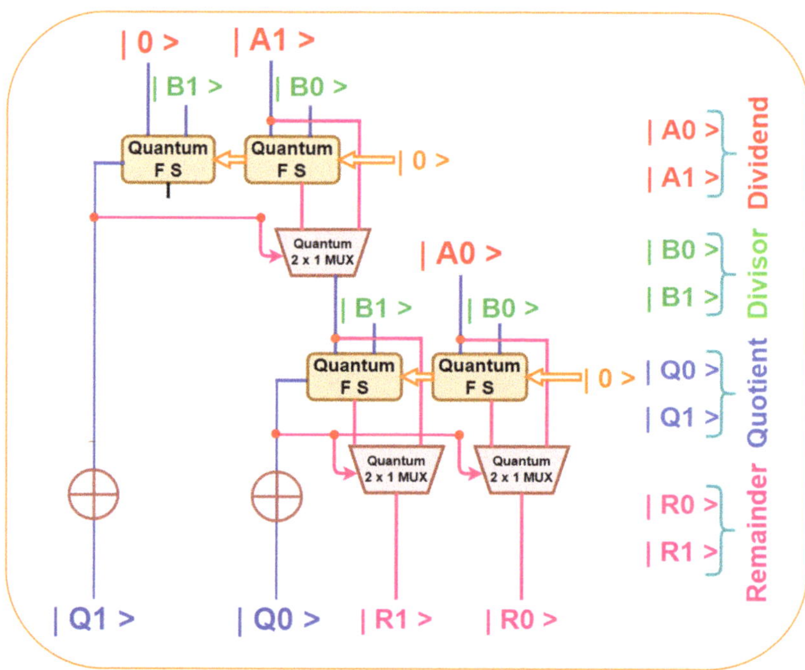

Figure 17.14. The quantum divider operation.

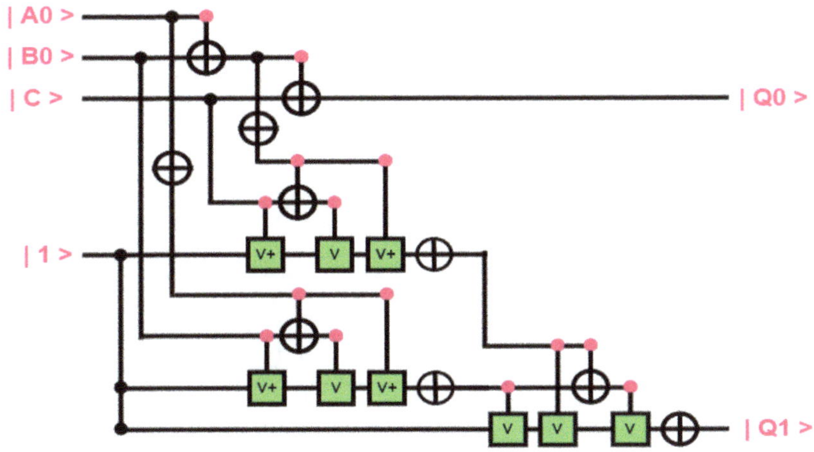

Figure 17.15. The quantum full subtractor operation.

to obtain the output of quantum AND operations with just three inputs. In order to obtain the precise output of $|Y_1\rangle$, the output of the quantum AND gate must then be connected with the input of the quantum OR gate. The fundamental characteristic of a quantum logic circuit is that the exact output of a quantum AND gate serves as a target qubit which is obtained by using the constant $|1\rangle$. For a maximum two-qubit output, two-bit divisor, and two-bit divisible are used to finish this circuit. Let $|A_0\rangle$ and $|A_1\rangle$ be $|1\rangle$ and $|0\rangle$, respectively, and let $|B_0\rangle$ and $|B_1\rangle$ be $|1\rangle$ and $|1\rangle$, respectively, where $|A_0\rangle|A_1\rangle$ is the divisor and $|B_0\rangle|B_1\rangle$ is the dividend. This is the two qubit quantum divider's fundamental mode of operation. For better understanding, the full quantum subtractor is given in figure 17.15.

17.4.7 The quantum accumulator

One quantum AND gate and two quantum D flip-flops make up a two-qubit accumulator with the help of quantum NAND gates. The inputs of the quantum AND gate are $|\text{LOAD}\rangle$ and $|\text{CLK}\rangle$, and its output will be used as the input for each quantum D flip-flop. Following quantum logical operations, the first and second quantum D flip-flops provide the first and second outputs, respectively. The circuit structure is shown in figure 17.16.

This is a quantum register which is called a quantum accumulator that serves as a temporary storage area and stores a value used as a bridge in logical and mathematical processes. In the operation '2 + 3 + 4', for instance, the accumulator will initially store the value 2, then the value 5, and finally the value 9.

In the event that the output of the quantum AND gate is $|1\rangle$, the data will be saved in the accumulator. Additionally, no information will be saved in the accumulator if the output of the quantum AND is $|0\rangle$.

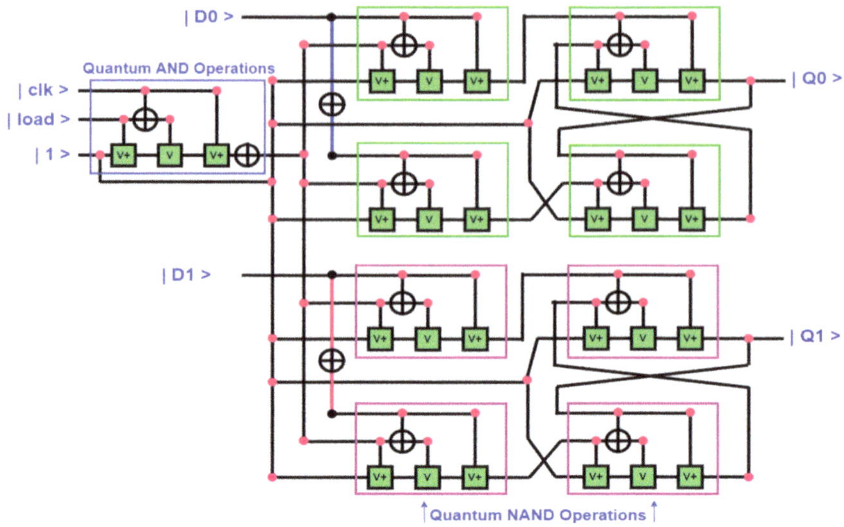

Figure 17.16. The two-qubit quantum accumulator.

17.5 Applications

It has already been established that quantum computing employs numerous methods to handle a variety of issues. Grover's database search method and Shor's integer factoring techniques are the two most well-known quantum algorithms. Additionally, Fourier transformations, error correction, and additionally, novel techniques that can move the state from one place to another include quantum teleportation and quantum key exchange. Such an algorithm, which can factor a huge number in polynomial time, was found by Peter Shor in 1994. He demonstrated how this approach can be used to defeat the RSA cryptosystem. This algorithm has a high possibility of providing the proper response, and by repeating the algorithm, it can also lower the likelihood that it will fail. Grover's algorithm is yet another crucial algorithm.

This method was created for searching databases that have not been sorted. This approach, for instance, helps speed up brute force key searches for symmetric key encryption systems such as AES. It takes $O(N)$ operations to search a size N unsorted array. Additionally, it requires O(N) for classical algorithms. It might be more accurate to refer to this algorithm as an inverting function. It may also be used to calculate the mean and median of a set of data as well as to solve the collision problems. $O(3/2)$ can be used to calculate the connectedness of an n-vertex graph. This has many benefits, one of which is the ability to execute parallel computations, which boosts system performance. The superposition state makes it simple to store the $2n$ inputs in n qubits. Arbitrary quantum computation can be carried out using n qubits as its input. Additionally, it can be utilized in computational chemistry, artificial intelligence, machine learning, pharmaceutical research and development, computer security and encryption, forecasting the weather, etc.

17.6 Summary

The goal of this chapter is to present potential architectural designs for the quantum processor. This chapter examines particular high-performance architecture concepts for quantum processors. Then, it describes designing a two-qubit central processing unit and its operating procedures. An efficient quantum processor with high system performance and a high error correction rate can be designed with the presented algorithms.

Critical thinking questions

1. Describe how the quantum control bus, quantum data bus, and quantum address bus are different from one another.
2. What are the typical values of a quantum RAM?
3. How many locations in memory can be accessed by a quantum RAM chip if it has *n* address input lines?
4. Describe the applications of a quantum RAM.
5. Describe the role of a quantum ALU in a quantum processor.
6. Using a circuit diagram, describe how a quantum decoder works.
7. How many enable lines are there in a quantum 2-to-4 decoder? Explain why a quantum decoder needs an enable line.
8. Using a quantum 2-to-1 MUX, construct a quantum 4-to-1 MUX.
9. What does the quantum instruction register contain?
10. What type of data is stored in a quantum instruction register?
11. Describe the uses of the quantum instruction register.

References

[1] Abrams D S and Lloyd S 1997 Simulation of many-body Fermi systems on a universal quantum computer *Phys. Rev. Lett.* **79** 4

[2] Arute F *et al* 2019 Quantum supremacy using a programmable superconducting processor *Nature* **574** 505–10

[3] Ernst D *et al* 2003 Razor: a low-power pipeline based on circuit-level timing speculation *Proceedings. 36th Annual IEEE/ACM Int. Symp. on Microarchitecture* p 7

[4] Ewald R H 2019 An introduction to quantum computing and its application *Quantum Technology and Optimization Problems: First Int. Workshop (Munich, 18 March)* (Berlin: Springer) pp 3–8

[5] Jain R 2019 *Introduction to Quantum Computing and its Applications to Cyber Security* (St. Louis County, MO: Washington University in Saint Louis)

[6] Kanamori Y and Yoo S-M 2020 Quantum computing: principles and applications *J. Int. Technol. Inf. Manag.* **29** 43–71

[7] Marchenkov A 2020 Reaction: can we grow a quantum processor? *Chem.* **6** 801–2

[8] Preskill J 1998 Computing: pro and con proceedings: mathematical *Phys. Eng. Sci.* **454** 469–86

[9] Schmidt F *et al* 2003 Realization of the Cirac–Zoller controlled-NOT quantum gate *Nature* **422** 408–11
[10] Tarasov V E 2009 Quantum nanotechnology *Int. J. Nanosci.* **8** 337–44
[11] Thomsen M K, Glück R and Axelsen H B 2010 Reversible arithmetic logic unit for quantum arithmetic *J. Phys. A: Math. Theor.* **43** 382002

IOP Publishing

Quantum Computing (Second Edition)
A pathway to quantum logic design
Hafiz Md Hasan Babu

Chapter 18

Applications of quantum computing technology

Learning objectives

- Demonstrate how to apply quantum computing to machine learning.
- Explain how to solve optimization issues.
- Discuss how to apply biomedical simulations using quantum computing technology.
- Learn how to employ quantum technologies in computational chemistry and financial services.
- Acquire knowledge about the uses of quantum technology in circuits, software, simulation of system faults, and weather forecasting.
- Discover how quantum technology can be used for scheduling, logistics, and cyber security.

There are many potential future uses for quantum computing technology. This is an important phenomenon that could change the total level of computing in the coming decade and the computers we know today will become museum pieces. In this chapter we discuss some real world applications of quantum computers/circuits. The basic structure of a quantum computer is shown in figure 18.1.

18.1 Optimization

Optimization problems are some of the most difficult problems to solve.

Imagine that you are building a house and have a list of things you need to have in your home. However, you cannot afford the cost of everything on your list since you have limited finances. What you truly need to work out is the combination of things which will give you the best value for your money.

This is a case of an optimization issue, where you are attempting to locate the best combination of things given a few requirements. While issues with only a couple of decisions are simple, as the number of decisions increases, they rapidly become difficult to settle ideally. With only 270 on/off switches, there are more conceivable combinations than there are molecules in the Universe! These types of advancement issues exist in various areas, for example framework structures, mission

doi:10.1088/978-0-7503-5683-1ch18

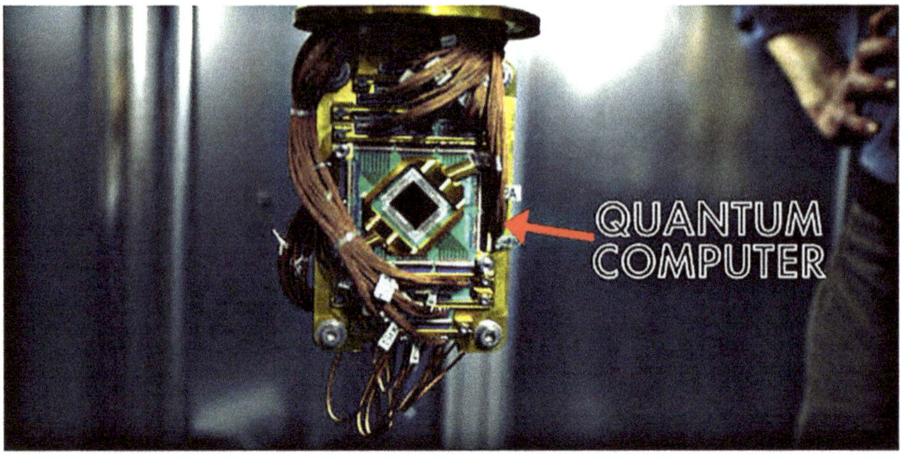

Figure 18.1. A quantum computer.

arrangements, aircraft planning, money related investigations, web searches, and malignant growth radiotherapy. These are some of the most complex issues on the planet, with tremendous potential advantages for organizations, individuals, and science if ideal solutions can be achieved quickly.

It is easily understandable how quantum computing can help in these types of scenarios. Some of these optimization problems are examined below, where quantum computers are playing a role in solving them.

18.1.1 The Roswell Park Cancer Institute

Enhancement techniques are a basic requirements to supply enough radiation to kill malignant cells while avoiding non-carcinogenic adjacent cells. Researchers at Roswell Park, which is shown in figure 18.2, utilized a D-Wave framework for use in quantum annealing for beamlet intensity optimization for intensity-modulated radiation therapy.

18.1.2 Volkswagen group

Volkswagen was the primary vehicle maker to utilize a quantum PC to determine traffic streams, as shown in figure 18.3. Their researchers carried out an explorative study of traffic stream improvement. Utilizing information from 10 000 taxis in Beijing, they modified a calculation to improve the movement times of taxis in the city.

18.1.3 Recruit Communications

Recruit Communications and D-Wave are working together to apply quantum processing to showcasing, promoting, and correspondence. The first task was to streamline the proficiency of coordinating notices to clients for web publicizing.

Figure 18.2. The Roswell Park Cancer Institute.

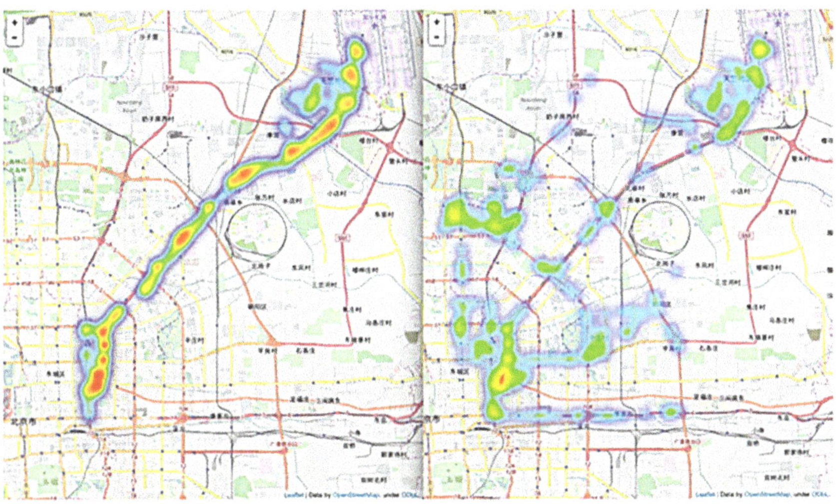

Figure 18.3. Non-optimized and optimized traffic flow research using quantum computers. Reproduced from [11]. CC BY 4.0.

18.2 Machine learning

When you look at a photograph it is simple for you to identify the distinct elements in the picture: trees, mountains, and so forth. This assignment is relatively easy for humans, however, it is a tremendously difficult job for PCs to accomplish. This is because software engineers do not know how to characterize the essence of a 'tree' in PC code.

Machine learning is the best way to deal with this issue, using which developers can devise algorithms that naturally determine how to perceive the 'substance' of objects by recognizing repeating designs in enormous quantities of information. In view of the amount of information involved in this procedure, and the vast number of potential combinations of information components, this is a computationally costly issue. Similarly to other streamlining issues, these types of tasks can be mapped to the local ability of the D-Wave quantum computer.

18.2.1 QxBranch

Predicting the result of an election is a challenge. Consolidating quantum processing with neural system innovations could enhance predictions, as indicated by the work carried out by QxBranch. They utilized such a system in a simple manner to predict the 2016 US presidential race. An example of embedding a problem into a quantum computation model is shown in figure 18.4.

18.2.2 Los Alamos National Laboratory

Utilizing a set of 2429 pictures of faces, Los Alamos National Laboratory researchers utilized unsupervised machine learning techniques by means of framework factorization to break down large datasets. This is one of the numerous quick reaction extensions that Los Alamos scientists have carried out on the D-Wave quantum computing framework, first introduced at Los Alamos, which is shown in figure 18.5.

18.2.3 NASA

Machine learning depends strongly on inspecting complex likelihood calculations. NASA researchers effectively prepared the D-Wave 2X framework for picture information collection in a generative unsupervised learning setting, which stands out as one of the most challenging ideal models in machine learning.

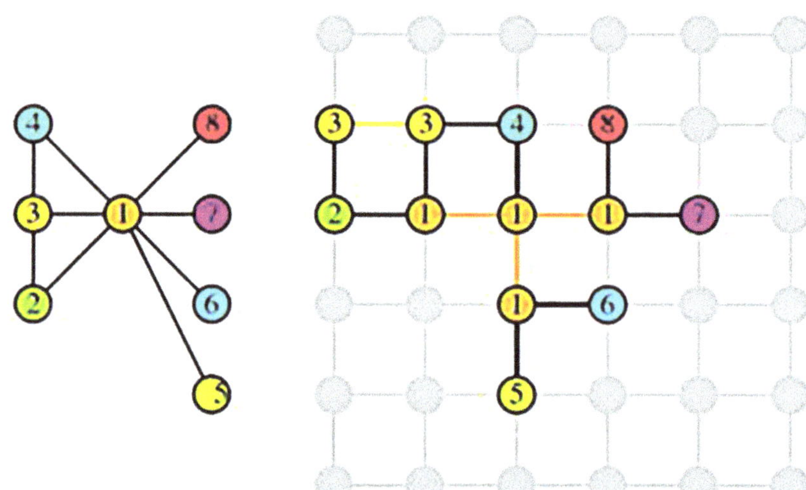

Figure 18.4. Example of embedding a problem into a quantum computation model.

Figure 18.5. The National Security Sciences Building at Los Alamos National Laboratory. Photo courtesy of Los Alamos National Laboratory.

18.3 Biomedical simulations

With quantum computers we can mimic and model sub-atomic structures. In 2012, analysts at Harvard University utilized a D-Wave One quantum PC, as shown in figure 18.6, to understand the riddle of how a few proteins overlap.

'The model comprised of scientific portrayals of amino acids in a cross section, associated by various communication qualities' wrote Geoffrey Brumfiel in a news article for *Nature* about the Harvard specialists' protein collapsing models, noting that '[t]he D-Wave PC found the least setups of amino acids and associations, which relates to the most efficient collapsing of the proteins.' While the innovation was not even close to complete—as the researchers wrote, '10 000 estimations utilizing an 81-qubit rendition of the examination gave the right answer only multiple times'—it had the ability to achieve a remarkable accomplishment by displaying the behavior of protein collapsing with some level of exactness.

18.4 Financial services

Quantum processing is now on its way. D-Wave, an organization supported by Goldman Sachs and Bezos Expeditions, among others, introduced its first business quantum computer, the D-Wave 2000Q, a quantum enhancing framework with 2000 qubits.

Notwithstanding their limitations, these computers are being used in specialist fields as demonstrated by the Harvard team's utilization of D-Wave's first model in 2012. The frameworks could be utilized for complex budgetary displays. Quantum computing could be utilized to discover 'better approaches to show budgetary

Figure 18.6. D-Wave quantum computers. Reproduced courtesy of [3].

Figure 18.7. IBM expands the quantum computing network.

information' and exclude 'key worldwide hazard factors', as per IBM. A development architecture of a quantum computing network by IBM is shown in figure 18.7.

'It is extraordinary to assemble frameworks to help Wall Street better oversee hazards utilizing this sort of innovation', D-Wave Systems President and CEO Vern Brownell told Bloomberg News: 'They spend a ton on figuring power to oversee risk.'

18.5 Computational chemistry

There are numerous issues in materials science that could accomplish tremendous results if we could simple locate the correct catalyst or procedure to construct a new material, or to produce a current material more productively. Currently, notable efforts are being made to utilize traditional PCs to mimic mixture interactions, yet much of the time the problems prove too difficult to solve traditionally. Thus the first thought, introduced by Richard Feynman, is why not utilize a quantum PC to reproduce the quantum mechanical procedures that occur. Here are only a few of the myriad of critical issues that could see significant solutions if only we could understand them:

1. Supplanting the Haber procedure to deliver ammonium nitrate for use in fertilizers.
2. Finding new materials that can be used to achieve a room-temperature superconductor.
3. Discovering a catalyst that can enhance the efficiency of carbon capture.
4. Devising a new battery science that can improve performance compared to current lithium-ion batteries.

18.6 Logistics and scheduling

Numerous standard improvements made in industry can be categorized under coordination and planning. Consider an aircraft coordination supervisor who needs to make sense of how to organize aircraft to achieve the best management at minimal cost, or a manufacturing plant supervisor who has a regularly changing combination of machines, stock, generation requests, and individuals, and requirements to limit cost and throughput times while improving yield.

Consider, on the other hand, a chief assessor at a car organization who needs to make sense of the optimal costs of a considerable variety of vehicle choices to enhance consumer loyalty and benefit. Although traditional calculations are utilized extensively to perform such tasks, some questions might be too difficult for traditional processing, while a quantum approach might have the capacity achieve such a task.

18.7 Cyber security

Digital security is becoming a more significant problem these days as threats around the globe are expanding their capacities and we are becoming progressively more powerless as we increase our reliance on advanced digital frameworks. Different strategies to combat digital security risks can be created utilizing quantum machine learning approaches to perceive the dangers ahead of time and alleviate the harm that they may cause.

18.8 Circuit, software, and system fault simulation

When one develops expansive programming with a huge number of lines of code or substantial ASIC chips that have billions of transistors, it can become very difficult and costly to assess them for correctness. There can be billions or trillions of different states and it is unthinkable for a traditional PC to check each and every one in simulation. In addition to wanting to comprehend what will happen when a framework is working normally, one also needs to understand what occurs if there is an equipment or other error. Will the framework identify this error and does it have a recovery system to alleviate any conceivable issues? The costs of a error can be high, as some of these frameworks may be utilized in situations where lives or a huge number of money may depend on their being error-free. By utilizing quantum computing to help in these simulations, one may be able to provide a vastly improved assessment of their performance in a much shorter time.

18.9 Weather forecasting

National Oceanic and Atmospheric Administration (NOAA) Chief Economist Rodney F Weiher has claimed that almost 30% of the US GDP ($6 trillion) is specifically or in an oblique way influenced by the weather, affecting food production, transportation, and retail, among others. The capacity to more accurately predict the weather would have huge advantages for numerous fields, while also providing better opportunities to protect people from extreme weather events.

While this has been an objective of researchers for quite some time, the conditions underlying weather and the climate include numerous factors, making traditional simulations protracted. As quantum specialist Seth Lloyd pointed out, '[u]tilizing a traditional PC to perform such investigation may take longer than it takes the real climate to advance!' This inspired Lloyd and partners at MIT to demonstrate that the conditions that determine the climate have a concealed wave nature which is amenable to be addressed by a quantum computer.

The executive of design at Google, Hartmut Neven, likewise noticed that quantum computers allowed better atmosphere models that could provide us with a greater understanding of how humans are affecting the natural world. These models are what we base our estimations of future warming on, and what will enable us to determine what steps need to be taken now to prevent future calamities. The United Kingdom's national weather service, the benefit Met Office, has recently started putting resources into such developments to meet the energy and adaptability requirements they will be facing from 2020 onward for larger scale computing.

Thus we can easily understand that quantum computers are the future of computing. All the types of quantum computers are very important for various types of computational problems. We only discuss some of its greater impacts, but there are many other areas where we will see the effects of the future of computing—quantum computing.

18.10 Summary

This chapter describes different applications of quantum computing. Quantum computing is leading to a paradigm shift in many ways. Quantum computing provides access to new levels of computational ability and it also inspires new ways of thinking. As we look at problems in new ways, this can, in turn, contribute new ideas for how we could advance classical computation as well. With more and more individuals conceiving problems from different angles, more and more ideas and results will emerge. In the world of quantum evolution, we will exercise the ability to think about problems in new ways, become familiar with programming quantum computers, and even simulate this work so that we can obtain useful quantum computers.

Critical thinking questions

1. Why does quantum computing technology need to be optimized? Explain using an example.
2. Give some examples of optimization problems that can be solved with quantum computing.
3. How is quantum computing applied in machine learning? Give some examples.
4. How does quantum computing technology affect cyber security?

References

[1] Jackson M 2017 6 things quantum computers will be incredibly useful for *SingularityHub* https://singularityhub.com/2017/06/25/6-things-quantum-computers-will-be-incredibly-useful-for/#sm.0000trlmgo7c0f9mwo3107yechmv7 (Accessed: 27 December 2018)

[2] Volkswagen 2018 Avoiding traffic jams and surviving tsunamis *Volkswagen* https://www.volkswagenag.com/en/news/stories/2018/06/avoiding-traffic-jams-and-surviving-tsunamis.html (Accessed: 27 December 2018)

[3] D-Wave Practical quantum computing *D-Wave* https://www.dwavesys.com/home (Accessed: 27 December 2018)

[4] D-Wave 2012 Harvard researchers use D-Wave quantum computer to fold proteins *D-Wave* https://www.dwavesys.com/news/harvard-researchers-use-d-wave-quantum-computer-fold-proteins (Accessed: 27 December 2018)

[5] NASA Quantum computing *NASA Advanced Supercomputing Division* https://www.nas.nasa.gov/projects/quantum.html (Accessed: 27 December 2018)

[6] Microsoft Quantum computing applications for innovation and impact https://www.microsoft.com/en-us/quantum/quantum-computing-applications (Accessed: 27 December 2018)

[7] LANL Quantum Institute *Los Alamos National Laboratory* https://www.lanl.gov/collaboration/research-opportunities/quantum-institute.php (Accessed: 27 December 2018)

[8] QxBranch QxBranch builds software and applications for quantum computers *QxBranch* https://www.qxbranch.com/quantum-computing/ (Accessed: 27 December 2018)

[9] Recruit Communications and D-Wave Collaborate to Apply Quantum Computing to Marketing, Advertising, and Communications Optimization https://www.globenewswire.com/en/news-release/2017/05/12/1202661/0/en/Recruit-Communications-and-D-Wave-Collaborate-to-Apply-Quantum-Computing-to-Marketing-Advertising-and-Communications-Optimization.html (Accessed 27 December 2018)

[10] Roswell Park https://www.roswellpark.org/ (Accessed: 27 December 2018)

[11] Neukart F, Compostella G, Seidel C, Dollen D V, Yarkoni S and Parney B 2017 Traffic flow optimization using a quantum annealer *Front. ICT* **4** 29

Part II

Quantum fault tolerance

An overview of quantum fault-tolerant circuits

A fault is an error that occurs in a system which forces the system to deviate from its normal behavior. The purpose of fault tolerance is to enable reliable quantum computations when the computer's basic components are unreliable. To achieve this, the qubits in the computer are encoded in blocks of a quantum error-correcting code, which allows the state to be corrected even when some qubits are wrong. A fault-tolerant protocol prevents catastrophic error propagation by ensuring that a single faulty gate or time step produces only a single error in each block of the quantum error-correcting code. The fault models vary according to the type of description that is being considered, which in turn varies according to the level of abstraction.

Fault tolerance is the property that enables a system to operate accurately in the presence of the failure of one or more of its components. Fault tolerance in a quantum circuit reflects the robustness of the system. Fault-tolerant systems are capable of detecting and correcting faults. If the logic circuit itself is made of fault-tolerant components, then the detection and correction of faults in the circuit become cheaper, easier, and more simple. To achieve fault tolerance, the first step is to identify the occurrence of a fault. To detect the occurrence of a fault, parity-preserving techniques are considered as one type of solution. Any fault that affects only one signal is detectable at the circuit's primary outputs in fault-tolerant quantum circuits.

A conventional circuit dissipates energy to reload missing information because of overlapped mapping between input and output vectors. Quantum computing recovers this energy loss and prevents bit errors by including fault-tolerant mechanisms. A fault-tolerant quantum circuit is capable of preventing error at the outputs. The goal of any program to produce a fault-tolerant logic gate is to implement a universal set of gates with the properties of fault tolerance. Essentially, the goal is to implement a set of gates that will allow any arbitrary quantum computation.

In practice, quantum computers are difficult to implement. The reason for this is that they require the qubits they are constructed of to be isolated from the environment and remain completely unaltered by any processes other than the logical gates which are applied to them. In reality, such isolation is extremely difficult to achieve. Undesired interactions with the qubits' surroundings, and even imperfect executions of quantum gates and measurements, all introduce the possibility of error. These errors are by nature random, so in the modeling it is considered that faults are probabilistic events that occur throughout the evolution of a quantum circuit. Here, the ramifications of the error rates and circuit designs are considered, which will inform the basis for fault-tolerant quantum computations.

Quantum computers appear to be capable, at least in principle, of solving certain problems far faster than any conceivable classical computer. In practice, quantum computing technology is still in its infancy. While a practical and useful quantum computer may eventually be constructed, at present we cannot envision clearly what the hardware of that machine will be like. Nevertheless, one can be quite confident that any practical quantum computer will incorporate some types of error correction in its operation. Quantum computers are far more susceptible to making errors than conventional digital computers, and some method of controlling and correcting such errors will be needed to prevent a quantum computer from crashing.

The basic idea of error-correcting codes is that, to store and transmit data without said data being compromised, it is necessary to build some kind of larger mathematical structure to serve as a shield for the structure containing the data. This shield must be effective, namely, it should be able to both survive one or more errors and should be able to correct those errors over time by restoring the original state.

This part of the book describes some background and preliminary studies of quantum fault-tolerant circuits in Chapter 19. The approach for designing a fault-tolerant quantum adder (full-adder) is given in this chapter. A design for a fault-tolerant multiplier (signed multiplier) is also shown. Finally, the design procedure for a quantum fault-tolerant integer division circuit is introduced in this chapter. In this design a restoring division algorithm as well as quantum restoring integer division circuitry are introduced.

IOP Publishing

Quantum Computing (Second Edition)
A pathway to quantum logic design
Hafiz Md Hasan Babu

Chapter 19

Quantum fault-tolerant circuits

Learning objectives

- Explain the significance of quantum fault-tolerant circuits.
- Discuss the restorative division algorithm and the quantum failure resilient integer divider.
- Become familiar with the fault-tolerant quantum full-adder.
- Describe the operational module for conditional addition.
- Construct a fault-tolerant multiplier in an appropriate manner.
- Demonstrate the circuitry for integer division in quantum restoration.
- Discuss a fault-tolerant signed multiplier with its design procedure.

The discovery of quantum error correction has greatly improved the long-term prospects for quantum computing technology. Encoded quantum information can be protected from the errors that arise due to uncontrolled interactions with the environment, or due to imperfect implementations of quantum logical operations. Recovery from errors can work effectively even if occasional mistakes occur during the recovery procedure.

19.1 The need for quantum fault-tolerant circuits

The discovery of quantum effects alone is not sufficient to ensure that a quantum computer can perform reliably. To carry out a quantum error-correction protocol we must first encode the quantum information we want to protect and then repeatedly perform recovery operations that reverse the errors that accumulate. However, encoding and recovery are themselves complex quantum computations, and errors will inevitably occur while we perform these operations. Thus we need to find methods for recovering from errors that are sufficiently robust to succeed with high reliability even when some errors are made during the recovery step.

Furthermore, to operate a quantum computer, we must process the information of two or more encoded qubits that interact with one another and if an error occurs in one qubit, then that qubit interacts with another quantum gate and the error is

doi:10.1088/978-0-7503-5683-1ch19

likely to propagate to the second qubit. We must design the gates to minimize the propagation of error.

Incorporating quantum error correction will surely complicate the operation of a quantum computer. To establish the redundancy needed to protect against errors, the number of elementary qubits will have to increase. The encoded information on performing gates and inserting periodic error-recovery steps will slow down the computation.

A device that works perfectly even when its primary components are imperfect is said to be fault-tolerant. With the development of fault-tolerant methods, it will be possible, in principle, for the operator of a quantum computer to actively intervene to stabilize the device against errors in a chaotic environment. In the long run, fault tolerance might be achieved in practical quantum computers by a rather different route—with intrinsically fault-tolerant hardware. Such hardware, designed to be impervious to localized influences, could be operated relatively unsupervised, yet could still store and process quantum information robustly.

Fault-tolerant quantum computing falls under the framework of proposed ideas that allow qubits to be protected from quantum errors introduced by the inadequate control of environmental interactions, called quantum error correction (QEC). Moreover, the appropriate innovations in quantum circuits will be implemented in both QEC and encoded logic operations in such a way as to avoid these errors cascading through quantum circuits. By avoiding a cascade of errors, there comes a point (when the fundamental accuracy of individual qubits is high enough), when the QEC is correcting more errors than are being created. Once this threshold has been achieved, the expansion of the size of the protective quantum will exponentially decrease the failure of the encoded information and allows us to achieve algorithms implemented with noisy devices.

The parity-preserving quantum circuits are a class of quantum logic with the additional property that the parity of the input is the same as the parity of the output. A quantum circuit will be parity-preserving if the Ex-OR of the inputs matches the EX-OR of the outputs, i.e. the parity of the input and the output remains the same.

19.2 The fault-tolerant quantum adder

A quantum adder is a quantum circuit that performs the addition operation. In a full-adder, two binary numbers and a carry or overflow bit might be the input to complete the addition operation. The output of the circuit will be the sum and another carry bit. Full-adders are commonly connected to each other to add bits to an arbitrary length of bits, such as 32 or 64 bits.

Fault tolerance is the ability of a system to hold its normal operation without failure when some part of the system fails to operate properly. It increases the wear-out time for any system at a cost of increased hardware. Fault-tolerant approaches must be incorporated in any safety-critical system for continued functioning without failure even if an error occurs in the system.

The adder is an important circuit in quantum circuit design. To design a fault-tolerant quantum full-adder some basic circuits are necessary, which are discussed in the next section.

19.2.1 The fault-tolerant full-adder

In this section, the conventional full-adder, as presented in figure 19.1, is discussed and a fault-tolerant full-adder is designed. Finally, the quantum fault-tolerant full-adder is presented. A classic full-adder, as shown in figure 19.1, will have two inputs and one carry or overflow bit and from the truth table (table 19.1) the following expressions can be achieved:

$$\text{Sum} = A \oplus B \oplus C \tag{19.1}$$

$$\text{Carry} = (A \oplus B)C \oplus AB. \tag{19.2}$$

Based on equations (19.1) and (19.2) a parity-preserving logic gate called the fault-tolerant full-adder (FTFA) is implemented and is presented in figure 19.2. In order to verify that the implemented gate has worked as a fault-tolerant quantum adder, the corresponding truth table of the gate is presented in table 19.2. It can be seen from the truth table that the input and output vectors show one-to-one mapping and at the same time the condition of the input parity $A \oplus B \oplus C \oplus D \oplus E$ is equal to the output parity $P \oplus Q \oplus R \oplus S \oplus T$.

From figure 19.2 it is discovered that if the inputs $D = 0$ and $E = 0$, the extended fault-tolerant full-adder (EFTFA) generates three garbage outputs, which is the minimum (see figure 19.3).

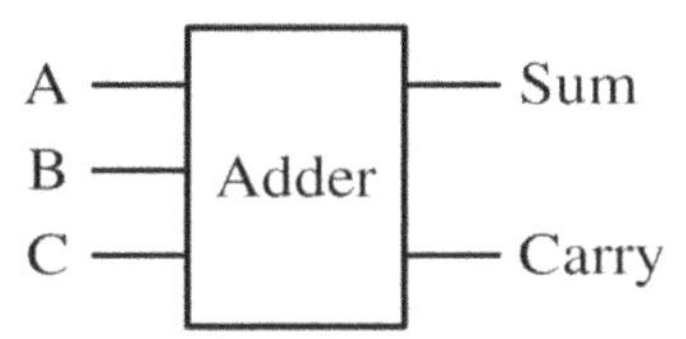

Figure 19.1. A conventional full-adder.

Table 19.1. The truth table of a conventional full-adder.

A	B	C	Sum	Carry
0	0	0	0	0
0	0	1	1	0
0	1	0	1	0
0	1	1	0	1
1	0	0	1	0
1	0	1	0	1
1	1	0	0	1
1	1	1	1	1

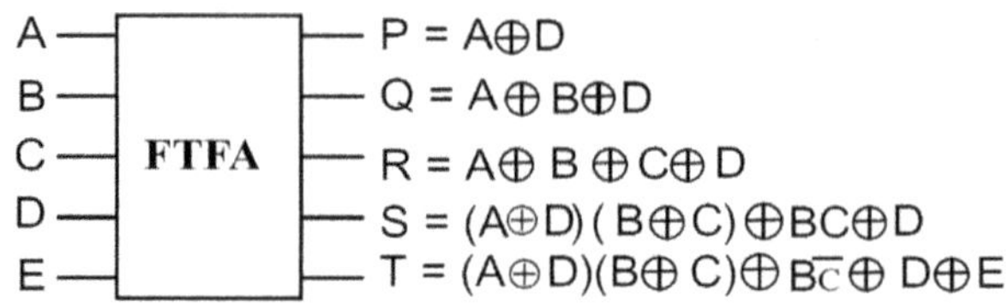

Figure 19.2. The block diagram of an FTFA.

Table 19.2. The truth table of a quantum FTFA.

Input					Output				
A	*B*	*C*	*D*	*E*	*P*	*Q*	*R*	*S*	*T*
0	0	0	0	0	0	0	0	0	0
0	0	0	0	1	0	0	0	0	1
0	0	0	1	0	1	1	1	1	1
0	0	0	1	1	1	1	1	1	0
0	0	1	0	0	0	0	1	0	0
0	0	1	0	1	0	0	1	0	1
0	0	1	1	0	1	1	0	0	0
0	0	1	1	1	1	1	0	0	1
0	1	0	0	0	0	1	1	0	1
0	1	0	0	1	0	1	1	0	0
0	1	0	1	0	1	0	0	0	1
0	1	0	1	1	1	0	0	0	0
0	1	1	0	0	0	1	0	1	0
0	1	1	0	1	0	1	0	1	1
0	1	1	1	0	1	0	1	0	1
0	1	1	1	1	1	0	1	0	0
1	0	0	0	0	1	1	1	0	0
1	0	0	0	1	1	1	1	0	1
1	0	0	1	0	0	0	0	1	1
1	0	0	1	1	0	0	0	1	0
1	0	1	0	0	1	1	0	1	1
1	0	1	0	1	1	1	0	1	0
1	0	1	1	0	0	0	1	1	1
1	0	1	1	1	0	0	1	1	0
1	1	0	0	0	1	0	0	1	0
1	1	0	0	1	1	0	0	1	1
1	1	0	1	0	0	1	1	1	0
1	1	0	1	1	0	1	1	1	1
1	1	1	0	0	1	0	1	1	0
1	1	1	0	1	1	0	1	1	1
1	1	1	1	0	0	1	0	0	1
1	1	1	1	1	0	1	0	0	0

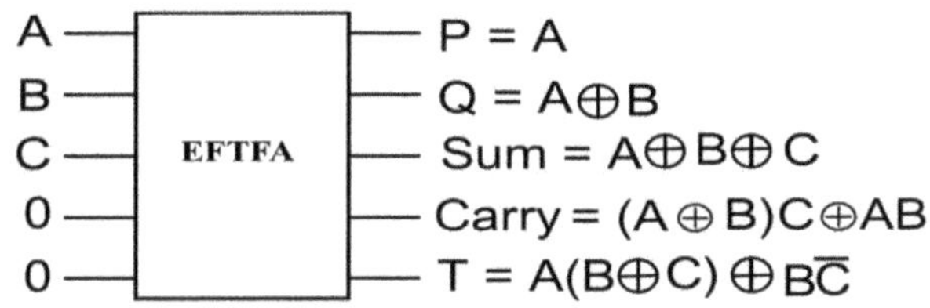

Figure 19.3. The block diagram of an EFTFA, when $D = 0$ and $E = 0$.

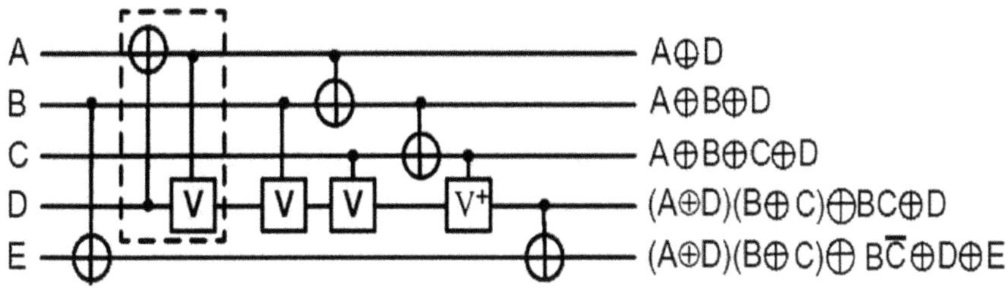

Figure 19.4. The quantum implementation of the EFTFA. Reproduced with permission from [9]. Copyright 2014 Taylor and Francis.

The equivalent quantum circuit of the EFTFA is illustrated in figure 19.4 and the dotted rectangle in the figure is equivalent to a 2 × 2 CNOT gate. Therefore, the quantum cost of the FTFA is 8.

19.3 The fault-tolerant multiplier

Fault-tolerant multiplier circuits have special importance because of the fact that they are the integral components of computer systems, mobile devices, and most audio/video components. To implement a fault-tolerant multiplier, quantum representations of the Ex-OR, NAND, and basic logic operations are needed. The parity-preserving quantum fault-tolerant operations are as follows:

A. *Quantum fault-tolerant Ex-OR operation*: figure 19.5 presents the block diagram (block A) and the quantum realization for the XOR operation, where the inputs are A, B, and C, and the outputs are $P = A$, $Q = A \oplus B$, and $R = A \oplus C$.
B. *Quantum fault-tolerant basic logic operations*: figure 19.6 presents the block diagram (block B) and the quantum realization for basic logic operations, where the inputs are A, B, and C, and the outputs are $P = A$, $Q = \bar{A}B \oplus AC$, and $R = \bar{A}C \oplus AB$.
C. *Quantum fault-tolerant NAND, Ex-OR, and basic logic operations*: figure 19.7 presents the block diagram (block C) and the quantum realization for the NAND, XOR, and basic logic operations, where the inputs are A, B, and C, and the outputs are $P = A\bar{C} \oplus BC$, $Q = A \oplus B$, and $R = A\bar{C} \oplus \bar{B}C$.
D. *Quantum fault-tolerant half-adder operation*: figure 19.8 presents the block diagram (block D) of half-adder operation and its quantum realization,

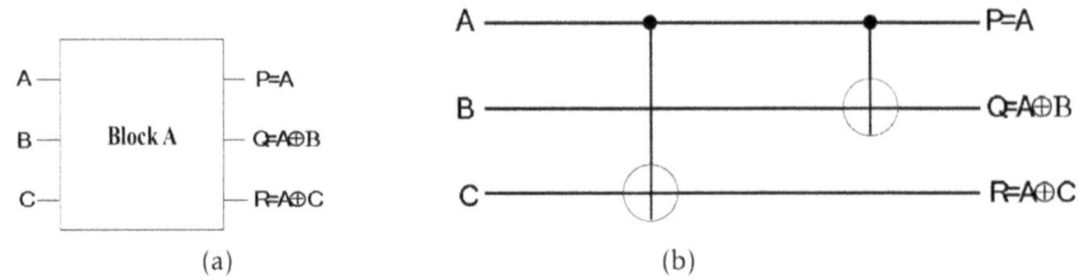

Figure 19.5. The fault-tolerant quantum realization of Ex-OR operation. (a) Block diagram. (b) Quantum implementation of block A.

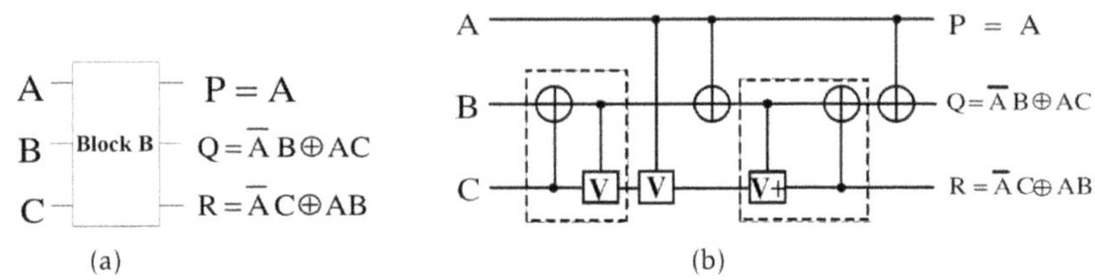

Figure 19.6. The fault-tolerant quantum realization of basic logic operations. (a) Block diagram. (b) Quantum implementation of block B. Reproduced with permission from [9]. Copyright 2014 Taylor and Francis.

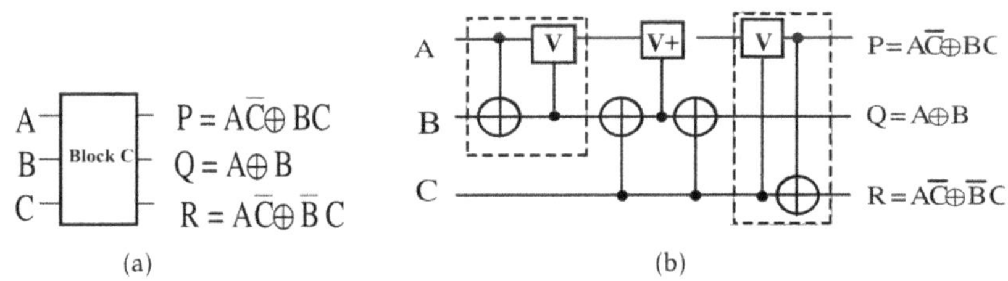

Figure 19.7. The fault-tolerant quantum realization of NAND, Ex-OR, and basic logic operations. (a) Block diagram. (b) Quantum implementation of block C.

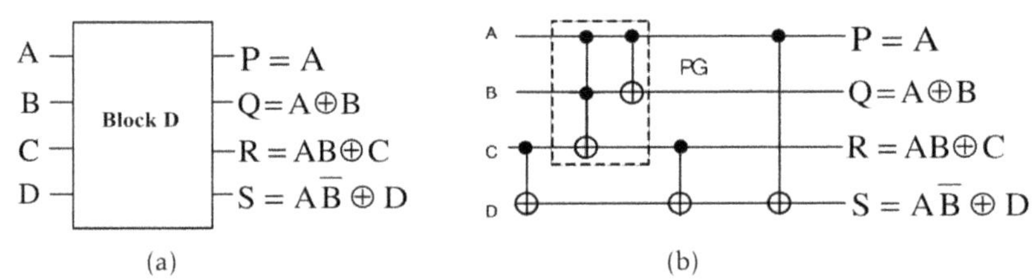

Figure 19.8. The fault-tolerant quantum realization of the half-adder operation. (a) Block diagram. (b) Quantum implementation of block D.

where the inputs are A, B, C, and D and the outputs are $P = A$, $Q = A \oplus B$, $R = AB \oplus C$, and $S = A\bar{B} \oplus D$.

E. *Quantum fault-tolerant full-adder operation*: figure 19.9 presents the block diagram (block E) of full-adder operation and the quantum realization,

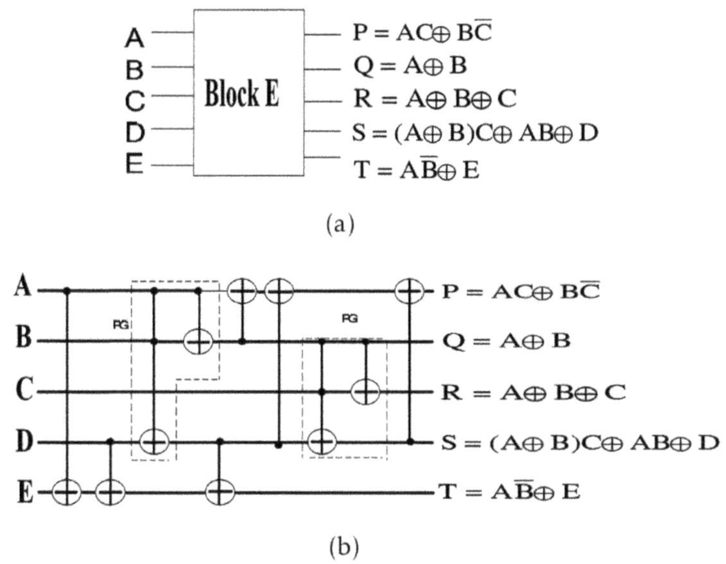

Figure 19.9. The fault-tolerant quantum realization of full-adder operation. (a) Block diagram. (b) Quantum implementation of block E.

where the inputs are A, B, C, D, and E and the outputs are $P = AC \oplus B\bar{C}$, $Q = A \oplus B$, $R = A \oplus B \oplus C$, $S = (A \oplus B)C \oplus AB \oplus D$, and $T = A\bar{B} \oplus E$. The working principles of block E are presented in table 19.3.

19.3.1 The fault-tolerant signed multiplier

The speed of the multiplier plays an important role in the performance of a system. In general the speed of a multiplier depends on the algorithm and the structure of the circuits. The Baugh–Wooley algorithm is adopted to perform the multiplication process, as shown in figure 19.10. Baugh–Wooley algorithms can be used directly in the complement multiplier, improving the speed of the multiplier. It only needs one kind of full-adder which requires supplements and it can be achieved as shown in figure 19.9(a).

In the signed multiplier, a Wallace tree structure is applied for implementing one for n operands. In order to reduce the carry transfer delay of the partial product in the adder array of the multiplier, it is necessary to design its Wallace tree. In other words, it improves the operating speed in the whole adder array. In the Wallace tree, the bits with the same weight in the corresponding partial products are added together rather than one by one, as shown in figure 19.11. Usually, a full-adder is used to accomplish the addition of bits with the same weight. In the layers of the Wallace tree, the number of partial product vectors can be decreased according to the proportion 3:2, e.g. every three figures produces one sum bit and one carry bit. The fault-tolerant signed multiplier is shown in figure 19.12.

The fault-tolerant quantum signed multiplier is composed of block D and block E, which are the quantum realizations of half-adder and full-adder circuits, respectively.

Table 19.3. The truth table for block E.

Input					Output				
A	*B*	*C*	*D*	*E*	*P*	*Q*	*R*	*S*	*T*
0	0	0	0	0	0	0	0	0	0
0	0	0	0	1	0	0	0	0	1
0	0	0	1	0	0	0	0	1	0
0	0	0	1	1	0	0	0	1	1
0	0	1	0	0	0	0	1	0	0
0	0	1	0	1	0	0	1	0	1
0	0	1	1	0	0	0	1	1	0
0	0	1	1	1	0	0	1	1	1
0	1	0	0	0	1	1	1	0	0
0	1	0	0	1	1	1	1	0	1
0	1	0	1	0	1	1	1	1	0
0	1	0	1	1	1	1	1	1	1
0	1	1	0	0	0	1	0	1	0
0	1	1	0	1	0	1	0	1	1
0	1	1	1	0	0	1	0	0	0
0	1	1	1	1	0	1	0	0	1
1	0	0	0	0	0	1	1	0	1
1	0	0	0	1	0	1	1	0	0
1	0	0	1	0	0	1	1	1	1
1	0	0	1	1	0	1	1	1	0
1	0	1	0	0	1	1	0	1	1
1	0	1	0	1	1	1	0	1	0
1	0	1	1	0	1	1	0	0	1
1	0	1	1	1	1	1	0	0	0
1	1	0	0	0	1	0	0	1	0
1	1	0	0	1	1	0	0	1	1
1	1	0	1	0	1	0	0	0	0
1	1	0	1	1	1	0	0	0	1
1	1	1	0	0	1	0	1	1	0
1	1	1	0	1	1	0	1	1	1
1	1	1	1	0	1	0	1	0	0
1	1	1	1	1	1	0	1	0	1

19.4 The quantum fault-tolerant integer divider

Dividers are one of the major computational units in quantum arithmetic. Integer division has applications in the circuit design of quantum algorithms, computation of power series, and trigonometric functions. Quantum computers of many qubits are extremely difficult to realize. Thus the number of qubits in quantum circuits needs to be minimized. The fabrication constraint of realizing quantum circuits with

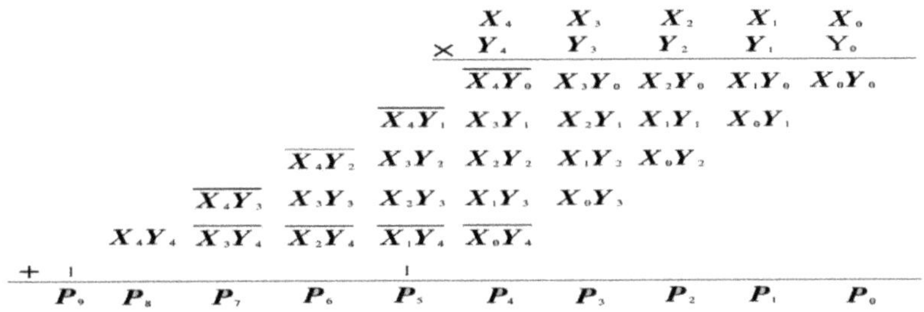

Figure 19.10. The modified Baugh–Wooley signed multiplier.

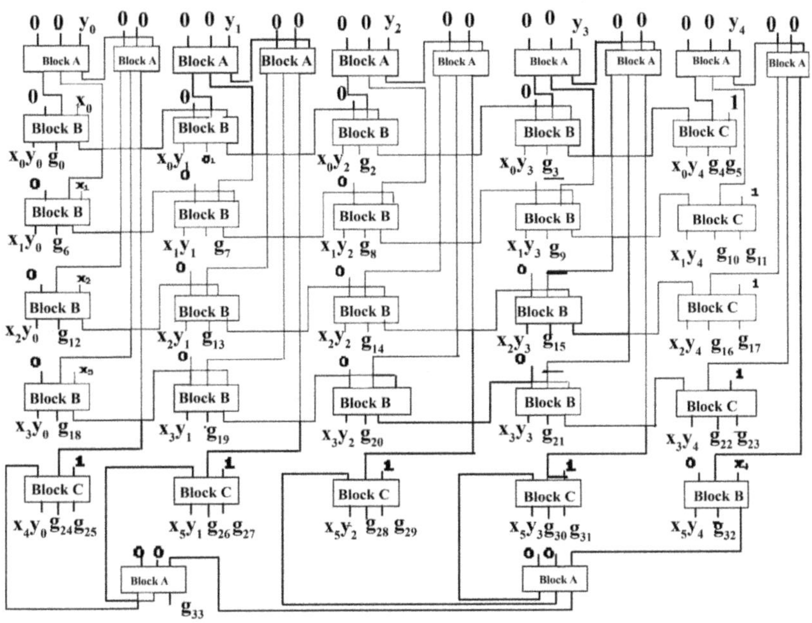

Figure 19.11. The block diagram of a fault-tolerant quantum circuit for partial product generation.

a large number of qubits has the objective of optimizing the number of ancillary qubits in a quantum circuit. Designing a scalable and reliable quantum computer is necessary now as well as in the future. Hence fault-tolerant quantum circuits are being explored based on Cliford and T gates.

Cliford+T gates based on a quantum circuit design of integer division are presented with n ancillary qubits, where n is the number of qubits in the operands. The design has no garbage outputs and the quantum circuit is based on the restoring division algorithm. It employs optimized quantum designs of conditional integer ADD operation and integer subtraction.

19.4.1 The restoring division algorithm

The restoring division algorithm for quantum circuits is shown in algorithm 19.1. In the algorithm, the inputs are: (i) $|Q_{[0:n-1]}\rangle$, n-qubit register in which the dividend is

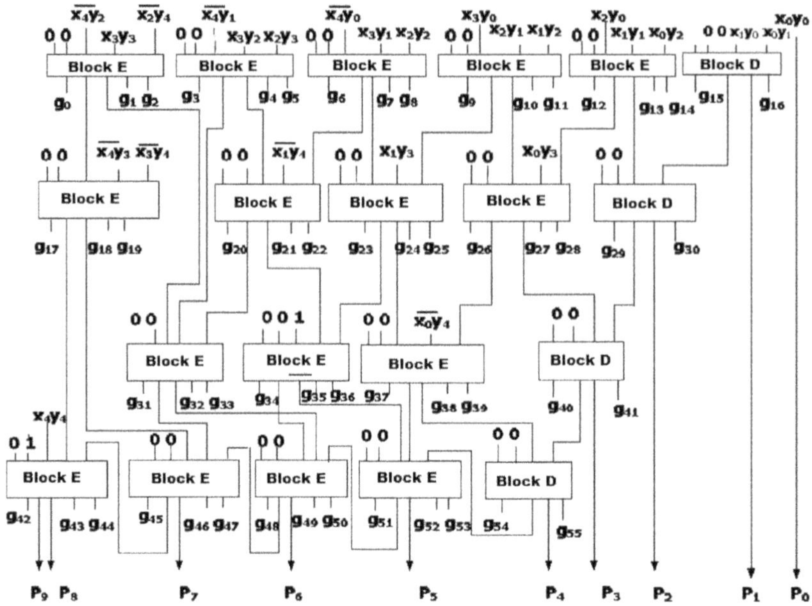

Figure 19.12. The block diagram of a fault-tolerant quantum circuit for signed multiplier generation.

stored; (ii) $|D_{[0:n-1]}\rangle$, n-qubit register in which the divisor is loaded; and (iii) $|R_{[0:n-1]}\rangle$, n-qubit remainder register which is initiated to 0 at the start. Therefore, for initiating $|R_{[0:n-1]}\rangle$, n ancillary qubits are required. The algorithm executes repeatedly and from the algorithm it can be observed that at the end of n iterations the quotient is $|Q_{[0:n-1]}\rangle$ and the remainder is $|R_{[0:n-1]}\rangle$. The divisor is retained at the output.

START
Restore $(|Q_n\rangle, |R_n\rangle, D_n\rangle)$
for i = 0 to n-*1* **do**
  $(|Q_{[1:n-1]}\rangle|R_{[0:n-1]}\rangle) = LEFTSHIFT(|Q_{[0:n-1]}\rangle), |R_{[0:n-1]}\rangle;$
  $|R - D_{[0:n-1]}\rangle = |R_{[0:n-1]}\rangle - |D_{[0:n-1]}\rangle;$
  if $(|R_{[0:n-1]}\rangle > 0)$ **then**
    $|Q_{[0]}\rangle$=1
    $R_{[0:n-1]}\rangle = |R - D_{[0:n-1]}\rangle;$
  else
    $|Q_{[0]}\rangle = 0;$
    $|R_{[0:n-1]}\rangle = |R - D_{0:n-1}\rangle + |D_{[0:n-1]}\rangle;$
  end if
end for
return R;
END

Algorithm 19.1. Restoring division algorithm.

The quantum circuits that are required to develop the hardware implementation of the restoring division algorithm are: (i) left shift operation circuitry; (ii) n-qubit quantum subtractor; and (iii) conditional ADD operation circuitry.

Combining $|R_{[0:n-2]}\rangle$ and $|Q_{[n-1]}\rangle$ forms an n-qubit register which is actually equal to performing a left shift operation. By aggregating the qubits in this way, a separate left shift operation circuitry is not necessary.

19.4.2 The subtractor module

Figure 19.13 shows the symbol and working principle of the quantum subtractor circuitry. The subtractor circuitry takes two n-qubit inputs $a_{[0:n-1]}$ and $b_{[0:n-1]}$. The input a is regenerated at the output. The n-qubit output $s_{[0:n-1]}$ has the result of the subtraction of b and a, i.e. $b - a$. Figure 19.14 presents the circuit design of an N-qubit subtractor based on an N-qubit quantum ripple carry adder. As demonstrated in figure 19.14, a quantum ripple carry adder is required to develop a quantum subtractor circuitry. The approach that is followed for developing the quantum subtractor circuitry is $b{-}a = \overline{(\overline{b} + a)}$. Both inputs are passed through the quantum ripple carry adder. The input qubits $b_{[0:n-1]}$ are complemented at the start and at the end. The qubits $a_{[0:n-1]}$ are just passed through the quantum ripple carry adder.

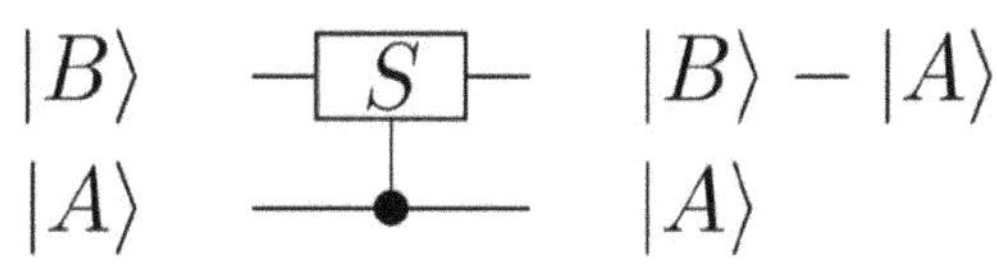

Figure 19.13. Graphical representation of a quantum subtractor.

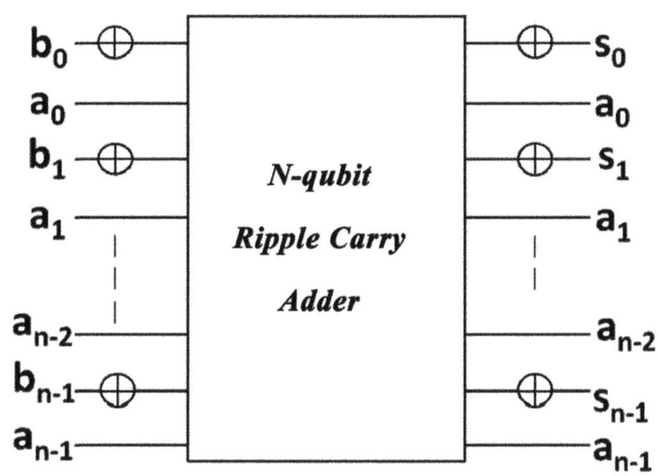

Figure 19.14. A quantum subtractor based on an N-qubit ripple carry adder.

19.4.3 The conditional addition operation module

Figure 19.15 presents the graphic symbol of the quantum conditional ADD operation circuit. The quantum controlled ADD operation circuitry operates as follows: (i) when the input labeled 'ctrl' is high, the circuit output is $|P\rangle = |B + A\rangle$, and (ii) when the 'ctrl' input is low, the circuit output is $|P\rangle = |B\rangle$. The complete working circuit of quantum conditional ADD operation circuitry is demonstrated in figure 19.16 for four-qubit operands. The quantum conditional ADD circuit uses a modified version of the ripple carry adder. The addition architecture uses Peres gates to perform the addition operation. The Peres gate can be decomposed into a quantum Feynman and a Toffoli gate. By replacing the Feynman gates with three input quantum Toffoli gates, it will be possible to use the control line (ctrl) to perform addition or no operation. It is possible to extend the four-qubit operands to any operand size.

19.4.4 Quantum restoring integer division circuitry

Figure 19.17 presents the quantum circuit of restoring division for one iteration. The details of the movement of the circuit are as follows:

Step 1. The $|D_{[0:n-1]}\rangle$ keeps the divisor, $|R_{[0:n-1]}\rangle$ initialized to zero, and $|Q_{[0:n-1]}\rangle$ holds the dividend.

Step 2. Let $|Q_{[n-1]}\rangle$ and $|R_{[0:n-2]}\rangle$ be one combined register. This allows us not to use a left shifting circuit thus saving quantum resources.

Step 3. The combined register mentioned above in step 2 and $|D_{[0:n-1]}\rangle$ are given as inputs to the quantum subtractor circuitry. Register $|D_{[0:n-1]}\rangle$ emerges unchanged. The combined register now holds the result of subtraction of the R and D registers. Let us call this result $|R - D_{[0:n-1]}\rangle$.

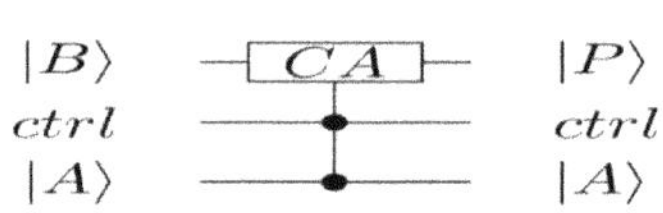

Figure 19.15. Graphical representation of the conditional ADD operation.

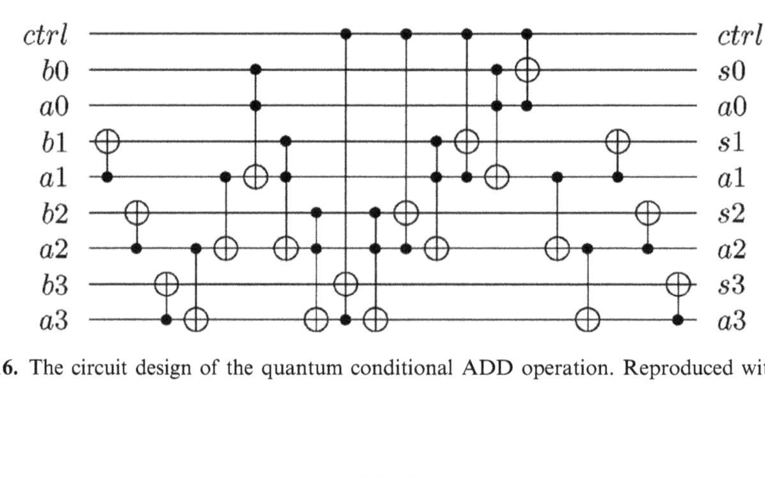

Figure 19.16. The circuit design of the quantum conditional ADD operation. Reproduced with permission from [8].

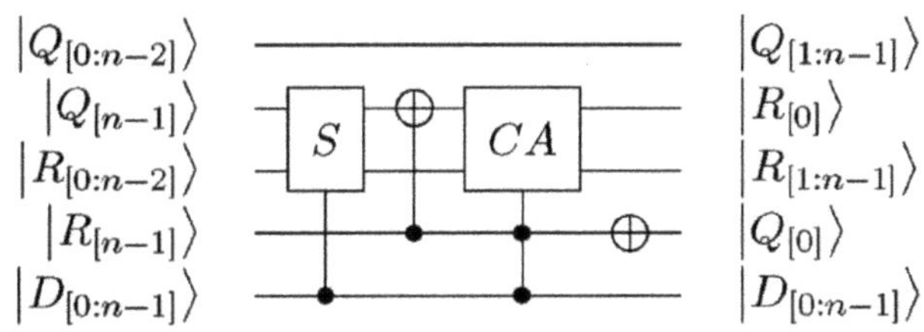

Figure 19.17. The quantum restoring integer divider circuitry design for a single iteration. Reproduced with permission from [8].

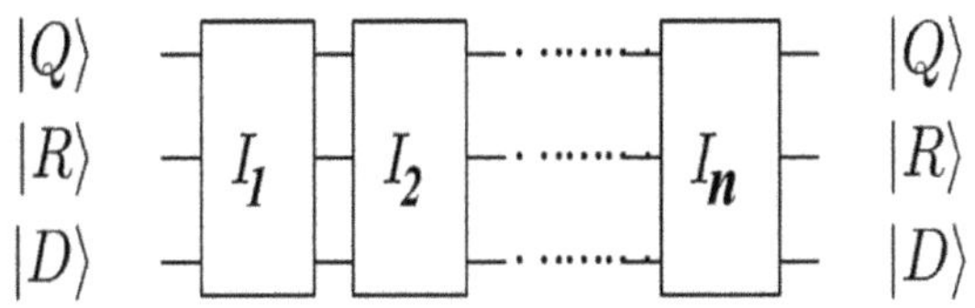

Figure 19.18. The quantum restoring integer divider circuitry design (for n iterations). Reproduced with permission from [8].

Step 4. Qubits $|R - D_{[n-1]}\rangle$ and $|R_{[n-1]}\rangle$ are supplied to a CNOT gate. $|R - D_{[n-1]}\rangle$ is the control qubit and $|R_{[n-1]}\rangle$ is the target qubit. The target now holds the value of $|R - D_{[n-1]}\rangle$ because $|R - D_{[n-1]}\rangle$ is always zero throughout the computation.

Step 5. The $|R - D_{[n-1]}\rangle$ computed in step 4 now acts as the control qubit to the conditional ADD circuit. $|R - D_{[0:n-1]}\rangle$ and $|D_{[0:n-1]}\rangle$ are two n-qubit inputs to the conditional ADD operation circuit. The outputs of conditional ADD operation are collected. $|R - D_{[n-1]}\rangle$ is complemented.

Step 6. All the above operations constitute the first iteration. From algorithm 19.1 it can be seen that the whole circuit is iterated n times. Hence the circuit in figure 19.17 is also iterated n times. This is done by using the outputs of the first iteration which will be used as inputs for the next iteration.

Step 7. This process continues for n iterations. In figure 19.18 I_1 represents the first iteration, I_2 represents the second iteration, and I_n represents the nth iteration. The steps 1 through 6 have to go through each iteration until it reaches n iterations.

Step 8. At the end of n iterations, the quotient is $|Q_{[0:n-1]}\rangle$, the remainder is $|R_{[0:n-1]}\rangle$, and the divisors are retained.

19.5 Summary

In this chapter fault-tolerant adder, multiplier, and division circuits are described with illustrations. The fault-tolerant quantum half-adder and full-adder are discussed and we show how circuits function as parity-preserving. The fault-tolerant quantum adder circuits, namely FTFA and EFTFA, are described and implemented using quantum gates, and it is shown that a fault-tolerant adder can be used for implementing all Boolean functions. In the fault-tolerant quantum multiplier design a Wallace tree structure is explained and demonstrated. A resource efficient design of the quantum circuitry of integer division is presented by optimizing the quantum circuit modules, and knitting them together efficiently. It is observed that a non-restoring division algorithm can be an attractive choice to design a quantum integer division circuit when minimizing the number of qubits is of primary concern. The designs can be integrated in larger data path subsystem designs to provide resource efficient implementation of quantum algorithms.

Critical thinking questions

1. Describe why quantum fault-tolerant circuits are needed.
2. Discuss some applications for quantum fault-tolerant circuits.
3. What distinguishes a quantum full-adder circuit from a quantum fault-tolerant full-adder circuit? Draw a diagram to explain.
4. Describe the restoring process for a quantum fault-tolerant integer divider.

References

[1] Amy M, Maslov D, Mosca M and Roetteler M 2013 A meet-in-the-middle algorithm for fast synthesis of depth-optimal quantum circuits *IEEE Trans. Comput.-Aided Des. Integr. Circuits Syst.* **32** 818–30

[2] Baugh C R and Wooley B A 1973 A two's complement parallel array multiplication algorithm *IEEE Trans. Comput.* **100** 1045–7

[3] Khosropour A, Aghababa H and Forouzandeh B 2011 Quantum division circuit based on restoring division algorithm *8th Int. Conf. on Information Technology: New Generations* (Piscataway, NJ: IEEE), pp 1037–40

[4] Kliuchnikov V, Maslov D and Mosca M 2012 Fast and efficient exact synthesis of single qubit unitaries generated by Clifford and T gates arXiv:1206.5236

[5] Li J, Peng X, Du J and Suter D 2012 An efficient exact quantum algorithm for the integer square-free decomposition problem *Sci. Rep.* **2** 260

[6] Qi X, Chen F, Zuo K, Guo L, Luo Y and Hu M 2012 Design of fast fault tolerant reversible signed multiplier *Int. J. Phys. Sci.* **7** 2506–14

[7] Thapliyal H and Ranganathan N 2013 Design of efficient reversible logic-based binary and BCD adder circuits *ACM J. Emerging Technol. Comput. Syst.* **9** 17

[8] Thapliyal H, Varun T S S and Munoz-Coreas E 2016 Quantum circuit design of integer division optimizing ancillary qubits and T-count arXiv:1609.01241

[9] Zhou R-G, Li Y-C and Zhang M-Q 2014 Novel designs for fault tolerant reversible binary coded decimal adders *Int. J. Electron.* **101** 1336–56

Part III

Quantum-dot cellular automata

An overview of quantum-dot cellular automata

Quantum-dot cellular automata (QCA) are not transistor-based, they are a type of nanotechnology. In QCA the basic element is a cell containing multiple quantum dots. The interaction between cells is purely Coulombic and there is no physical transport of charge. QCA are very promising for a number of reasons, among which their low power consumption and high density are two important features. Therefore, they seem to be appropriate for designing circuits for general-purpose computation as well as for embedded applications. Digital design in QCA is the focus of this part of the book.

QCA technology was introduced in 1993. It consists of an array of cells. Each cell contains several quantum dots, each of which is a nanoparticle or a crystal made of semiconducting materials such as silicon, cadmium, selenide, etc. A cell typically contains four quantum dots at the corners of a square and one central dot. Due to Coulomb interactions, electrons may occupy two 'diametrically' opposite quantum dots (or two antipodal sites). Labeling the quantum dots at the corners from 1 to 4 (in order), it can be observed that electrons may occupy dots 1 and 3 (or 2 and 4). Each cell has bi-stable behavior, facilitating its use in large scale cellular arrays. The physical interaction between neighboring cells is used to implement various logic functions. The key aspect of QCA is that the interaction between cells is purely Coulombic and there is no transport of charge between cells. The notion of cellular automata (CA) is due to the fact that the state of a given cell at a particular time depends on the state of its neighbors during the previous clock cycle.

Arrays can, in general, be regular or irregular and can lead to realization of various logic functions. Early efforts were created on a single QCA device using different approaches. Semiconductor, molecular, and magnetic implementations have been pursued. One of the earliest logic functions realized was the three-input majority gate. This has been used for computation of other basic logic functions, such as AND and OR. Combining a majority gate with an inverter has facilitated the realization of NAND, NOR, and other logic elements. All circuits in QCA technology are clocked. Clocking serves as a means of synchronization. Further, circuits for the realization of arithmetic functions may involve some types of wires crossing in the plane.

Contemporary microprocessors and application-specific integrated circuits are largely based on the complementary metal-oxide semiconductor (CMOS) technology. It is believed that the performance of various circuits in current CMOS-based architectures is close to reaching the limit. When the feature size of transistors is reduced to a nanometer, quantum effects such as tunneling take place. Further, when device scaling takes place, the interconnections do not scale automatically due to the effects of wire resistance and capacitance. This part of the book introduces the reader to QCA, an emerging nanotechnology and an alternative to CMOS. We also examine the problem of designing efficient circuits using QCA.

The changes that the QCA devices have experienced have been evolutionary, but the most advanced chips still use the same computing paradigm as their predecessors. There is a lot of expectation that new paradigms will be developed for the processing of information. CMOS technology uses current switching, whereas QCA deal with encoding the binary information as they are arrays of cells and every cell has quantum dots. The interaction of the quantum mechanics with the Coulomb force in every cell determines the cell state.

In the near future, the era of 'beyond CMOS' will begin, as the scaling of the current CMOS technology will reach the fundamental limit. QCA are a transistor-less computation paradigm and a viable candidate for 'beyond CMOS' device technology. The continued and fast dimensional scaling of CMOS will eventually approach the fundamental limit. Also, short channel effects, high power dissipation, and quantum effects are limiting the further scaling of current CMOS technology devices. Emerging device technology could overcome the scaling limitation in the current CMOS technology. Single electron transistors (SETs), QCA, and resonant tunneling diodes (RTDs) are some of the 'beyond CMOS' technologies. Among these evolving nanotechnologies, QCA are the most favorable technology. QCA can achieve a device density of 1012 devices/cm^2 and an operating speed of THz. The QCA device paradigm can replace field-effect transistor (FET) based logic and exploits the quantum effects of small size. QCA are a means of representing binary information on cells through which no current flows, and achieving device performance by the coupling of those cells.

This part of the book starts with some fundamentals of QCA circuits, which are provided in Chapter 20. Many of the QCA gates are included in this chapter. The designs for a QCA cell and QCA clock are described here. Chapter 21 presents the characteristics and design procedure of QCA adder and subtractor circuits. The designs of a QCA half-adder and -subtractor as well as a QCA full-adder and -subtractor are also provided here. The designs for a QCA multiplier and divider circuit are shown in Chapter 22. In this chapter, the technique of the non-restoring binary divider is described. The QCA asynchronous and synchronous counters are introduced in Chapter 23. The design of a dual-edge triggered J–K flip-flop is also discussed in this chapter. The design procedures of the QCA decoder and encoder circuit are introduced in Chapter 24. In this chapter, a QCA 2-to-4 decoder and QCA 3-to-8 decoder are shown. In addition, the design of an encoder with single-feedback and multi-feedback is also given in this chapter. Chapter 25 describes the design for a QCA multiplexer and demultiplexer circuit. In this chapter, a QCA 2-to-1 multiplexer and QCA 4-to-1 multiplexer are presented. In addition, a QCA 1-to-2 demultiplexer and QCA 1-to-4 demultiplexer are provided. Chapter 26 describes QCA flip-flops and Chapter 27 covers QCA programmable logic devices. Chapter 28 presents the design for a QCA random access memory (RAM) cell, read only memory (ROM), and cache memory. In addition, the design procedures for instruction memory, data memory, arithmetic logic units (ALUs), and integrated processors are also described in this chapter. Chapter 29 describes the QCA processor circuit, and Chapter 30 provides examples of applications of QCA technology.

IOP Publishing

Quantum Computing (Second Edition)
A pathway to quantum logic design
Hafiz Md Hasan Babu

Chapter 20

Quantum-dot cellular automata

Learning objectives

- Learn the significance of quantum-dot cellular automata (QCA).
- Learn the fundamental QCA components and gates.
- Discuss the fundamentals of QCA circuits.
- Discuss the QCA clock and its zone.
- Describe the structure of a QCA cell.
- Become familiar with QCA wires, unique cell configurations, and symmetric cells.
- Acquire knowledge about information and data propagation through QCA.

Quantum-dot cellular automata (QCA) are a new computational paradigm which encode binary information by charge configuration within a cell instead of using the conventional current switches. As there is no current flow, the Coulombic interaction is sufficient for computation. This revolutionary paradigm provides a possible solution for transistorless computation at the nanoscale. QCA allow operating frequencies in a high range with a small feature size and ultra-low power consumption. Currently, devices are becoming smaller and consume low power with a higher frequency of operation. For this reason, optimizing the overall cost is becoming more difficult day by day.

For circuit layout and functionality checking, a simulation tool for QCA circuits, QCA Designer, is used. This tool allows users to create a custom layout and then verify QCA circuit functionality using simulations. It includes two different simulation engines such as a bi-stable approximation and a coherence vector.

20.1 Fundamentals of QCA circuits

The basic properties of QCA circuits are area, delay, power, and overall cost. The descriptions of basic components are given below.

doi:10.1088/978-0-7503-5683-1ch20

20.1.1 Area

The area of a QCA circuit is the area of the base rectangular block on which the circuit is plotted. The area of a QCA circuit is determined by multiplying the vertical length and horizontal length of the circuit. Thus, the area (A) of a QCA circuit is calculated as

$$A = C_{\mathrm{h}} \times C_{\mathrm{w}} \times L^2, \tag{20.1.1}$$

where C_{h} = maximum number of cells in height, C_{w} = maximum number of cells in width, and L = length of a cell. The units to measure the area are m^2, nm^2, and $\mu\mathrm{m}^2$.

Example 20.1. If a circuit has 22 cells in the maximum horizontal chain, 25 cells in the maximum vertical chain, and the length of a cell is 20 $\mu\mathrm{m}^2$, then the area of the circuit is $A = 0.22\ \mu\mathrm{m}^2$.

20.1.2 Delay

Delay represents the number of clock cycles to generate the output from an input. For a QCA circuit, the delay is the value with the maximum clock cycle among all clock cycles of different input paths to the same output of the QCA circuit. The delay (D) is determined as

$$D = \max{}_{D_i \in D_a}(D_i), \tag{20.1.2}$$

where D_a = the set of minimum clock cycles of each path, D_i= the ith element of the set D_a, the measuring unit of delay is the clock cycle.

Example 20.2. If a QCA circuit contains three different input paths such as D_1, D_2, and D_3 to the same output and the paths need 12, 11, and 13 clock cycles to reach the same corresponding output, then the delay of the QCA circuit is 13 clock cycles.

20.1.3 Kink energy

Kink energy E_{kink} represents the energy difference between two horizontally or vertically adjacent polarized cells with the same or the opposite polarization and electrostatics (Coulomb's law) gives the means to calculate its value. According to Coulomb's law the following equation is obtained:

$$E_{\mathrm{kink}} = F_{\mathrm{q}} \times d, \tag{20.1.3}$$

where q = the charge of an electron and d = the distance between two repulsive electrons.

20.1.4 Power

It is known that in a tightly coupled environment and for an irreversible operation, each cell will dissipate E_{kink} for each cycle. If the delay for each cycle is constant, then the power dissipation is proportional to E_{kink}. The power dissipation of any QCA circuit is directly related to the number of cells needed to design the circuit. Thus the power (P) is measured as

$$P = N \times P_{\text{r}}, \qquad (20.1.4)$$

where N = the number of cells of the circuit and P_{r} = the relative power of a cell. The measuring unit of power is a relative value. Thus it has no unit. The power needed for a four-dot QCA is taken as 1 unit.

If a circuit needs 2345 cells and the relative power of the QCA cell is 0.4, then the power of the QCA cell is $P = 938$.

20.1.5 Overall cost

Cost is the parameter of a circuit which defines its efficiency. Cost in a QCA circuit is the measurement in terms of area, power, and delay. The overall cost of a circuit can be formulated in different ways. The overall cost of a circuit can be calculated as

$$\text{Cost} = N \times P_{\text{r}} \times A \times D, \qquad (20.1.5)$$

where N = the number of cells (i.e. complexity implies to power), P_{r} = relative power of a cell (which equals 1 for a standard four-dot QCA cell), A = the area of the circuit, and D = the delay of the circuit.

Thus the derivation of the overall cost for a standard four-dot QCA cell is Cost = Area × Delay × Power.

20.2 The QCA cell

In general, an electronics-based transistor functions through the transportation of electrons, whereas QCA apply the adjustment of electrons in a fixed area of tiny square nanometers. QCA are implemented in a small quadratic region which is called the QCA cell. Each small square cell contains exactly four potential dots to hold the extra electrons. Electrons can only reside in the potential dots. The QCA innovation depends on the association of bi-stable QCA cell outlines from four quantum dots. A schematic chart of a basic cell structure is shown in figure 20.1. The

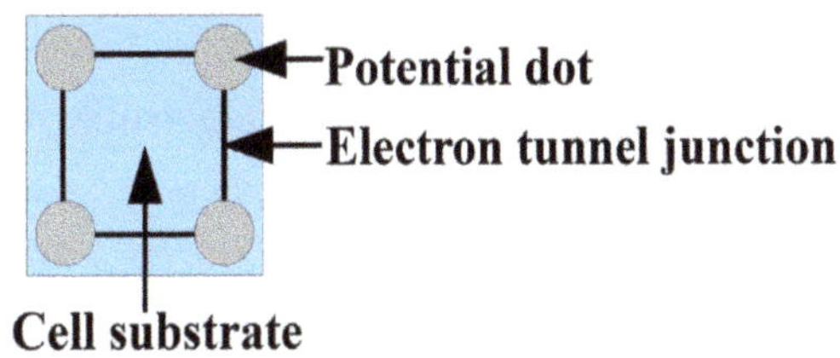

Figure 20.1. The structure of a QCA cell.

cell is charged by two free electrons by burrowing between nearby spots. These electrons have a tendency to possess antipodal locales as an after-effect of their shared electrostatic Coulombic repulsion.

The potential dots are connected with electron tunnel junctions. They can be opened for the electrons to travel through them under a particular condition by a clock signal. Without any interaction from outside, the two electrons will try to separate from each other as far as possible due to the Coulomb force that acts between them. As a result they will reside in diagonally located potential dots, because the diagonal is the largest possible distance for them to reside (figure 20.2).

There are two diagonals in a square, which means the electrons can reside in exactly two possible variants in the QCA cell. Regarding these two arrangements, they are interpreted as a binary 0 and binary 1, i.e. each cell can be in two states. The state 0 and the state 1 are shown in figure 20.2. A binary system is something familiar, as Boolean logic is used already in today's computers. There, a high voltage is often interpreted as binary 1 and a low voltage as binary 0.

20.3 Information and data propagation

If two QCA cells are placed next to each other, it is possible to exchange their states, i.e. the adjustments of the electrons in them. The QCA cell that should transfer its state to a neighboring cell must have its tunnel junctions closed, and the tunnel junctions in the neighboring cell have to be opened to allow the electrons to travel through the tunnel junctions between the potential wells. As soon as they open, the electrons in the neighboring cell are pushed by the Coulomb force of the original cell as far away as possible. As they are also pushed away from each other, they will travel into the same potential wells as in the original cell. As soon as the tunnel junctions are closed again, the transfer of the state is completed.

The state of a cell can also be transferred to multiple neighboring cells. This works in the same way as for a single neighboring cell, but the tunnel junctions of all the sequentially neighboring cells should be opened at the same time, which makes the transfer much faster than transferring the state cell by cell. This allows us to build 'wires' made of QCA cells to transport information over larger distances.

20.4 Basic QCA elements and gates

So far we know how to interpret and transport information with QCA cells, as shown in figure 20.3, but yet we lack the possibility of computations. For QCA cells,

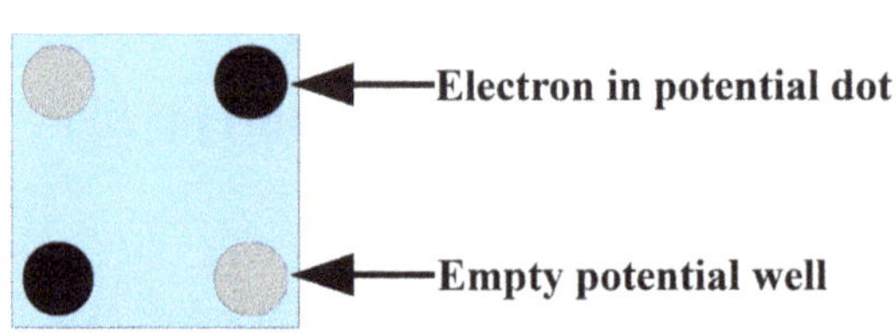

Figure 20.2. Electrons in potential dots.

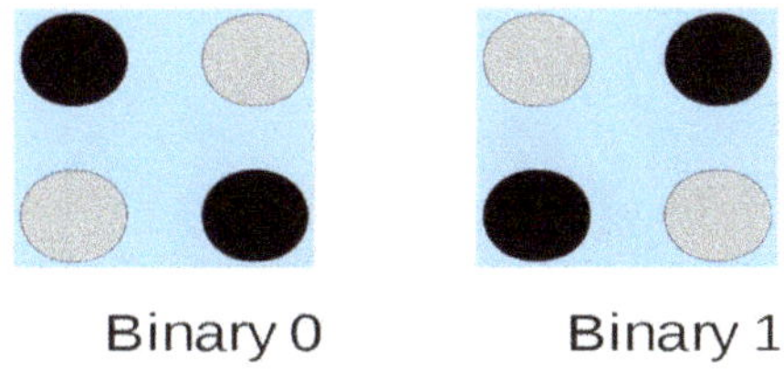

Figure 20.3. Binary interpretation of adjustments.

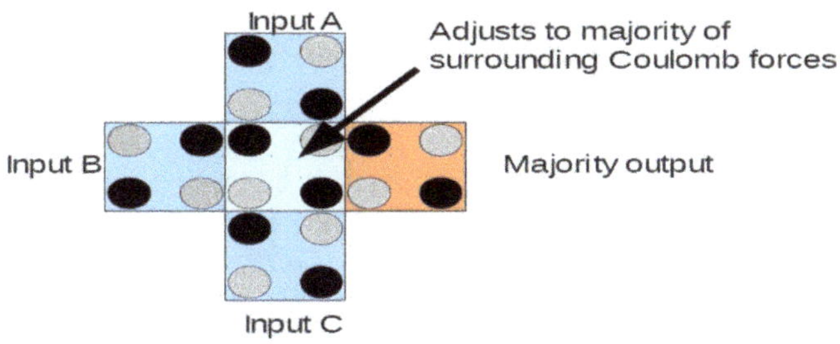

Figure 20.4. The QCA majority voter.

the basic gate is a three-input majority voter. It is built from five cells, arranged as a cross.

20.4.1 The QCA majority voter

It is known that the Coulomb forces of several electrons are summed up and the majority voter takes advantage of this effect. The cells at the top, at the left, and at the bottom work as input connection cells. As the Coulomb forces of the electrons of all input cells sum up, the middle cell adjusts to the majority of adjustments of the input connection cells. Finally the output cell adjusts to the middle cell and the resulting state of the majority vote, as shown in figure 20.4, can be obtained from the output cell.

20.4.2 The QCA AND gate

As we work in the field of QCA with known binary representation, it is preferable to have further logic gates with which we are already familiar. With a slight modification, it is possible to turn the majority voter into an AND gate.

The Boolean AND outputs are 1 if all inputs are 1, otherwise 0. Regarding two inputs of the majority voter, as the inputs of an AND gate and the voter should not output 1 if only one of the two inputs is 1, a fixed cell is added as the third input that is always in the state 0. If both AND inputs are 1, the two ones sum up to a stronger Coulomb force than the single fixed 0 cell and the majority voter is turned into a two-input AND gate, as shown in figure 20.5. The fixed cell can be obtained by setting it to the 0 state and never opening the electron tunnel junctions.

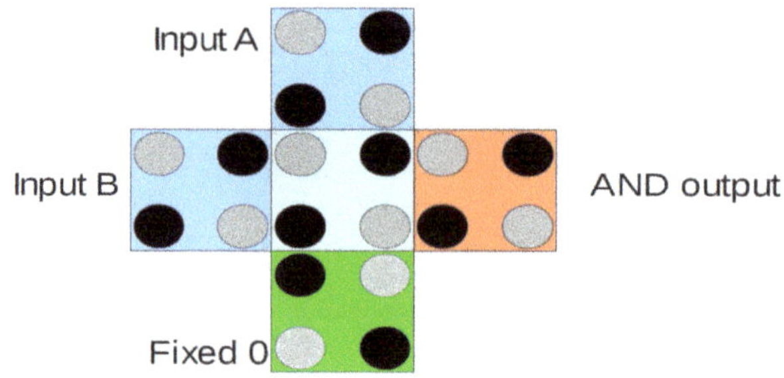

Figure 20.5. The QCA AND gate.

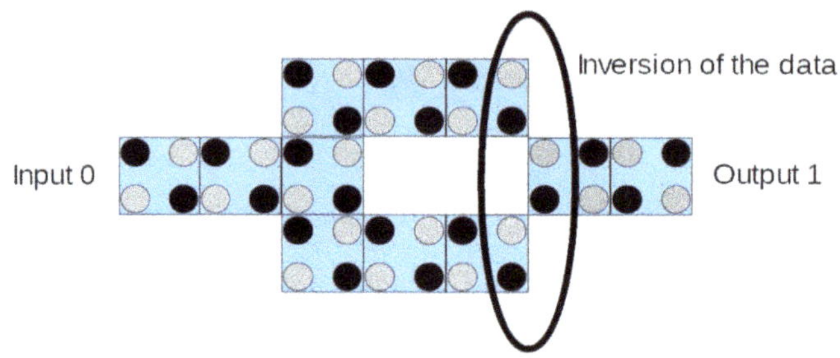

Figure 20.6. The QCA NOT gate.

20.4.3 The QCA OR gate

The OR gate is built almost exactly like the AND gate, but instead of a fixed 0, a fixed 1 QCA cell must be attached as one input. The fixed 1 cell sums up to a stronger Coulomb force with a single other input being adjusted to 1, so that the OR gate will output 1, if one of the free inputs is 1.

20.4.4 The QCA NOT gate

It is also possible to build a QCA NOT gate, as shown in figure 20.6. The implementation in QCA takes advantage of the geometry of cell adjustments. One QCA wire is forked to two wires. The switch of the cell adjustment takes place by putting the output cell next to the forked wires so that only the corners touch. Since only cell corners are touching the right of the fork and the cells at the end of the fork will have the same adjustment, and the cells of the right fork will not adjust with an electron close to an electron at the corner at the end of the fork, the adjustment on the right of the fork will be inverted. This makes a 1 at the input and a 0 at the output and vice versa.

20.4.5 The QCA wire

A string of QCA cells acts like a wire. A representation of a QCA wire is shown in figure 20.7. Within each clock cycle, half of the wire is active for signal transmission, while the other half remains steady. In the middle of the following clock cycle, half of the past active clock zone is deactivated and the remaining active zone cells trigger

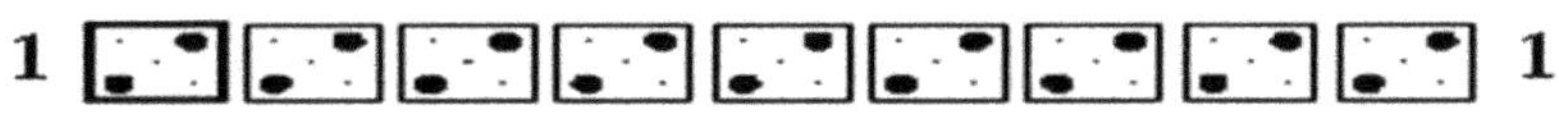

Figure 20.7. The QCA wire that propagates the signal across the circuit.

the recently initiated cells to be polarized. In this way the signal propagates from one clock zone to the next.

20.5 The QCA clock

A QCA cell has four clock zones, where each clock zone has four stages: switch, hold, release, and relax. Figure 20.8 demonstrates its operation process. In the switch stage the QCA cells start as unpolarized and their inter-dot potential boundaries are low. The obstructions are then raised in the middle stage and the QCA cells can be polarized as per the condition of the driver of the QCA gate (i.e. the information cell). It is in this clock stage that the real calculation (or exchanging) happens. Before the end of this clock stage, boundaries are sufficiently high to stifle any electron burrowing and cell states are settled. In the hold stage boundaries are held high, so the output of the sub-exhibit can be utilized as inputs to the following stage. In the release stage barriers are imposed and cells are permitted down to an unpolarized state. Finally, in the fourth clock stage, the relax stage, cell barriers are brought down and cells stay in an unpolarized state. Figure 20.8 demonstrates every clock zone flag and shows the pipeline component. All cells in a sure zone are controlled by the same QCA clock signal. Cells in every zone perform a particular count and the condition of a zone is then settled. Thus, they can serve as data signs to the following zone and data moves in a pipelined manner.

20.5.1 Special cell arrangements and symmetric cells

During the design of a QCA circuit, the situation is likely to occur that QCA wires have to be crossed. In contrast to classic transistor technology where wires can only cross by inserting another layer, QCA wires can be crossed in the same layer. This works by introducing a QCA cell type where the four potential wells are not in the corners of the cell but in the middle of the edges, as shown in figure 20.9.

If several of these QCA cells are put together to form a wire, the adjustment of the succeeding cell is inverted to its predecessor and so on. The advantage of this type of QCA cell originates in its symmetric effect of Coulomb force on regular cells. Although the electrons do interact with electrons in neighboring regular QCA cells, through the symmetry they do not push the electrons in regular QCA cells to a particular adjustment. In the other direction electrons in a regular cell do not push the electrons in a symmetric cell into a particular potential well. This allows the construction of wire crossings, as shown in figure 20.10, of these QCA cells with regular ones. A crossing is built using a continuous wire of special cells, building a gap in across a wire of regular cells.

Of course it has to be possible to connect symmetric QCA cells to regular ones and vice versa. This works by putting a regular cell as a neighbor of two symmetric

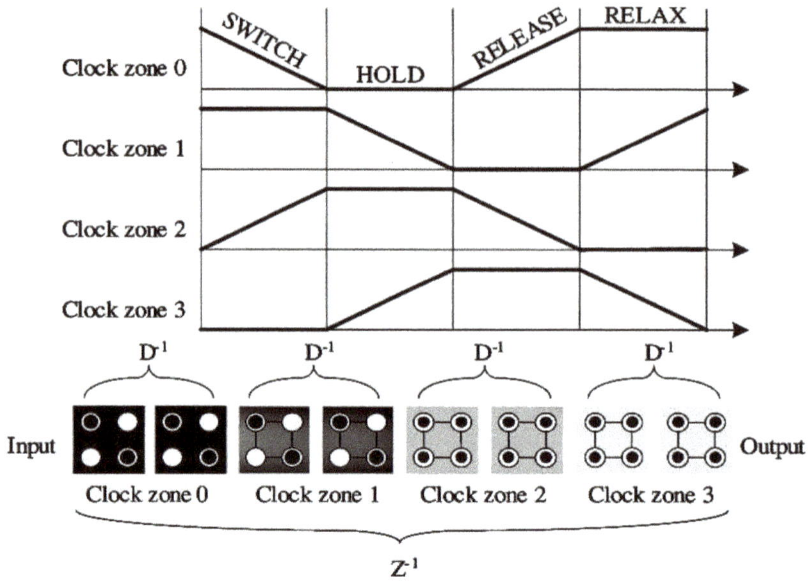

Figure 20.8. The four phases of a QCA clocking mechanism.

Figure 20.9. The symmetric adjustable QCA cell.

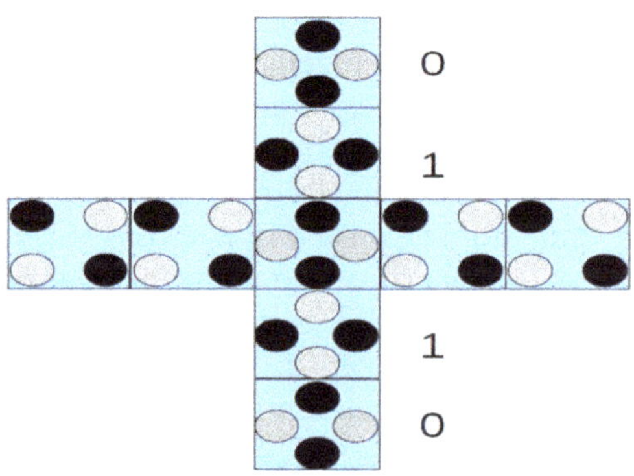

Figure 20.10. The QCA wire crossing.

cells near the beginning or end of a wire of symmetric cells. One has to take care as to which two symmetric cells are chosen, as neighbors of this type are in the inverted adjustment depending on the necessary adjustment, the original or the inverted one, as shown in figure 20.11.

20.5.2 The NOT gate clock zones

Since the NOT gate has a slightly critical zone of arranged QCA cells, it is important to put the cells in the proper clock zones to avoid randomly flipped adjustments of electrons near the forking wire.

The clock zone of the input should end at the beginning of the 'fork'. The complete fork itself, i.e. the U-shaped wire, should be in the subsequent clock zone of the input and the output should be in the subsequent clock zone of the fork. The QCA NOT gate clock zones are shown in figure 20.12.

20.5.3 The majority voter clock zones

For the majority voter it is important that all cells are in the same clock zone. Putting some cells in different clock zones can lead to incorrect results. The center cell and the three input cells plus the output cell have to be in the same clock zone. The QCA majority voter clock zones are presented in figure 20.13.

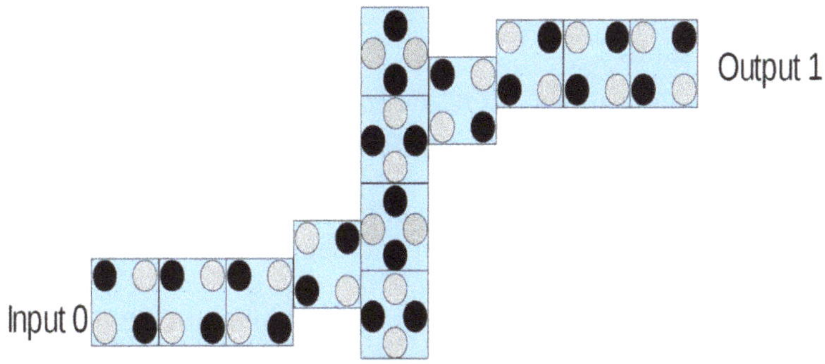

Figure 20.11. The connections between symmetric and regular QCA cells.

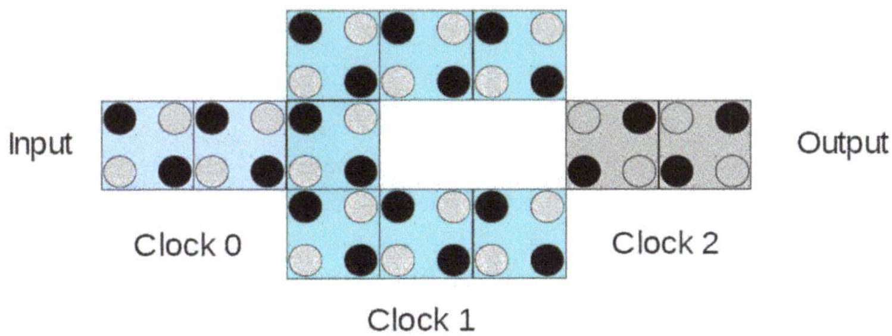

Figure 20.12. The QCA NOT gate clock zones.

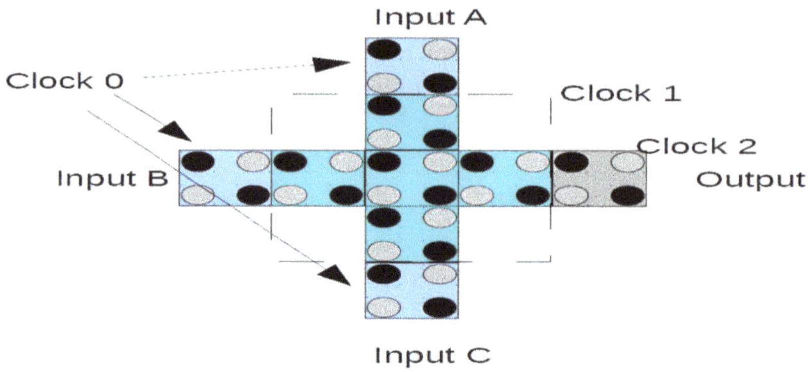

Figure 20.13. The QCA majority voter clock zones.

20.6 Summary

This chapter presents the basic components of quantum-dot cellular automata. The QCA circuit's overall cost in terms of the cell area, delay, kink energy, and power is described here. This chapter explains the QCA clock and its zone, QCA wires and their crossing, and the connection between symmetric and regular QCA cells. The ideas in this chapter will help the reader to understand the different implementations of QCA logic.

Critical thinking questions

1. Describe the applications of QCA circuits.
2. What interpretation should be given to negative results from the necessary QCA criteria?
3. Find the area of the circuit if a circuit has 20 cells in the maximum horizontal chain, 15 cells in the maximum vertical chain, and the length of a cell that is 12 μm.
4. Find the delay of the QCA circuit if a QCA circuit contains three different input paths such as D_1, D_2, and D_3 to the same output and need 15, 12, and 17 clock cycles to reach the same corresponding output.
5. Find the power of the QCA cell if a circuit needs 3325 cells. Please note that the relative power of the QCA cell is 1.4.
6. Describe some disadvantages of the QCA circuits.

References

[1] Nano-Arch Online 2012 Quantum-dot cellular automata (QCA) *University of Erlagen* http://www.cs3.tf.uni-erlangen.de/Research/KOMINA/QCA_slides.pdf (Accessed: 6 December 2018)

[2] Gladshtein M 2011 Quantum-dot cellular automata serial decimal adder *IEEE Trans. Nanotechnol.* **10** 1377–82

[3] Liu M and Lent C S 2005 Power dissipation in clocked quantum-dot cellular automata circuits *63rd Device Research Conf. Digest* vol 1 (Piscataway, NJ: IEEE) pp 123–4

[4] Liu M 2006 Robustness and power dissipation in quantum-dot cellular automata *PhD Thesis* University of Notre Dame, IN https://www3.nd.edu/~lent/pdf/nd/Robustness_and_Power_Dissipation_in_Quantum-Dot_Cellular_Automata.pdf

[5] Momenzadeh M, Huang J, Tahoori M B and Lombardi F 2005 Characterization, test, and logic synthesis of and-or-inverter (AOI) gate design for QCA implementation *IEEE Trans. Comput.-Aided Des. Integr. Circuits Syst.* **24** 1881–93

[6] Tahoori M B, Huang J, Momenzadeh M and Lombardi F 2004 Testing of quantum cellular automata *IEEE Trans. Nanotechnol.* **3** 432–42

[7] Timler J and Lent C S 2002 Power gain and dissipation in quantum-dot cellular automata *J. Appl. Phys.* **91** 823–31

IOP Publishing

Quantum Computing (Second Edition)
A pathway to quantum logic design
Hafiz Md Hasan Babu

Chapter 21

The QCA adder and subtractor

Learning objectives

- Discuss the QCA Ex-OR gate with its proper circuit.
- Describe the construction procedure of a full-subtractor using QCA.
- Design the QCA half-adder circuit.
- Discuss how to build a QCA full-adder.
- Construct a half-subtractor using QCA logic.
- Mention the uses of QCA full-subtractors.

A quantum dot is a nano-crystal made of semiconductor material that exhibits quantum mechanical properties. The flow of information in the quantum-dot cellular automaton (QCA) cell is based on the interaction of switching cells. It encodes the binary information as the electronic configuration of a cell. In this chapter the adder and subtractor circuits are discussed using the QCA implementation of the exclusive-OR (Ex-OR) gate. This model has the potential for combinational circuits that are compatible with QCA gates within the nanoscale.

The basic logic gates, i.e. Ex-OR and exclusive-NOR (Ex-NOR) gates, are used in the design of digital circuits. These gates have special functions and applications. They are particularly useful in arithmetic operations as well as code generators. The Ex-OR and Ex-NOR gates are usually found as two-input gates. No multiple-input Ex-OR/Ex-NOR gates are available since they are complex to fabricate using current hardware. The design has a complexity of cells and coplanar cross-overs or multiple layers that need to be implemented.

21.1 The Ex-OR gate

It is possible to implement all combinational and sequential logic functions by properly arranging cells so that the polarization of one cell can adjust the polarization of a nearby cell. A schematic representation of an Ex-OR gate in terms of a QCA majority gate and the truth table of this gate are shown in figure 21.1 and

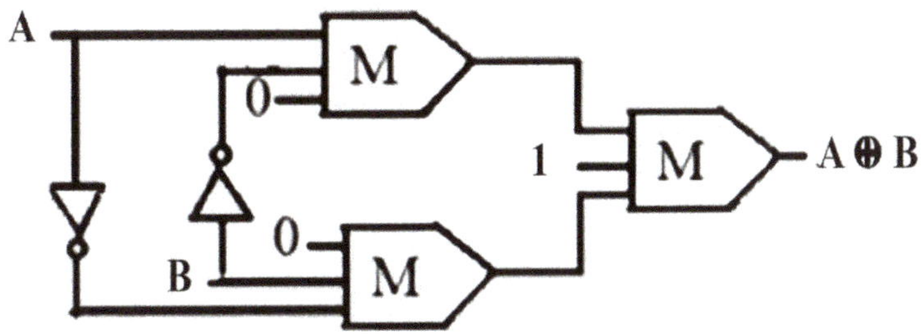

Figure 21.1. Schematic of the Ex-OR gate.

Table 21.1. The truth table of the Ex-OR gate.

A	B	$A \oplus B$
0	0	0
0	1	1
1	0	1
1	1	1

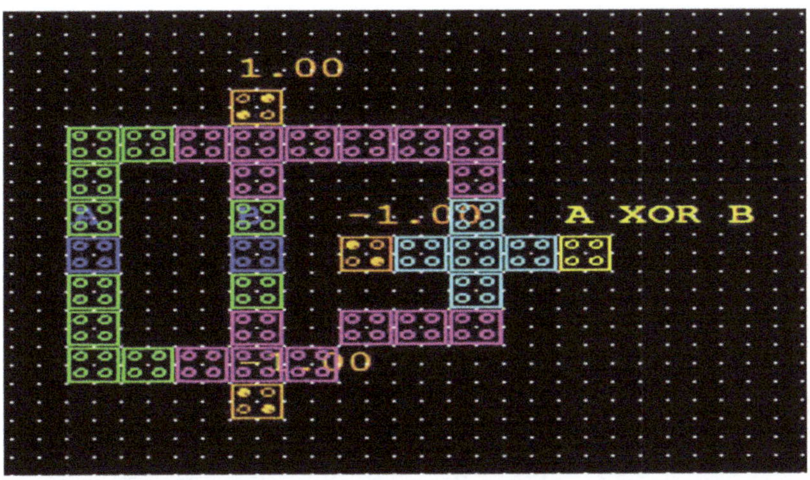

Figure 21.2. The QCA Ex-OR gate.

table 21.1, respectively. A representation of an Ex-OR gate using the QCA Designer tool is shown in figure 21.2.

21.2 The QCA half-adder and -subtractor

A half-adder is implemented by using the combination of an Ex-OR gate and an AND gate. The half-adder circuits could be implemented using different logics. The schematic of a half-adder is shown in figure 21.3 with its truth table shown in table 21.2, and the QCA implementation of figure 21.3 is presented in figure 21.4.

Figure 21.3. The design of a half-adder circuit.

Table 21.2. The truth table of a half-adder.

Input		Output	
A	*B*	Sum	Carry
0	0	0	0
0	1	1	0
1	0	1	0
1	1	0	1

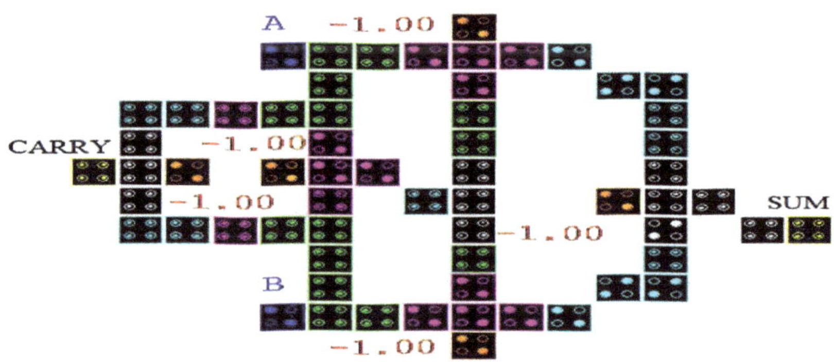

Figure 21.4. The design of a QCA half-adder circuit.

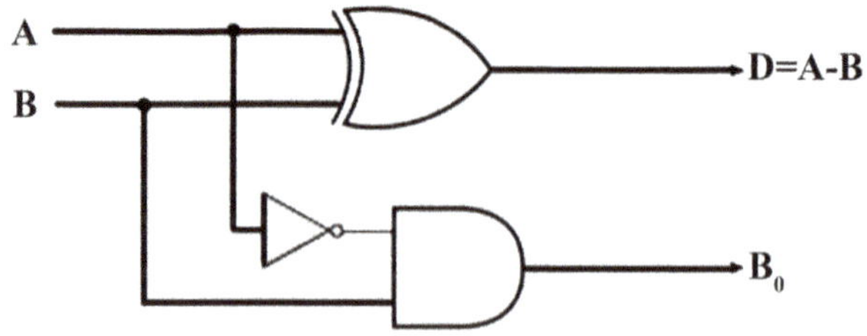

Figure 21.5. The half-subtractor.

The half-adder circuit might be used to make the subtractor circuit with the addition of one NOT gate. The schematic diagram of a subtractor circuit is shown in figure 21.5 with its truth table in table 21.3.

Table 21.3. The truth table of the half-subtractor.

Input		Output	
A	*B*	Difference	Borrow
0	0	0	0
0	1	1	0
1	0	1	0
1	1	0	1

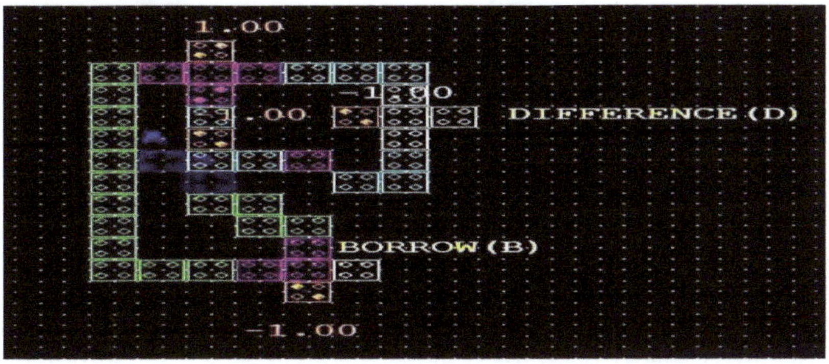

Figure 21.6. The QCA half-subtractor.

Figure 21.6 represents the QCA implementation of the half-subtractor circuit.

21.3 The QCA full-adder and full-subtractor

Imagine a full-adder with three inputs (*A*, *B*, and *C*). The *C* input will act as the carry bit which is transferred from the previous stage. The implementation function of a full-adder is

$$\text{Sum} = A \oplus B \oplus C, \tag{21.1}$$

$$\text{Carry} = AB + AC + BC. \tag{21.2}$$

As can be seen from equations (21.1) and (21.2), each full-adder has two outputs (sum and carry) and the sum output is the result of Ex-OR of circuit inputs. According to equation (21.2) it should also be noted that carry can be defined as an output of the three-input majority gate (with *A*, *B*, and *C* as its inputs). Hence, it is possible to design an optimal full-adder in QCA technology with this simple reason. The implementation of this full-adder is shown in figure 21.7. This figure shows that the space occupied and the number of cells used in the proposed circuit are smaller than in the circuits using only three-input and five-input majority gates.

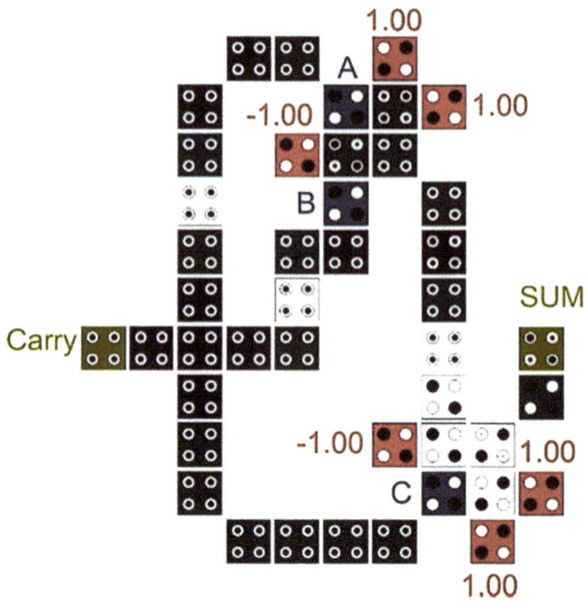

Figure 21.7. Implementation of the full-adder in QCA technology. Reproduced with permission from [9]. CC BY 4.0.

As can be seen from figure 21.7, the design can be generalized to higher bit number full-adder design using the coplanar method. This is the advantage of this full-adder.

21.3.1 Implementation of the full-adder and full-subtractor

A full-subtractor circuit is composed of three inputs like the full-adder. The sub and borrow outputs of the full-subtractor are as follows:

$$\text{Sub} = A \oplus B \oplus C, \tag{21.3}$$

$$\text{Borrow} = AB + AC + BC. \tag{21.4}$$

Since the sub output of the full-subtractor is the same as sum in the full-adder (see equations (21.3) and (21.4)), implementing full-adder and full-subtractor circuits in one integrated structure is reasonable. Again it should be mentioned that from equations (21.3) and (21.4), to generate the carry and borrow a majority gate is needed. Since both circuits have the same inputs and use the majority and similar Ex-OR gates, it is possible to design an integrated circuit for both operations. Therefore, a new full-adder and full-subtractor structure with the minimum number of quantum cells will be introduced.

Figure 21.8 shows the implementation of the proposed circuit. As it is obvious in the figure, this circuit makes use of all the gates in the adder. Although an inverter gate and one majority gate are added to the circuit, it is more efficient in comparison to other circuits in terms of the number of cells and area. As can be seen, the full-

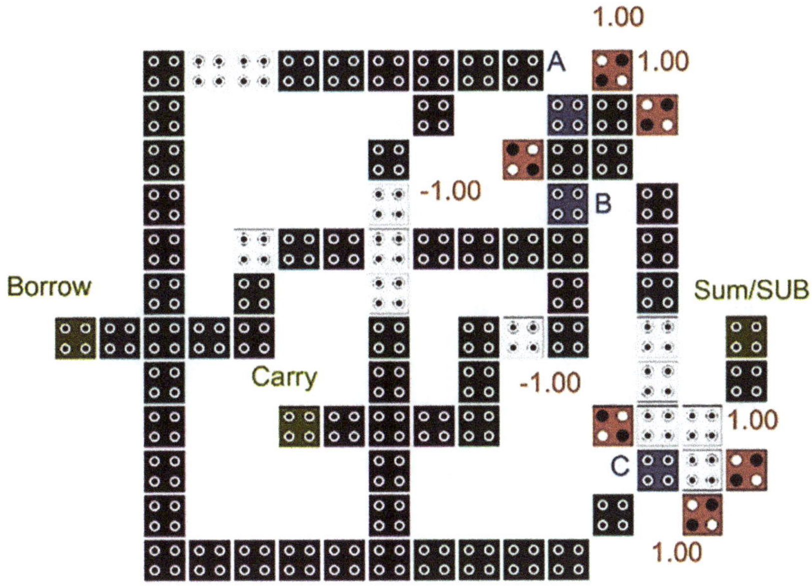

Figure 21.8. Implementation of the full-adder and full-subtractor in QCA technology. Reproduced with permission from [9]. CC BY 4.0.

adder and full-subtractor have fewer cells and smaller area with good delay performance.

21.4 Summary

This chapter presents the QCA adder and subtractor circuits. The half-adder and half-subtractor circuits are designed from the Ex-OR gate and the schematic and QCA implementation of the circuits are described. Moreover, the full-adder and full-subtractor are presented together in the QCA implementation.

Critical thinking questions

1. Describe a method to construct a compact QCA adder.
2. Describe the uses of QCA subtractors.
3. Is it possible to use a QCA adder circuit as a QCA subtractor? Explain using an appropriate figure.
4. What do the terms minuend and subtrahend denote in a QCA subtractor?
5. Let the inputs of a QCA adder be A and B. What will the output be if $A = B$?

References

[1] Ahmad F, Bhat G M and Ahmad P Z 2014 Novel adder circuits based on quantum-dot cellular automata (QCA) *Circuits Syst.* **5** 142

[2] Hayati M and Rezaei A 2015 Design of novel efficient adder and subtractor for quantum-dot cellular automata *Int. J. Circuit Theory Appl.* **43** 1446–54

[3] Lakshmi S K 2010 Efficient design of logical structures and functions using nanotechnology based quantum dot cellular automata design *Int. J. Comp. Appl.* **3** 35–42

[4] Polisetti S and Santhosh K 2008 QCA based multiplexing of 16 arithmetic and logical subsystem—a paradigm for nano computing *3rd Annual IEEE-Int. Conf. on Nano/Micro Engineering Molecular System, Hainan Island, China* (Piscataway, NJ: IEEE)

[5] Rani T S, Karmakar S, Metri A A and Sharan P 2015 Comparative study of half adder and subtractor circuits based on quantum-dot cellular automata (QCA) *Int. Conf. on Advances in Computer and Communication Engineering* vol 3 pp 61–7

[6] Shahidinejad A and Selamat A 2012 Design of first adder/subtractor using quantum-dot cellular automata *Adv. Mater. Res.* **403** 3392–7

[7] Tougaw P D and Lent C S 1999 Logical devices implementation using quantum dot cellular automata *J. Appl. Phys.* **75** 1818

[8] Walus K, Dysart T J, Jullien G A and Budiman R A 2004 QCADesigner: a rapid design and simulation tool for quantum-dot cellular automata *IEEE Trans. Nanotechnol.* **3** 26–31

[9] Zoka S and Gholami M 2019 A novel efficient full adder–subtractor in QCA nanotechnology *Int. Nano Lett.* **9** 51–4

IOP Publishing

Quantum Computing (Second Edition)
A pathway to quantum logic design
Hafiz Md Hasan Babu

Chapter 22

The QCA multiplier and divider

Learning objectives

- Explain the QCA multiplier.
- Calculate the bit product matrix for four-bit multiplication using QCA.
- Understand the multiplication network.
- Define the QCA divider.
- Discuss QCA multiplication networks with an example.
- Construct a QCA divider.
- Design a multiplier for the QCA.

In this chapter the QCA multiplier and divider circuits are discussed with schematic diagrams and QCA layouts. For the multiplier circuit, the filter network is explained and illustrated and finally the QCA implementation is presented. For the divider circuit, non-restoring division techniques are explained using a schematic and the QCA layout.

22.1 The QCA multiplier

The multiplier has important functions in signal processing and various other applications. With advancements in technology, high speed, low power consumption, regularity of layout and hence low area, or even a combination of these, can be applied in one multiplier. This makes them suitable for various high speed, low power, and compact implementations. In general, the multiplication method uses add and shift algorithms.

To perform multiplication of two numbers, we need to start with the filter networks, i.e. the finite impulse response (FIR) filter, and the output of the network is defined by

$$y_i = \sum_{k=0}^{N-1} b_k x_{i-k}. \tag{22.1}$$

doi:10.1088/978-0-7503-5683-1ch22

Let Z^{-1} be the one cycle delay operator such that $Z^{-1}x_i = x_{i-1}$. Z^o is defined by the unit operator and Z^{-n} is defined by $Z^{-n} = Z^{-1}Z^{-n+1}$. The characteristics are the following:

- $Z^{-n}x_i = x_{i-n}$.
- $Z^{-1}F(x) = F(Z^{-1}x)$.
- If C is the time invariant, $Z^{-1}C = C$.

$$\begin{aligned} y_i &= \sum_{k=0}^{N-1} b_k x_{i-k} \\ &= \sum_{k=0}^{N-1} b_k Z^{-k} x_i \\ &= \left(\sum_{k=0}^{N-1} b_k Z^{-k} \right) x_i. \end{aligned} \tag{22.2}$$

Equation (22.2) can be implemented by the network presented in figure 22.1. The circles in the figure with b_i represent multiplication by the constant written inside them and $\oplus$ entails the addition of two inputs. The data x_i, b_i, and y_i are words of arbitrary size.

In the pipeline design both upper and lower signal lines take the same additional delay units. It is assumed that $Z^{-\frac{1}{4}}$ comprises the possibility and application of the $Z^{-\frac{1}{2}}$ delay element to each section with upper and lower lines. Equation (22.3) proves that figure 22.2 provides the correct filter output result with $N/2$ cycle delays.

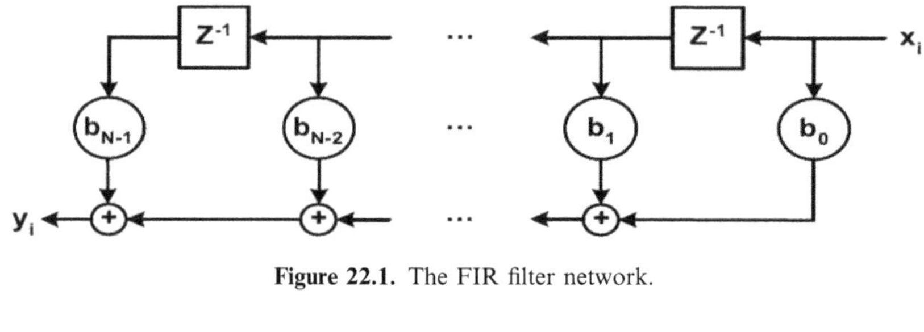

Figure 22.1. The FIR filter network.

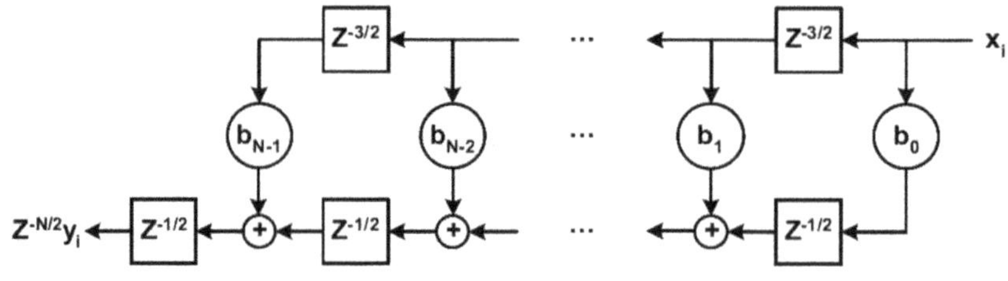

Figure 22.2. The pipelined FIR filter network.

The pipelined FIR filter outputs are the following:

$$
\begin{aligned}
&= Z^{-\frac{1}{2}}\left(b_{N-1}Z^{-\frac{3}{2}(N-1)} + Z^{-\frac{1}{2}}\left(b_{N-2}Z^{-\frac{3}{2}(N-2)} + \cdots + Z^{-\frac{1}{2}}(b_0 Z^0)\right)\right)x_i \\
&= Z^{-\frac{N}{2}}b_{N-1}Z^{-(N-1)}x_i + Z^{-\frac{N}{2}b_{N-2}}Z^{-(N-2)}x_i + \cdots + Z^{-\frac{N}{2}}b_0 x_i \\
&Z^{-\frac{N}{2}}\left(\sum_{k=0}^{N-1} b_k Z^{-k}\right)x_i \\
&= Z^{-\frac{N}{2}}y_i.
\end{aligned}
\tag{22.3}
$$

22.1.1 Multiplication networks

The above relations of a channel system can be connected for the duplication procedure. Let us consider an example, where an unsigned number framework is imagined; figure 22.3 shows the bit item matrix for an unsigned multiplication. For this situation only single digit calculations are utilized in the network, so all additions and multiplications utilize binary calculations. A one-digit multiplication is addressed by a logical AND and a one-digit addition presents to a full-adder. The fundamental difference between the FIR filter and the multiplication network is the treatment of the carry-out of the full-adder. The channel organizes an inside use carry flow stream, yet the multiplication arrangement needs distinct signal streams, so the system for multiplication ought to be changed in a similar manner to the filter channel.

Let $(a_i; b_i)$ be the multiplicand and multiplier pair and p_i be the product sum of the bit position i. Bits a_i and p_i correspond to the words x_i and y_i of the filter example. The position i is considered as the input applied at time i. Define the sum and carry-out of the full-adder at the ith time and jth location as (s_{ij}, c_{ij}) when $0 \leqslant i \leqslant 2N - 1$ and $0 \leqslant j \leqslant N - 1$, where j is numbered from right to left. Assume that the sum generation takes at least $Z^{-\frac{1}{2}}$ and the carry generation takes at least $Z^{-\frac{1}{4}}$. Although figures 22.1 and 22.2 ignored the zeroth full-adder, the derivation includes that adder. The implementation can be performed in two ways:

						a_{N-1}	a_{N-2}	…	a_1	a_0
					X	b_{N-1}	b_{N-2}	…	b_1	b_0
						$a_{N-1}b_0$	$a_{N-2}b_0$	…	a_1b_0	a_0b_0
					$a_{N-1}b_1$	$a_{N-2}b_1$	…	a_1b_1	a_0b_1	
				⋮	⋮	…	⋮	⋮		
		$a_{N-1}b_{N-2}$	$a_{N-2}b_{N-2}$	…	a_1b_{N-2}	a_0b_{N-2}				
	$a_{N-1}b_{N-1}$	$a_{N-2}b_{N-1}$	…	a_1b_{N-1}	a_0b_{N-1}					
p_{2N-1}	p_{2N-2}	p_{2N-3}	…	p_N	p_{N-1}	p_{N-2}	…	p_2	p_1	p_0

Figure 22.3. The bit product matrix for unsigned multiplication.

$$\begin{aligned}(s_{ij},\, c_{ij}) &= \mathrm{Add}\left(b_j Z^{-\frac{7}{4}j} a_i,\; Z^{-\frac{3}{4}} s_{i(j-1)},\; Z^{-\frac{1}{4}} c_{i(j+1)}\right)\\ &= \mathrm{Add}\left(b_j a_{i-\frac{7}{4}j},\; s_{(i-\frac{3}{4})(j-1)},\; c_{(i-\frac{1}{4})(j+1)}\right)\end{aligned} \tag{22.4}$$

$$\begin{aligned}(s_{ij},\, c_{ij}) &= \mathrm{Add}\left(b_j Z^{-\frac{3}{2}j} a_i,\; Z^{-\frac{1}{2}} s_{i(j-1)},\; Z^{-1} c_{ij}\right)\\ &= \mathrm{Add}\left(b_j a_{i-\frac{3}{2}},\; s_{(i-\frac{1}{2})(j-1)},\; c_{(i-1)j}\right).\end{aligned} \tag{22.5}$$

Equation (22.4) sends the carry-out in a backward direction with minimum delay. Equation (22.5) instead uses a feedback loop to the adder itself using a one clock delay unit. Both handle the carry-out correctly, carrying it to the higher bit calculation. They are called a carry shift multiplication (CSM) and a carry delay multiplication (CDM), respectively. Figures 22.4 and 22.5 present the network diagrams based on equations (22.4) and (22.5). From the stream direction of the information in and out, they are called right-to-left (RtL) systems. The calculated yield line of a full-adder is the carry-out. The CSM configuration has the advantage of the least delay for the convey shift while the CDM configuration minimizes the latency of the yield.

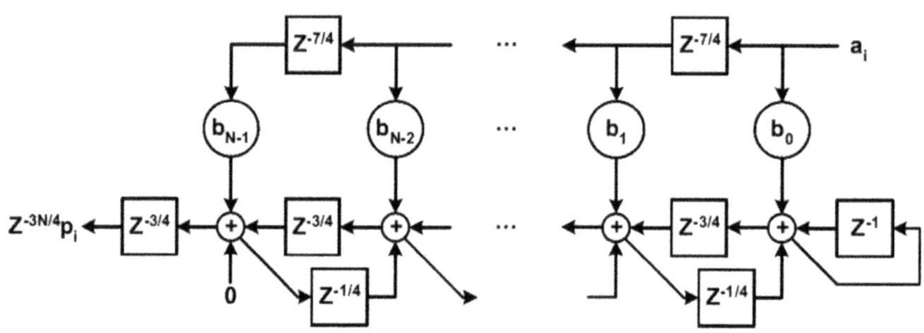

Figure 22.4. The RtL carry shift multiplication network.

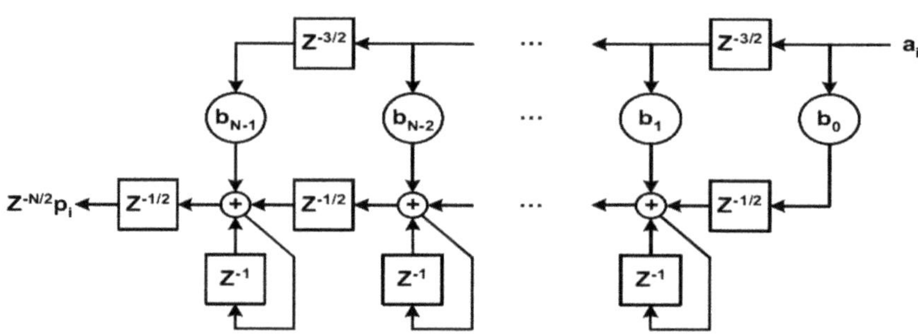

Figure 22.5. The RtL carry delay multiplication network.

In this realization, the minimum latency from the first input to first output is either $3N/4$ or $N/2$ cycles. From figure 22.1 the redirected output to the right side, which is the same side to the input in figure 22.5 is present for the redirection graph and for a pipeline design, can be redrawn as shown in figure 22.6 using the following equation:

$$\begin{aligned} Z^{-\frac{1}{2}}y_i &= Z^{-\frac{1}{2}}\left(\sum_{k=0}^{N-1} b_k Z^{-k}\right)x_i \\ &= Z^{-\frac{1}{2}}\left(\sum_{k=0}^{N-1} b_k Z^{-\frac{k}{2}} Z^{-\frac{k}{2}}\right)x_i \\ &= Z^{-\frac{1}{2}}\left(\sum_{k=0}^{N-1} Z^{-\frac{k}{2}} b_k Z^{-\frac{k}{2}}\right)x_i. \end{aligned} \tag{22.6}$$

Finally, figure 22.6 is a network design which is comparable to figure 22.2. The main difference is that there exists a much smaller latency from the first input to the first output. Figure 22.7 presents the block diagram of a redirected pipelined FIR filter network which can easily be derived from figure 22.6.

Based on figure 22.6, the multiplication networks are represented by equations (22.7) and (22.8) depending on the carry-out handling. Figures 22.8 and 22.9 show the respective network implementations. Similarly, the networks are called a right-to-right (RtR) networks and they are distinguished by the carry flow as either CSM

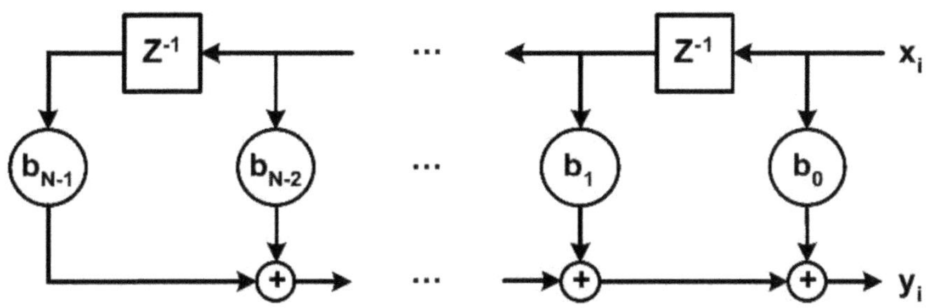

Figure 22.6. The redirected FIR filter network.

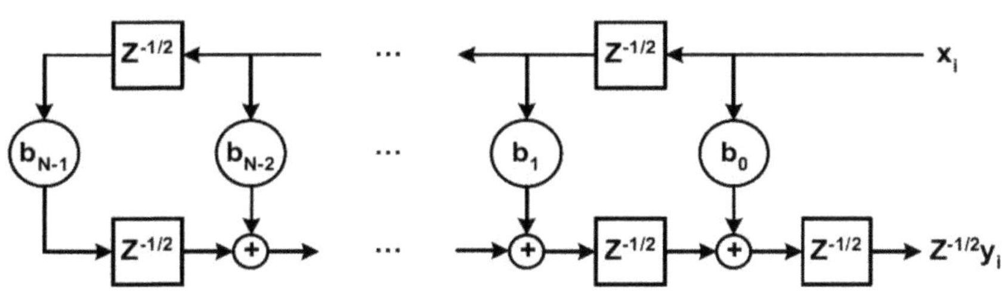

Figure 22.7. The redirected pipelined FIR filter network.

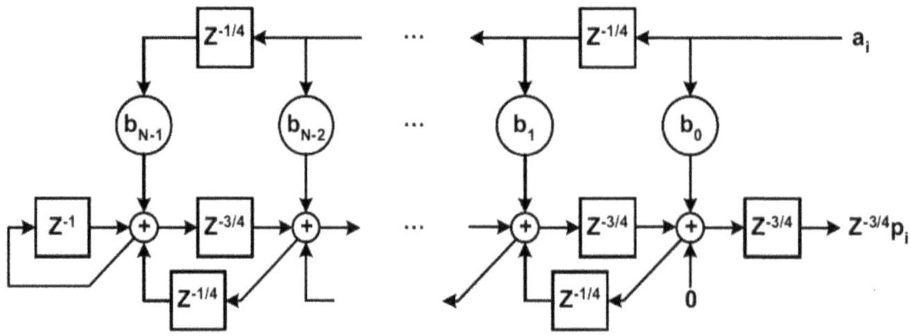

Figure 22.8. The RtR carry shift multiplication network.

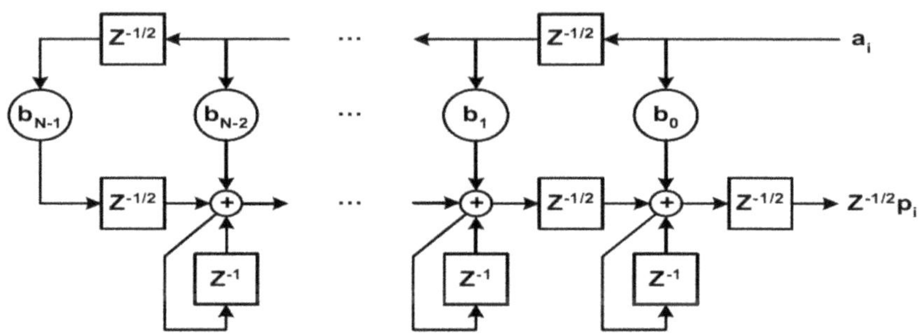

Figure 22.9. The RtR carry delay multiplication network.

or CDM networks. The CSM design has the minimum delay of the carry shift and the CDM design has the minimum latency to the output:

$$
\begin{aligned}
(s_{ij},\ c_{ij}) &= \mathrm{Add}\left(b_j Z^{-\frac{1}{4}j} a_i,\ Z^{-\frac{3}{4}} s_{i(j+1)},\ Z^{-\frac{1}{4}} c_{i(j-1)}\right) \\
&= \mathrm{Add}\left(b_j a_{i-\frac{1}{4}j},\ s_{(i-\frac{3}{4})(j+1)},\ c_{(i-\frac{1}{4})(j-1)}\right)
\end{aligned} \tag{22.7}
$$

$$
\begin{aligned}
(s_{ij},\ c_{ij}) &= \mathrm{Add}\left(b_j Z^{-\frac{1}{2}j} a_i,\ Z^{-\frac{1}{2}} s_{i(j+1)},\ Z^{-1} c_{ij}\right) \\
&= \mathrm{Add}\left(b_j a_{i-\frac{1}{2}},\ s_{(i-\frac{1}{2})(j+1)},\ c_{(i-1)j}\right).
\end{aligned} \tag{22.8}
$$

22.1.2 QCA multiplication networks

The QCA circuit has a four-phase clock and the circuit areas are divided into four clock zones. One clock zone delay is denoted by the D^{-1} operator which corresponds to a quarter cycle. That is, $D^{-}4 = Z^{-1}$. Based on the QCA circuit characteristics, one clock zone delays a quarter clock, so that the delay matches one D^{-1} operation. Assume a logical AND operation delay by one D^{-1} operation and one full-adder sum and carry-out computation from the input delay by D^{-2} and D^{-1} operations, respectively. Wires also cause some delay based on the wire length. After incorporating these characteristics the filter networks are redrawn as figures 22.10 and 22.11. The delay amounts in the upper and lower signal flows in either case are chosen to make one clock cycle difference between the adjacent paths.

From the filter network examples, the multiplier networks for QCA are drawn. Based on equations (22.4)–(22.5) and (22.7)–(22.8), equations (22.9)–(22.10) and (22.11)–(22.12) are reorganized for QCA multiplication. The former figures are revised by the carry flow routes. The serial multiplier can be realized with four distinct possibilities, as shown in figures 22.12, 22.13, 22.14, and 22.15. The

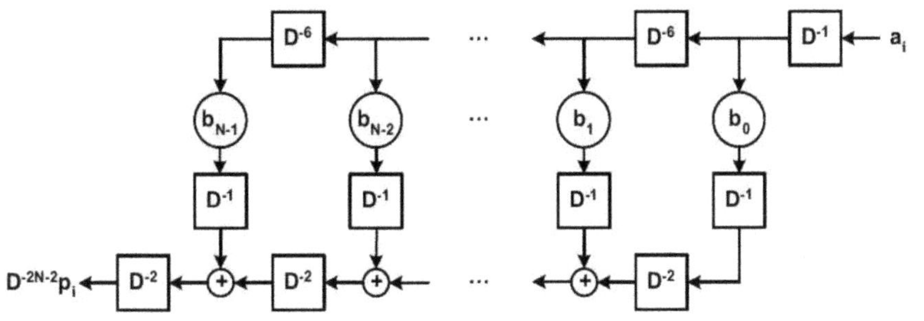

Figure 22.10. The FIR filter network for QCA.

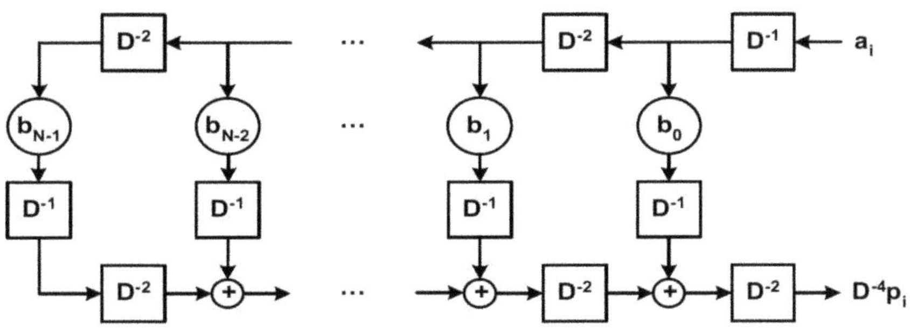

Figure 22.11. The redirected FIR filter network for QCA.

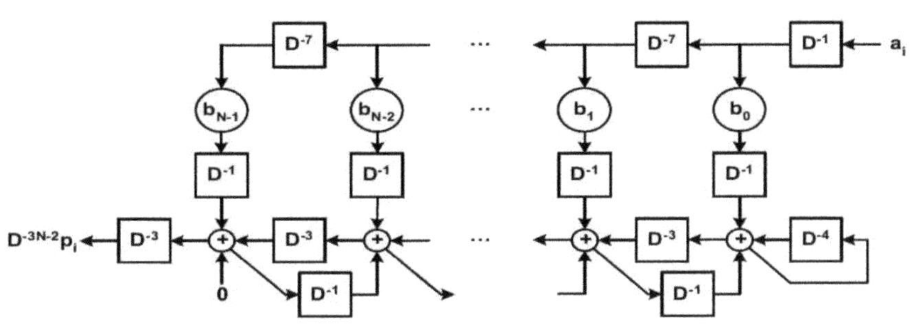

Figure 22.12. The RtL CSM network for QCA.

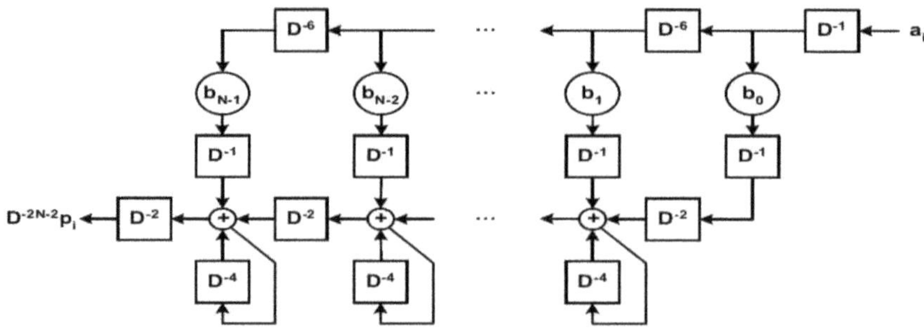

Figure 22.13. The RtL CDM network for QCA.

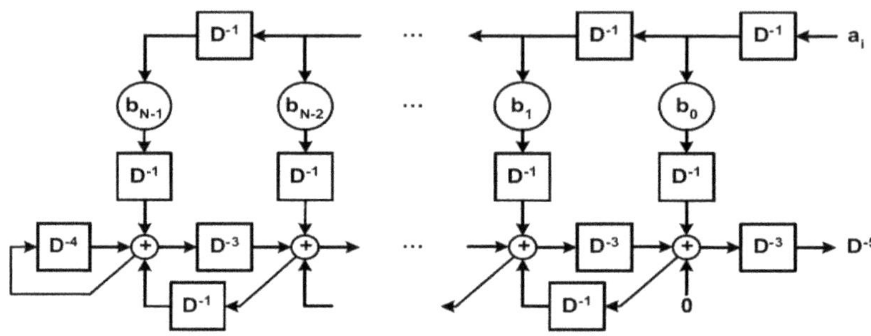

Figure 22.14. The RtL CSM network for QCA.

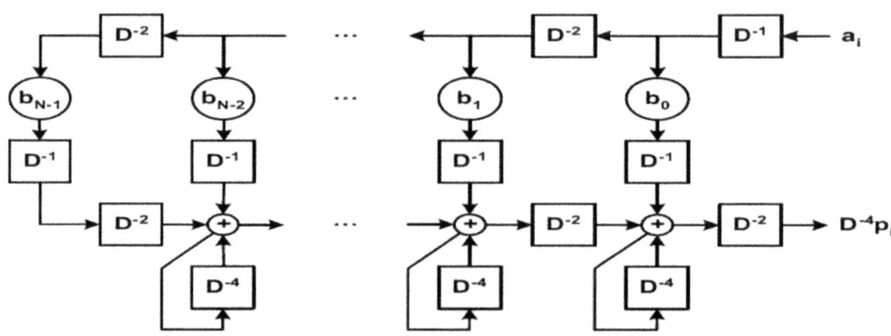

Figure 22.15. The RtR CDM network for QCA.

multiplier block diagram and the modified multiplier block diagram are presented in figures 22.16 and 22.17, respectively.

$$\begin{aligned}(s_{ij},\, c_{ij}) &= \mathrm{Add}\big(b_j D^{-7j-2} a_i,\ D^{-3} s_{i(j-1)},\ D^{-1} c_{i(j+1)}\big)\\ &= \mathrm{Add}\big(b_j a_{i-7j-2},\ s_{(i-3)(j-1)},\ c_{(i-1)(j+1)}\big)\end{aligned} \tag{22.9}$$

$$\begin{aligned}(s_{ij},\, c_{ij}) &= \mathrm{Add}\big(b_j D^{-6j-2} a_i,\ D^{-2} s_{i(j-1)},\ D^{-4} c_{ij}\big)\\ &= \mathrm{Add}\big(b_j a_{i-6j-2},\ s_{(i-2)(j-1)},\ c_{(i-4)j}\big)\end{aligned} \tag{22.10}$$

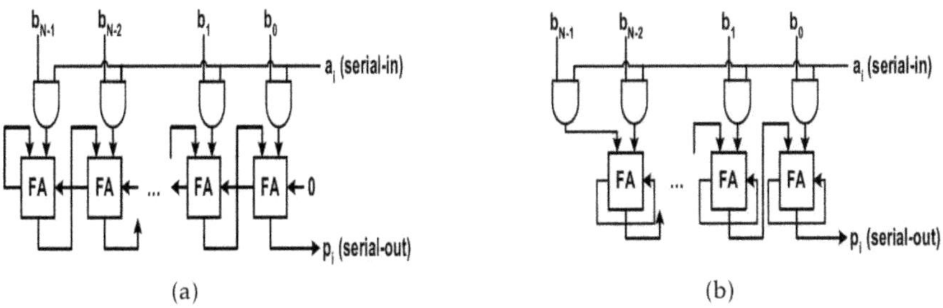

Figure 22.16. Multiplier block diagrams. (a) RtR CSM. (b) RtR CDM.

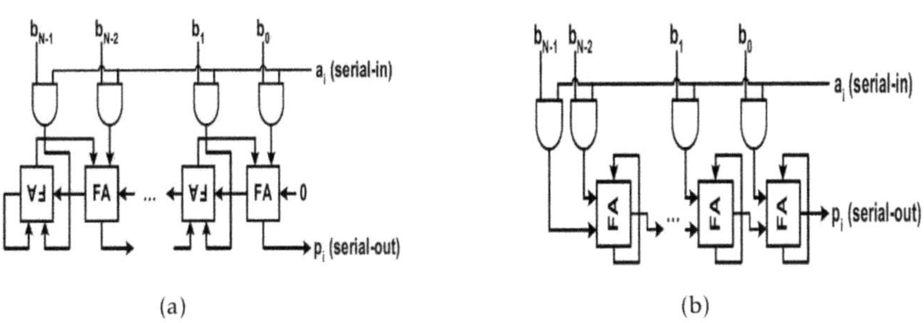

Figure 22.17. Modified multiplier block diagrams. (a) RtR CSM. (b) RtR CDM.

$$\begin{aligned}(s_{ij},\ c_{ij}) &= \mathrm{Add}\left(b_j D^{-j-2} a_i,\ D^{-3} s_{i(j+1)},\ D^{-1} c_{i(j-1)}\right)\\ &= \mathrm{Add}\left(b_j a_{i-j-2},\ s_{(i-3)(j+1)},\ c_{(i-1)(j-1)}\right)\end{aligned} \tag{22.11}$$

$$\begin{aligned}(s_{ij},\ c_{ij}) &= \mathrm{Add}\left(b_j D^{-2j-2} a_i,\ D^{-2} s_{i(j+1),\ D^{-4} c_{ij}}\right)\\ &= \mathrm{Add}\left(b_j a_{i-2j-2},\ s_{(i-2)(j+1)},\ c_{(i-4)j}\right).\end{aligned} \tag{22.12}$$

22.1.3 Multiplier design

The multiplication track is spread over a structure like figure 22.5 except multiplying $Z^{-\frac{1}{2}}$ to all the horizontal stream delay units. This design intrinsically experiences a lengthy delay and has an redundant right-most full-adder. An RtR structure is nominated for QCA execution since it is know to reduce the inactivity of the first input and output. The final proposal has two substitutes, as depicted in figures 22.12 and 22.13. Figure 22.14 presents the general block diagrams of QCA multipliers and figure 22.15 shows the block diagrams of the improved designs for QCA circuits.

22.1.3.1 QCA implementation

Figure 22.18 describes the bit product matrix for four-bit multiplication.

Full-adders are used for the carry shift multiplier and bit-serial addresses are used for the carry delay multiplier. Those addresses are made using the carry flow adder

X				a_3	a_2	a_1	a_0
				b_3	b_2	b_1	b_0
				a_3b_0	a_2b_0	a_1b_0	a_0b_0
			a_3b_1	a_2b_1	a_1b_1	a_0b_1	
		a_3b_2	a_2b_2	a_1b_2	a_0b_2		
	a_3b_3	a_2b_3	a_1b_3	a_0b_3			
p_7	p_6	p_5	p_4	p_3	p_2	p_1	p_0

Figure 22.18. The bit product matrix for four-bit multiplication.

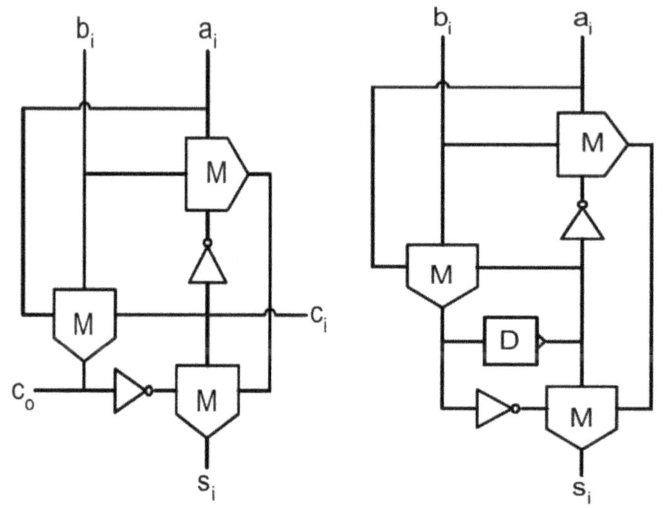

(a) (b)

Figure 22.19. One-bit adder schematic. (a) Full-adder. (b) Bit-serial adder.

(CFA), which is a layout optimized ripple carry adder. The bit-serial adder is similar to the full-adder except that the carry-in and carry-out are connected internally with a one clock delay. Figures 22.19 and 22.20 show the schematics and layouts of both adders.

Using the above mechanism, the four-bit RtR CSM and CDM are designed as shown in figure 22.17. These block diagrams are used to implement the multiplier circuits, as shown in figures 22.21 and 22.22. Multipliers for larger word sizes can also be implemented easily by adding additional bit slices, i.e. 32-bit CSM and CDM are shown in figures 22.23 and 22.24.

22.2 The QCA divider

The divider is the most entangled and tedious circuit among all the math units. However, for the advancement of exact instrumentation, i.e. for spaceflight and radar innovations, the divider is vital. Numerous calculations are utilized for the

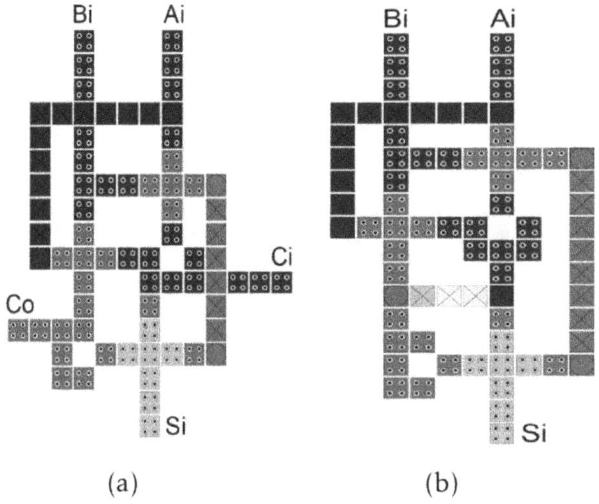

Figure 22.20. QCA one-bit adder layouts. (a) Full-adder. (b) Bit-serial adder. Reproduced with permission from [2]. Copyright 2007 IEEE.

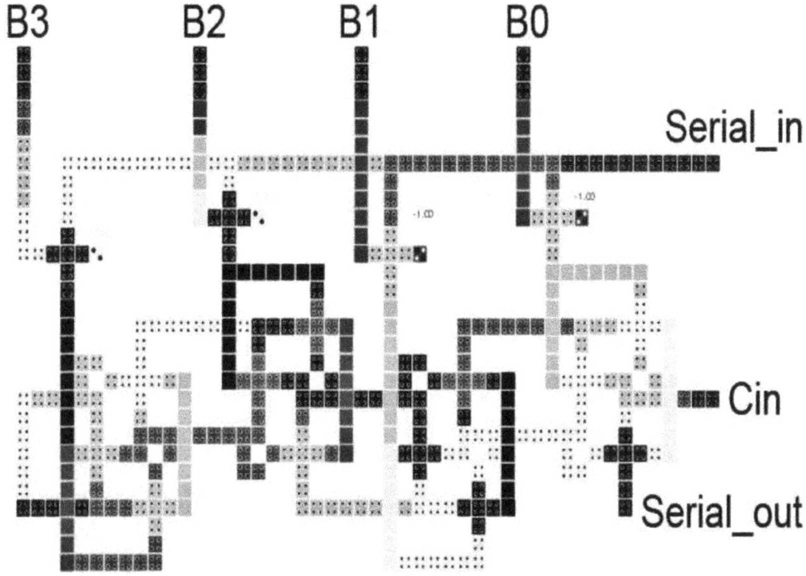

Figure 22.21. The layout of a four-bit QCA carry shift multiplier. Reproduced with permission from [2]. Copyright 2007 IEEE.

execution of the divider, specifically re-establishing, non-re-establishing, Sweeney, Robertson and Tocher techniques, and the Newton iterative calculation. Re-establishing and non-re-establishing calculations contain expansion and subtraction, which are appropriate for coordinated circuit usage. The generally utilized

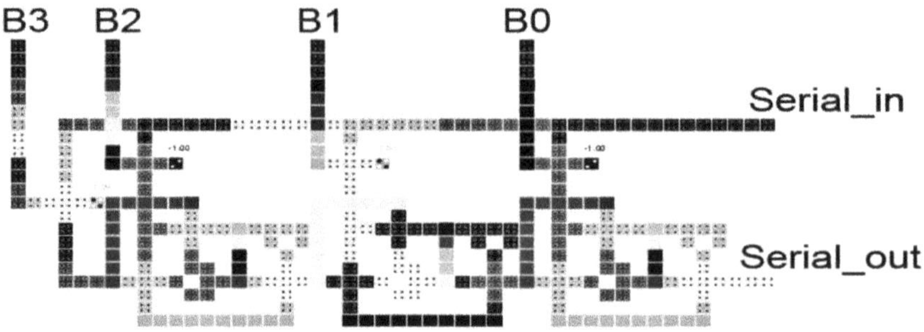

Figure 22.22. The layout of a four-bit QCA carry delay multiplier. Reproduced with permission from [2]. Copyright 2007 IEEE.

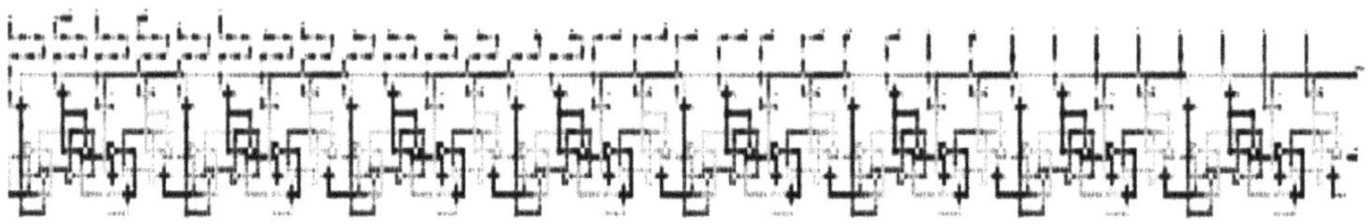

Figure 22.23. The layout of a 32-bit QCA carry shift multiplier. Reproduced with permission from [2]. Copyright 2007 IEEE.

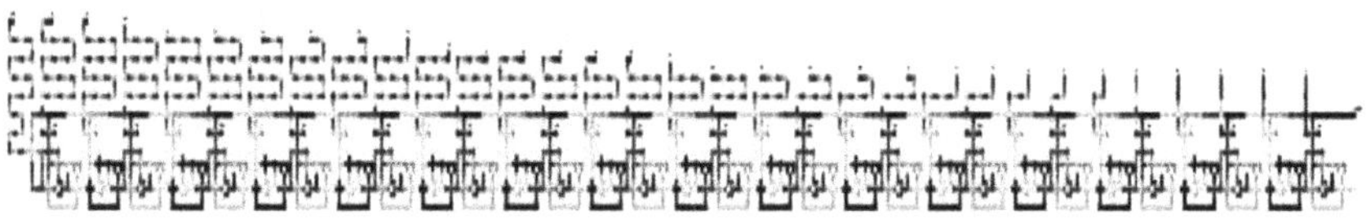

Figure 22.24. The layout of a 32-bit QCA carry delay multiplier. Reproduced with permission from [2]. Copyright 2007 IEEE.

calculation in the divider is the non-re-establishing division, and Huanqing displayed the structure of a non-re-establishing binary array divider in QCA.

22.2.1 The non-restoring binary divider

To understand the non-restoring division, it is necessary to explain the restoring division algorithm. The following notation will be used in the divider section:

N: numerator or dividend
Y: denominator or divisor
R_i: partial remainder after ith iterations, i = number of iterations
n: number of bits
q: quotient set for the algorithm
Q: quotient for the division

The restoring algorithm is the basic method for division operation, and the details of the method are described by the following set of equations:

$$q_{i+1} = \begin{cases} 1, & \text{if } 2R_i > Y \\ 0, & \text{if } 2R_i < Y, \end{cases} \tag{22.13}$$

$$R_{i+1} = 2R_i - q_{i+1}.\ Y. \tag{22.14}$$

The partial remainder is going through a left shift of R_i and a subtraction: $R_{i+1} = 2R_i - 1$. If $R_i + 1$ is positive, then $q_{i+1} = 1$, else $q_{i+1} = 0$ and a restoring addition is needed, as shown in equation (22.15). The addition is used to restore the proper partial remainder, $R_{i+1} = R_{i+1} + Y = 2R_i$, as shown in equation (22.17).

To overcome the problems of restoring division, we undertake an alternative method of binary division, the non-restoring division also known as the add–subtraction alternate algorithm, and the relevant divider is called non-restoring division (NRD). In this algorithm if a negative partial remainder is produced at the end of a certain cycle, no restoration is executed.

The non-restoring division operation is performed by the following formulas:

$$q_{i+1} = \begin{cases} 1, \text{if } R_i > 0 \\ 0, \text{if } R_i < 0, \end{cases} \tag{22.15}$$

$$R_{i+1} = \begin{cases} 2R_i - Y, \text{if } R_i > 0 \\ 2R_i + Y, \text{if } R_i < 0, \end{cases} \tag{22.16}$$

$$r = \begin{cases} 2^{-n}.\ R_n, \text{if } R_n > 0 \\ 2^{-n}.\ (R_n + Y) \text{ if } R_n < 0, \end{cases} \tag{22.17}$$

where $i = 0, 1, \cdots, n - 1$ is the recursion index. The initial partial remainder R_0 equals the dividend, and r is the final remainder.

In this manner, from equations (22.15)–(22.17) above it is anything but difficult to see that each time the negative partial remainder portion shows up, the non-re-establishing calculation makes one expansion stride not exactly as the re-establishing strategy. Thus there will be a very critical reduction of the quantity of clock cycles over the whole division process.

In binary non-restoring division, partial remainders are obtained by a subtraction or an addition between the dividend and the successively right-shifted versions of the divisor. The quotient bit is determined by the sign of the partial remainder which also decides whether to add or subtract the shifted divisor in the next cycle.

A two-dimensional array of pipelined two's-complement adder/subtractor cells (CAS cells) can be used to implement the non-restoring algorithm. Fundamentally the array consists of columns of carry propagate adders with one controlled input bit P. Figures 22.25(a) and (b) illustrate an array divider with a two-bit divisor ($0.\ y_1y_2$) and an four-bit dividend ($0.\ x_1x_2x_3x_4$). The first bit of both the divisor and dividend are their signs, it is used to represent a positive number. Each logic cell is made up of a full-adder and an Ex-OR gate. The Ex-OR gate provides the divisor input to the full-adder. The control signal P determines the function of the CAS cell, namely

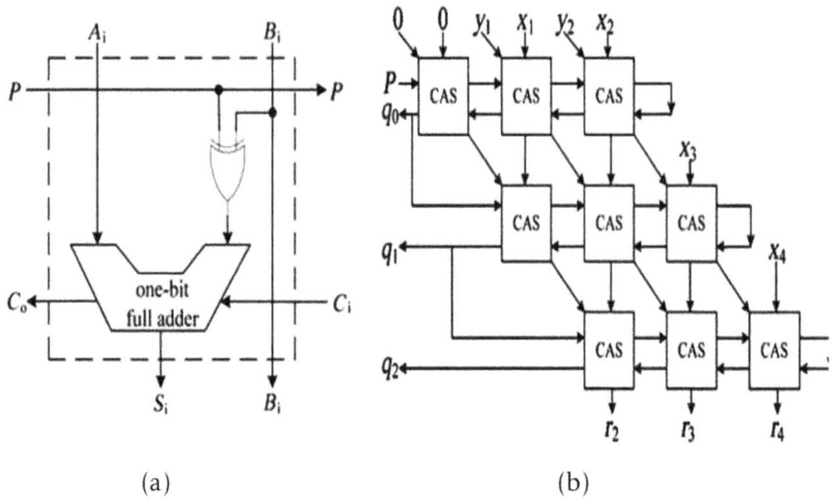

Figure 22.25. The binary array divider. (a) CAS cell. (b) Four-bit dividend, two-bit divisor.

whether an addition or a subtraction is to be executed. The two-bit divisor and the double-length four-bit dividend are imported at the top and right edges of the array, respectively. At first, a three-bit quotient is produced at the left side of the array. Then every quotient bit is transported to the next row as the control signal *P* and the low carry of the right-most cell. The three-bit final reminder appears at the bottom of the array.

In this section, the presented QCA multiplication circuit depends on pipeline engineering. The sequential parallel multiplier is a conceivable answer for this requirement, and the standard QCA pipeline configuration has the advantage of reducing wire delays.

22.2.2 Divider implementation

The non-restoring array divider is composed of CAS cells that have one full-adder and one Ex-OR gate. Figure 22.26 shows the CAS cell layout. The Ex-OR gate is demonstrated in the dashed box and it can be designed using only four majority gates and one inverter, its delay is a full clock cycle. The rest of the cell is a one-bit full-adder; it can be constructed using only three majority gates and two inverters. The full-adder takes three inputs, (Ex-OR$_O UT$, A, and C_i) and gives two outputs ($S_i = A \oplus B \oplus C_i$ and $C_o = MAJ(A, B, C_i)$). Both the outputs have a latency of $1\frac{1}{4}$ clock cycles. The different colors in the layout are used to distinguish different clock zones.

An n-bit divider is formed by combining n^2 CAS cells to the regular array shown in figure 22.27; only $n = 3$ is presented here for simplicity. In the figure the total number of QCA cells is 3742 and the area of the divider layout is 6.22 μm^2. Compare this to the eight-bit NRD implemented by 45 nm complementary metal-oxide semiconductor (CMOS) technology which has an area of 2015.64 μm^2. It is very

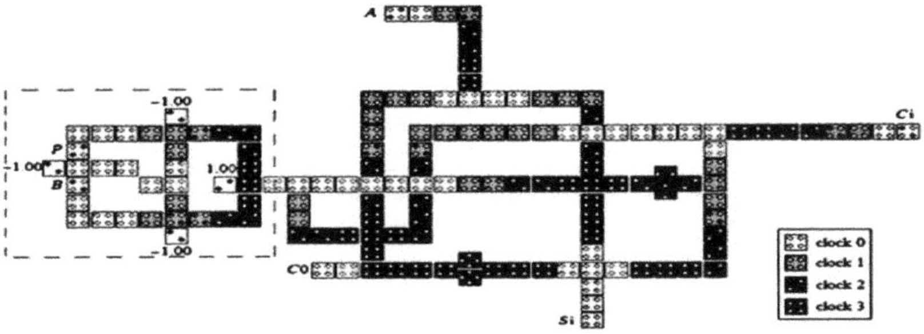

Figure 22.26. The QCA layout of the CAS cell.

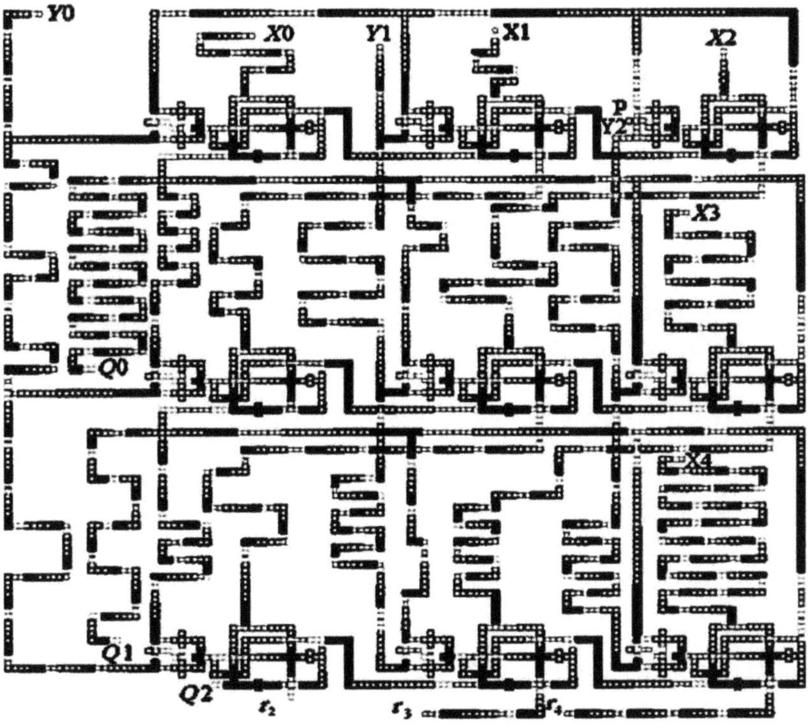

Figure 22.27. The QCA representation of the 3 × 3 non-restoring array divider.

easy to see that the QCA NRD is much more area efficient than the normal divider because the area of the eight-bit QCA NRD is 56.98 μm^2, which also shows the advantage of the QCA device over traditional transistors.

22.3 Summary

QCA multiplier and divider circuits are structured with detailed calculations. For the multiplier, a four-bit and 32-bit carry shift multiplier, and a four-bit and 32-bit carry delay multiplier are described. For the divider, a non-re-establishing binary array divider is discussed using QCA technology. It is built using CAS cell blocks.

Critical thinking questions

1. Describe the applications of the QCA multiplier with appropriate examples.
2. What kind of network is used in multiplication networks? Describe the network using a diagram.
3. How many types of distinct possibilities can be realized by the serial QCA multiplier? Describe each category in detail.
4. Describe the differences between the QCA multiplier and the QCA divider.

References

[1] Cappa M and Hamacher V C 1973 An augmented iterative array for high-speed binary division *IEEE Trans. Comput.* **C-22** 172–5

[2] Cho H *et al* 2007 Serial parallel multiplier design in quantum-dot cellular automata *18th IEEE Symp. on Computer Arithmetic* (Piscataway, NJ: IEEE) pp 7–15

[3] Cho H and Swartzlander E E 2006 Modular design of conditional sum adders using quantum-dot cellular automata *6th IEEE Conf. on Nanotechnology* vol 1 (Piscataway, NJ: IEEE) pp 363–6

[4] Cohen D 1987 A mathematical approach to computational network design *Systolic Signal Processing Systems* (New York: Marcel Dekker) pp 1–29

[5] Cui H, Cai L, Yang X, Feng C and Qin T 2014 Design of non-restoring binary array divider in quantum-dot cellular automata *IET Micro Nano Lett.* **9** 464–7

[6] Shin S-H, Jeon J-C and Yoo K-Y 2013 Wire-crossing technique on quantum-dot cellular automata *NGCIT2013, the 2nd Int. Conf. on Next Generation Computer and Information Technology* vol 27 pp 52–7

[7] Walus K, Dysart T J, Jullien G A and Budiman R A 2004 QCADesigner: a rapid design and simulation tool for quantum-dot cellular automata *IEEE Trans. Nanotechnol.* **3** 26–31

[8] Wang Y and Lieberman M 2004 Thermodynamic behavior of molecular-scale quantum-dot cellular automata (QCA) wires and logic devices *IEEE Trans. Nanotechnol.* **3** 368–76

Chapter 23

QCA asynchronous and synchronous counters

Learning objectives

- Define an asynchronous counter.
- Explain the asynchronous backward counter.
- Construct a dual-edge triggered J–K flip-flop using QCA.
- Design a three-bit asynchronous backward counter and a two-bit counter circuit in QCA.
- Discuss the synchronous counter in detail.
- Construct one-bit and three-bit synchronous counter circuits using QCA.
- Acquire knowledge of QCA synchronous counters.

In this chapter, quantum-dot cellular automaton (QCA) asynchronous and synchronous counters are described. In the asynchronous counter section, a dual-edge triggered J–K flip-flop is presented with its schematic and QCA implementation. In the synchronous counter section, J–K-type and T-type flip-flops are explained with their graphic symbols, schematics, and QCA layouts.

23.1 The asynchronous counter

In this section a falling-edge triggered J–K flip-flop is designed based on a QCA implementation. A dual-edge triggered J–K flip-flop is designed using QCA by arranging special clock zones precisely and serially. The dual-edge triggered flip-flop requires half of the clock compared to the falling-edge triggered J–K flip-flop and the clock power consumption can be reduced significantly. Based on dual-edge triggered J–K flip-flop an asynchronous backward counter design is explored.

23.1.1 The dual-edge triggered J–K flip-flop

A dual-edge triggered J–K flip-flop is designed with a dual-edge triggered structure. The QCA cells have quantum dots 18 nm in width and 5 nm in height. They are placed on a grid with a cell center-to-center distance of 20 nm. Figure 23.1 shows the layout of the dual-edge triggered structure. Its input is a clock pulse (CP) and the

doi:10.1088/978-0-7503-5683-1ch23

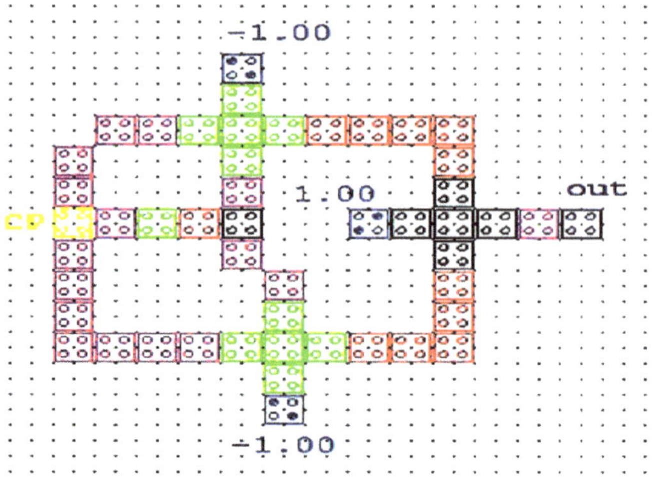

Figure 23.1. Layout of a dual-edge triggered structure.

Figure 23.2. A level-triggered J–K flip-flop description.

Table 23.1. The truth table of figure 23.1.

CP	$\overline{CP}$	CP_{old}	$\overline{CP_{old}}$	AND 1 = $\overline{CP} \cup CP_{old}$	AND 2 = $CP \cup \overline{CP_{old}}$	OR
1	0	1	0	0	0	0
1 $n \to 0$	1	1	0	1	0	1
0	1	0	1	0	0	0
0 $n \to 1$	0	0	1	0	1	1

output will only be high level when the CP is at its rising and falling edge. The output is added as the level input for level-triggered J–K flip-flop, as shown in figure 23.2.

It is necessary to take some precaution with the CP, because it is different to the QCA clock zone signal. It is just an input of a flip-flop while the QCA clock zone signal controls signal flow. Here the different shades of gray represent the different clock zones (clock zones 0, 1, 2, and 3). The truth table of the dual-edge triggered structure is given in table 23.1.

23.1.2 The design of dual-edge triggered J–K flip-flop

A dual-edge triggered J–K flip-flop is constructed using the aforementioned dual-edge triggered structure. The dual-edge triggered J–K flip-flop comprises QCA logic gates, such as AND gates, OR gates, and inverters. A schematic diagram of the dual-edge triggered J–K flip-flop is shown in figure 23.3, in which the dual-edge triggered structure is shown by the dotted box. The state transfer equation of the dual-edge triggered J–K flip-flop is

$$Q^{n+1} = \overline{\text{level input}}\, Q^n + (\text{level input})(J\overline{Q^n} + \overline{K}\, Q^n). \tag{23.1}$$

The layout of the dual-edge triggered J–K flip-flop is shown in figure 23.4. The three yellow boxes denote the J, K, and CP inputs, respectively. From the layout it can be observed that there is a clock cycle delay at the output of $(J\overline{Q}^n + \overline{K}\, Q^n)$ to ensure timing and delay matching. This makes sure that the CP can capture the values of J, K at the front edge when the CP is falling or rising. When the edge of the CP is applied, the level input becomes 1. Then the level-triggered J–K flip-flop carries through a different operation according to J and K. The dual-edge triggered

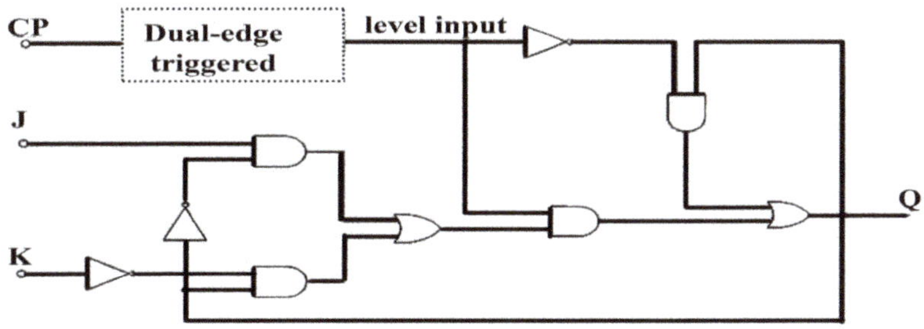

Figure 23.3. Schematic diagram of the dual-edge triggered J–K flip-flop.

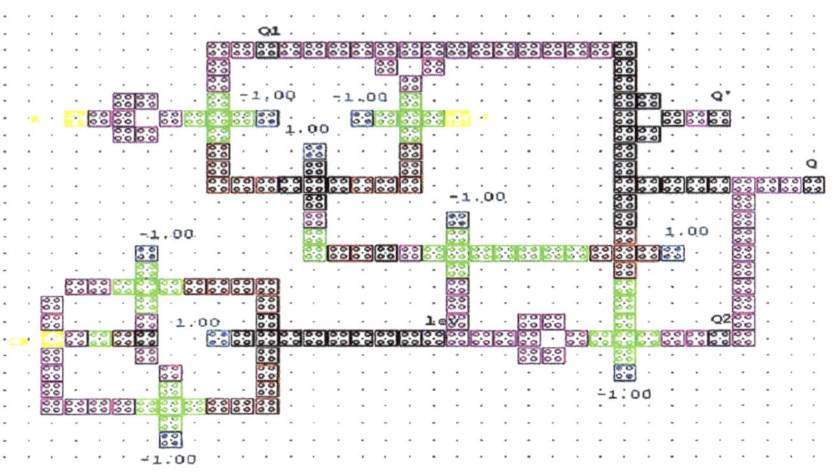

Figure 23.4. Layout of the dual-edge triggered J–K flip-flop.

structure and the level-triggered J–K flip-flop compose a dual-edge triggered J–K flip-flop. The detailed functional description of this dual-edge triggered J–K flip-flop is as follows:

1. If $J = 0$, $K = 0$, and level input = 0, the outputs of the dual-edge triggered J–K flip-flop do not change. If the falling or rising edge of the clock pulse is applied, level input = 1 and, according to equation (23.1), the output at the end of the edge is $Q^{n+1} = Q^n$. This is called a holding state.
2. If $J = 0$, $K = 1$, and level input = 0, the output of the dual-edge triggered J–K flip-flop does not change. If the falling or rising edge of the clock pulse is applied, level input = 1 and, according to equation (23.1), the output at the end of the edge is $Q^{n+1} = 0$. This is called set 0 in J–K flip-flop.
3. If $J = 1$, $K = 0$, and level input = 0, the output of the J–K flip-flop does not change. If the falling or rising edge of the clock pulse is applied, level input = 1 and, according to equation (23.1), the output at the end of the edge is $Q^{n+1} = 1$. This is called set 1 in the J–K flip-flop.
4. If $J = 1$, $K = 1$, and level input = 0, the output of the J–K flip-flop does not change. If the falling edge of clock pulse is applied, level input = 1 and, according to equation (23.1), the output at the end of the edge is $Q^{n+1} = Q^n$. At this time J–K flip-flop will act as a one-bit counter.

23.1.3 The asynchronous backward counter

Any sequential logic circuits can be designed using the J–K flip-flop and additional combinational circuits. Counters are widely used sequential circuits in digital systems. In this section the QCA implementation of a mod-8 asynchronous backward counter is described using the dual-edge triggered J–K flip-flop.

The equation of the n-bit asynchronous backward counter is

$$\begin{aligned} J_i &= K_i = 1,\ i = 1 \\ J_i &= K_i = \bar{Q}_{i-1},\ i = 1, 2, 3, \cdots, n \\ \mathrm{CP}_i &= \left\{ \frac{\mathrm{CP},\ i = 1, 2}{Q_{i-1,\ i=3,\ 4,\ \cdots,\ n}} \right\}. \end{aligned} \tag{23.2}$$

A counter circuit has n flip-flops and it includes 2^n possible states. A mod-8 counter can count from 0 to 7. On the other hand, the backward counter, which is from 7 to 0, that is the state $Q_3Q_2Q_1$ is 111 → 110 → 101 → 100 → 011 → 010 → 001 → 000, as presented in equation (23.2). Large modular size counters can be implemented by simply adding additional bit slices of dual-edge triggered J–K flip-flops. The schematic diagram of a three-bit asynchronous backward counter is shown in figure 23.5(a) and its QCA implementation is shown in figure 23.5(b). In this design, three dual-edge triggered J–K flip-flops are applied. The first two dual-edge triggered J–K flip-flops are clocked by CP, and the last dual-edge triggered J–K flip-flop is clocked by the inversion of Q_2. It can be seen from the figure that the inputs of J and K of the first dual-edge triggered J–K flip-flop are logic 1 and the output Q_1 will change state for every CP's rising and falling edges. Then the CP is delayed the same number of clock cycles as the

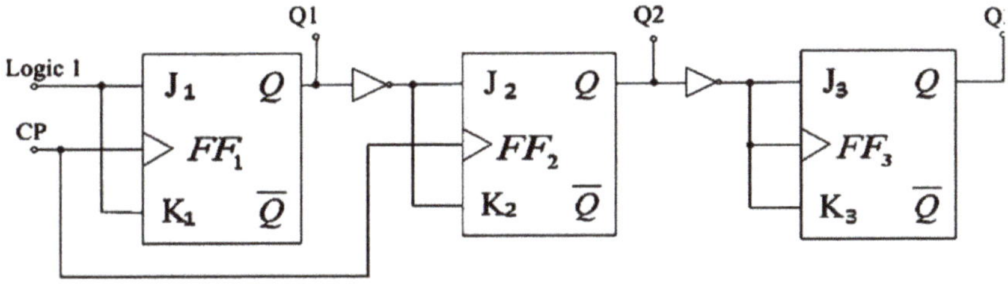

(a)

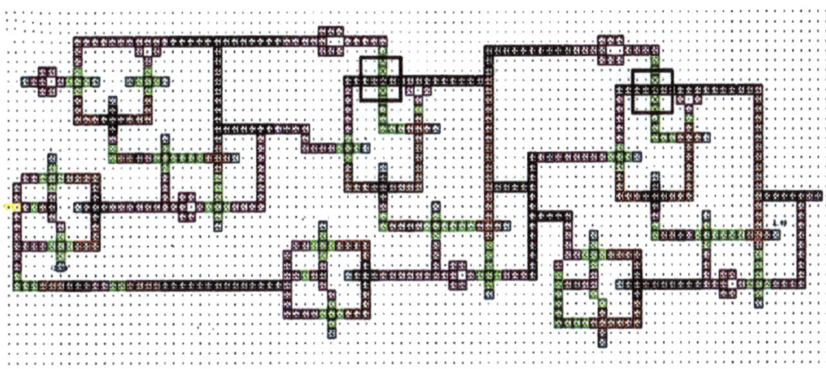

(b)

Figure 23.5. The QCA encoder with multi-feedback. (a) Block diagram. (b) QCA layout.

first dual-edge triggered J–K flip-flop is delayed. Then the delayed CP and Q_1 are imported to the second dual-edge triggered J–K flip-flop simultaneously. The Q_2 will change its state depending on the value of Q_1. When $Q_1 = 0$, Q_2 will change its state, and when $Q_1 = 1$, Q_2 will hold its state. The mechanism of the last dual-edge triggered J–K flip-flop is the same as the previous flip-flops. The only difference is that it is driven by Q_2. Similarly, a QCA encoder with multi-feedback is presented in figure 23.6.

23.2 The synchronous counter

The synchronous counter circuit can be constructed using any types of flip-flops. In QCA technology, synchronous counters have been implemented based on the J–K-type flip-flop (J–K-FF) and T-type flip-flop (T-FF). However, to-date there has been no implementation of synchronous counters using the D-type flip-flop (D-FF) in QCA technology. Here, efficient QCA synchronous counters are designed based on D-FFs. A QCA falling-edge triggered cascaded design is discussed in this section which employs well-optimized level-sensitive D-FFs in parallel with an 'edge-to-level' converter to realize synchronous counters with different bit sizes.

An innovative QCA design of a level-sensitive D-FF is explained here with the goal of building a high performance constructive model for implementing different sequential circuits in QCA technology, in particular QCA counters. The D-FF is a memory element with two inputs (D and CLK) and an output (Q) whose graphic symbol is shown in figure 23.7(a). It implements the following functionality:

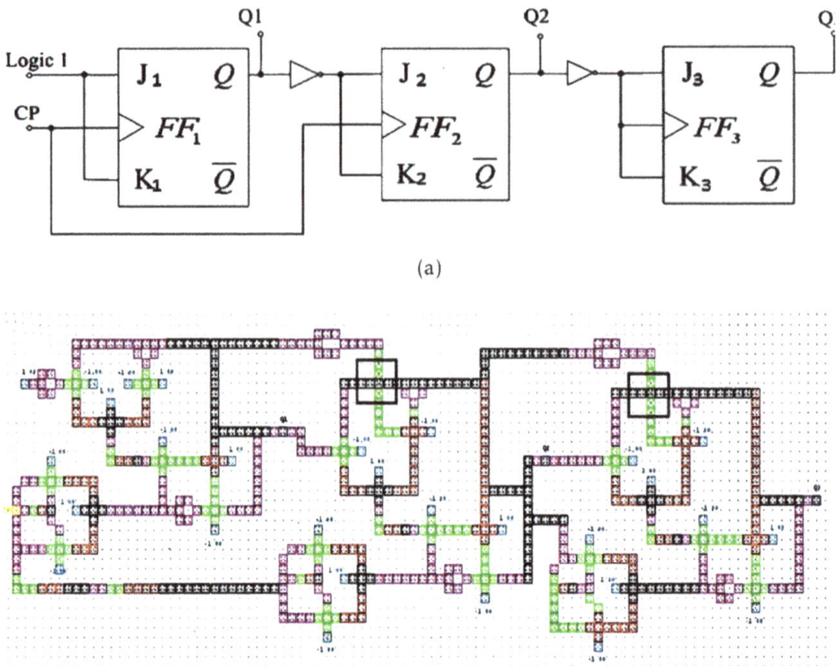

(a)

(b)

Figure 23.6. The QCA three-bit asynchronous backward counter. (a) Block diagram. (b) QCA layout.

transparent (Q follows input D) and hold (Q remains unchanged). The logic function of the level-sensitive D-FF design can be expressed using equation (23.3).

The operation of this D-FF is shown in table 23.2. According to this table when the clock (CLK) signal is activated by 1, the value of data input (D) is stored in the output (Q) and when CLK signal is deactivated by 0, the output is not changed. Due to the fact that majority gates are key components in designing QCA circuits, the corresponding design equation can also be represented based on majority voter gates, as shown in equation (23.4):

$$Q_t = \text{CLK}.\ D + \overline{\text{CLK}}.\ Q_{t-1} \tag{23.3}$$

$$Q_t = \text{Maj}(\text{Maj}(\text{CLK},\ D,\ \text{'0'}),\ \text{Maj}(\overline{\text{CLK}},\ Q_{t-1},\ \text{'0'})). \tag{23.4}$$

The corresponding schematic diagram of the level-sensitive D-FF is illustrated in figure 23.7(b). According to this figure, the design requires three majority gates connected together in two successive gate-levels. Figure 23.7(c) depicts the QCA layout of a D-FF with regular clock zones. It is implemented using the 4 × 4 USE grid with square dimensions of 5 × 5 QCA cells. As shown in figure 23.7(c), when the CLK signal is in 'high', the input bit D is transferred into the output Q and stored in the closed loop. On the other hand, when the CLK signal is in 'low', the stored value is preserved in the loop due to a QCA pipelined process. It consists of 74 cells in an area of 0.1 μm^2 and a latency of 1.5 QCA clocking cycles from input to output.

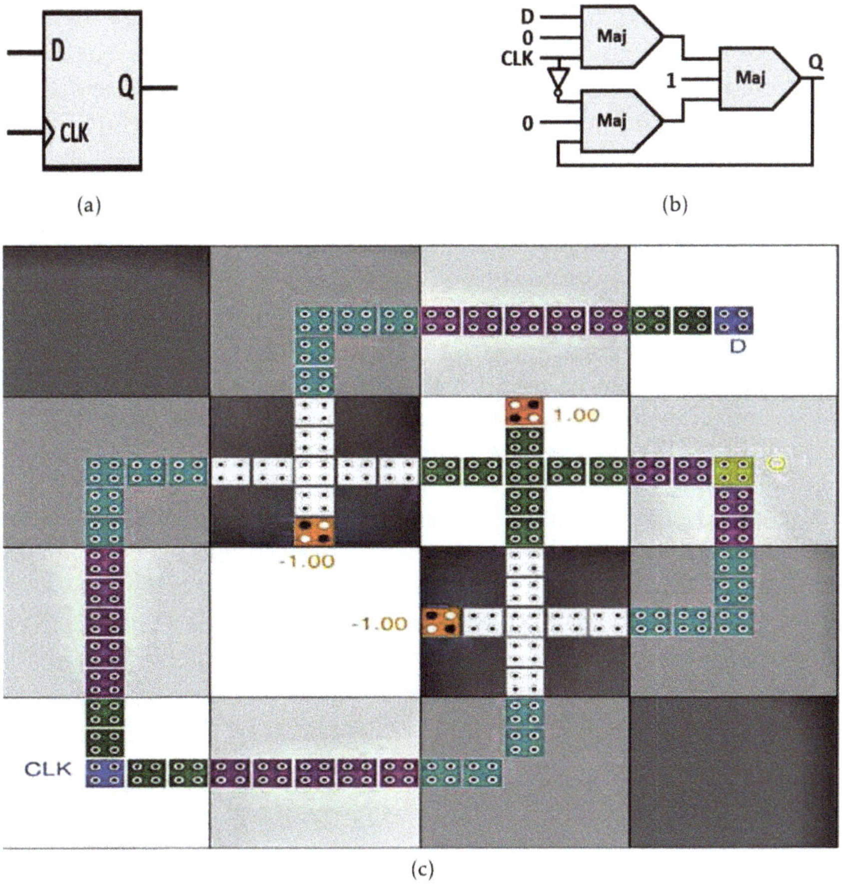

Figure 23.7. The QCA sensitive D-flip-flop. (a) Graphic symbol. (b) Schematic diagram. (c) QCA implementation. Reproduced with permission from [1]. Copyright 2017 Elsevier.

Table 23.2. The operation table of the level-sensitive D-FF.

$CLK_{in}(t-1)$	$CLK_{in}(t)$	CLK_{out}
0	0	0
0	1	0
1	0	1
1	1	0

23.2.1 QCA synchronous counters

In this subsection a well-optimized QCA design of n-bit synchronous counter is described. A counter design comprises n D-FFs and other simple combinational circuits to generate 2^n ascending count states. The level-sensitive D-FF is susceptible to noise if the period of the positive level is long in the clock signal. In this case, the output of D-FF would have a race-round condition. In order to avoid this unstable phenomenon, an 'edge-to-level' converter is essential to enable the edge detecting technique in the D-FF. The edge-to-level converter is used to create a narrow pulse with each falling transition of the input signal. The input and output signals of a falling-edge converter are shown in figure 23.8(a). The edge-to-level converter is constructed in QCA technology and its layout is shown in figure 23.8(b). This converter utilizes the inherent capability of QCA clocking zones. Apparently, an AND logic operation between the inversion value of the current clock signal and its delayed version (by one QCA clock cycle) are used to generate a narrow pulse at each negative transition (high-to-low) of the input signal. This converter operates based on the operation table demonstrated in table 23.3.

The sensitive D-FF can be used for designing a one-bit counter by attaching a falling-edge converter at the clock terminal, as shown in figure 23.8(a). We can see from figure 23.8(a) that a one-bit counter can be constructed using a simple and robust D-FF feeding back the inversed output value directly to the input D at each falling edge of the clock signal. Moreover, an AND gate is placed at the converter output to add the control mechanism to the counter through the count enable (CE) terminal. As a result, the output Q_0 changes to its opposite value on each CLK falling edge when the CE signal is actively high. The corresponding QCA layout is explored in figure 23.9(b). The QCA layout is achieved by employing the sensitive

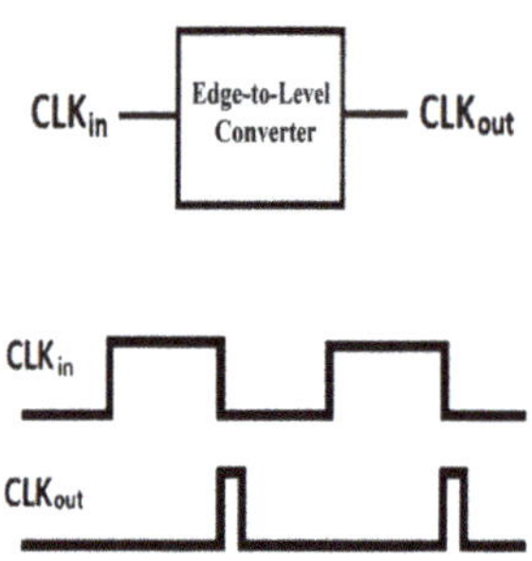

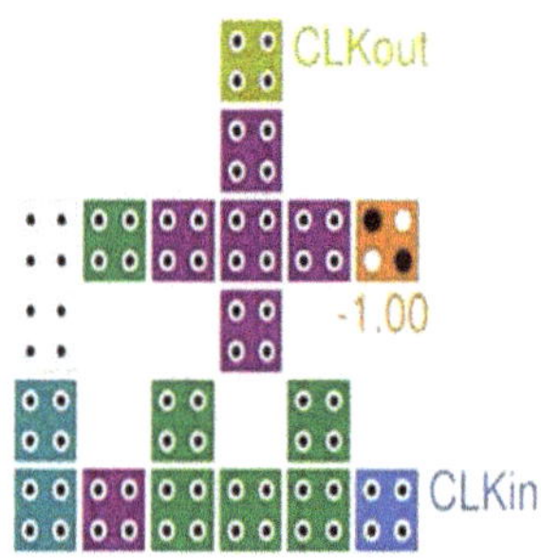

Figure 23.8. The falling-edge converter. (a) Signals. (b) QCA layout. Reproduced with permission from [1]. Copyright 2017 Elsevier.

Table 23.3. The operation table of the falling-edge converter.

CLK	D	Q_t	State
0	0	Q_{t-1}	Hold
0	1	Q_{t-1}	
1	0	0	Transparent
1	1	1	

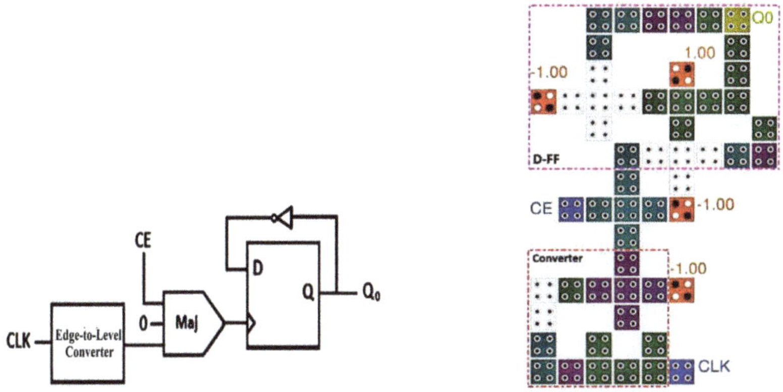

(a) (b)

Figure 23.9. The one-bit counter. (a) Block diagram. (b) QCA layout. Reproduced with permission from [1]. Copyright 2017 Elsevier.

D-FF and falling-edge converter, as shown by the dashed rectangular boxes in the figure.

The one-bit counter in figure 23.9 can be expanded for building *n*-bit synchronous counters. As an example, the three-bit synchronous counter is designed, as shown in figure 23.10, by employing the same building block that is depicted by a solid red rectangular box. As illustrated in figure 23.10(a), this counter is implemented by cascading three level-sensitive D-FFs. It is worth mentioning that to create a QCA falling-edge design, an edge-to-level converter is connected to the clock terminal of each D-FF. Count enable signals are evaluated as $CE_0 = 1$, $CE_1 = Q_0$, and $CE_2 = Q_0Q_1$. As a result, the three outputs of D-FFs ($Q_2Q_1Q_0$) count through 000–001–010–011–100–101–110–111 (from decimal 0 to 7, returning 0 again). The corresponding efficient QCA layout is shown in figure 23.10(a). The QCA layout is achieved by employing three level-sensitive D-FFs that are clocked by a falling-edge converter simultaneously. The counter is a high-speed QCA structure where the input-to-output delay is only two QCA clocking cycles.

Generally, an *n*-bit synchronous counter can be implemented, as shown in figure 23.11, in single-layer by involving a cascade structure of *n* level-sensitive D-FFs that are clocked in parallel using a falling-edge converter and controlled by a logical AND of the previous outputs.

23.3 Summary

In this chapter asynchronous and synchronous counters are explained with the block diagrams and QCA implementations. A three-bit asynchronous backward counter and two-bit counter circuits are implemented in QCA. For the synchronous counter, one-bit and three-bit counter circuits are designed with block diagrams and QCA implementations.

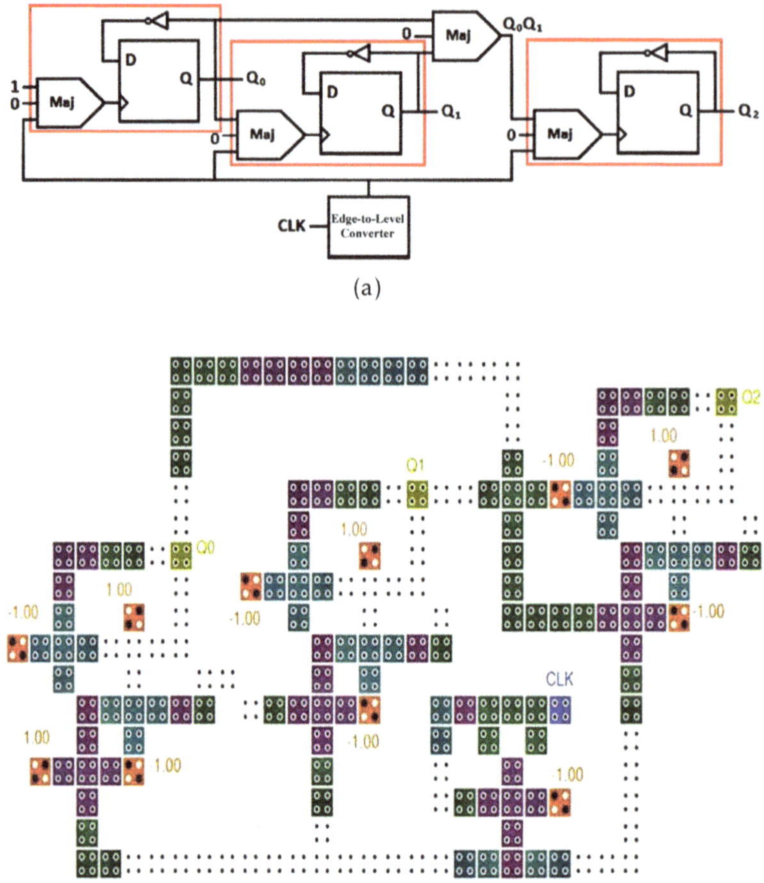

Figure 23.10. The three-bit counter. (a) Block diagram. (b) QCA layout. Reproduced with permission from [1]. Copyright 2017 Elsevier.

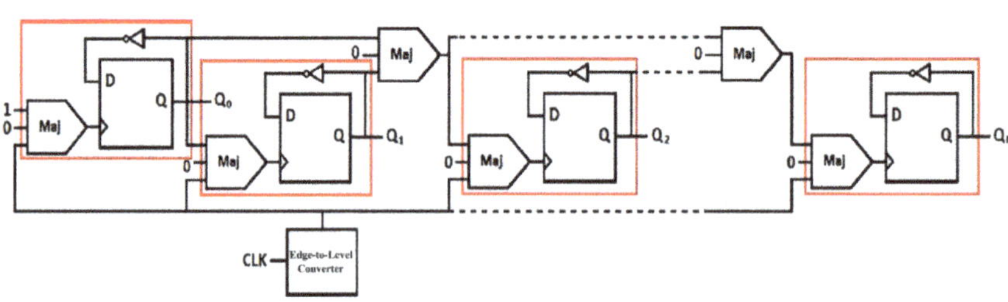

Figure 23.11. The n-bit synchronous counter. Reproduced with permission from [1]. Copyright 2017 Elsevier.

Critical thinking questions

1. What are the uses of QCA asynchronous counters in daily life?
2. What are the main drawbacks of QCA asynchronous counters?

3. Which QCA counter is capable of eliminating the internal propagation delay of the QCA asynchronous counter? Describe in detail.
4. Why are QCA asynchronous counters called QCA ripple counters?
5. Which drawbacks does a QCA synchronous counter overcome?
6. Why do QCA synchronous counters operate more quickly? Explain in detail.
7. What is the propagation delay in QCA synchronous counter circuits?

References

[1] Abutaleb M M 2017 Robust and efficient quantum-dot cellular automata synchronous counters *Microelectron. J.* **61** 6–14

[2] Angizi S, Sayedsalehi S, Roohi A, Bagherzadeh N and Navi K 2015 Design and verification of new *n*-bit quantum-dot synchronous counters using majority function-based JK flip-flops *J. Circuits Syst. Comput.* **24** 1550153

[3] Nelson V and Nagle H 1995 *Digital Logic Circuit Analysis and Design* (Englewood Cliffs, NJ: Prentice-Hall)

[4] Sheikhfaal S, Navi K, Angizi S and Navin A H 2015 Designing high speed sequential circuits by quantum-dot cellular automata: memory cell and counter study *Quantum Matter* **4** 190–7

[5] Wu C-B, Xie G-J, Xiang Y-L and Lv H-J 2014 Design and simulation of dual-edge triggered sequential circuits in quantum-dot cellular automata *J. Comput. Theor. Nanosci.* **11** 1620–6

[6] Yang X, Cai L, Zhao X and Zhang N 2010 Design and simulation of sequential circuits in quantum-dot cellular automata: falling edge-triggered flip-flop and counter study *Microelectron. J.* **41** 56–63

Chapter 24

The QCA decoder and encoder

Learning objectives

- Define the QCA decoder.
- Construct the QCA turbo encoder.
- Explain the QCA 2-to-4 decoder with its circuit.
- Learn the resistor–capacitor (RC) encoder with single-feedback.
- Construct a QCA 3-to-8 decoder.
- Acquire knowledge of an RC encoder with multi-feedback.
- Define the QCA encoder.

In modern computing, the encoder and decoder are used to convert data from one form to another. The encoder and decoder are used frequently in communication systems, namely telecommunication, networking, and end-to-end transfer of data. They are also used for the easy transmission of data, which are replaced with codes and then transmitted. At the end of the receiver, the coded data are collected from the code and then processed for display. The encoder and decoder have a wide variety of applications domains, such as the fast synchronization of multiple motors in industries, military applications, flying robots with night vision flying cameras, robotic vehicles with metal detectors, radiofrequency based home automation systems, and automatic health monitoring systems.

24.1 The QCA decoder

The Ex-OR and decoder are the most valuable functions in the Boolean circuit family and have been of central concern in the QCA field for quite a while. One type of QCA Ex-OR gate is investigated in this section which will be utilized to design a decoder circuit. The truth table of the Ex-OR gate is given in table 24.1. The investigation covers the way that the general circuit can be streamlined if the significant yields as $\bar{A}B$ and $A\bar{B}$ can be adjusted. Along these lines, one single settled cell is utilized with polarization 1 in the larger part functions to play out the AND activities in both of the cases, i.e. $\bar{A}B$ and $A\bar{B}$. Figures 24.1(a) and (b) suggest this concept.

doi:10.1088/978-0-7503-5683-1ch24

Table 24.1. The truth table of the Ex-OR logic.

Inputs		Output
A	B	$Y = A \oplus B$
0	0	0
0	1	1
1	0	1
1	1	0

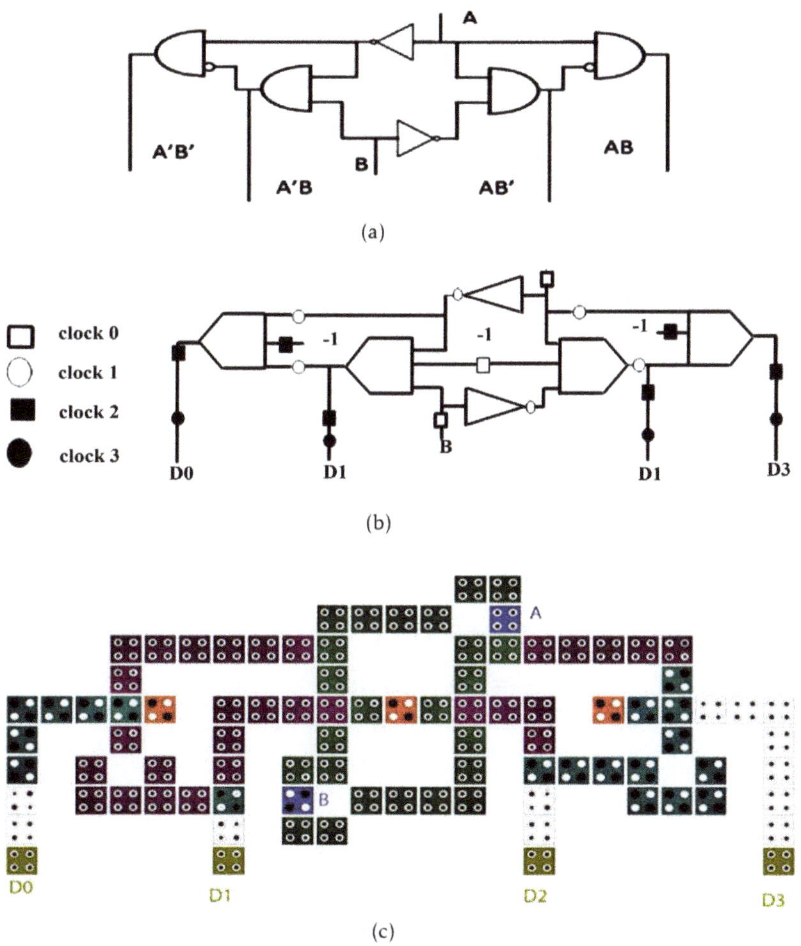

Figure 24.1. The QCA 2-to-4 decoder. (a) Schematic for the 2-to-4 decoder circuit. (b) Three-input majority gate implementation of the 2-to-4 QCA decoder. (c) QCA layout. Reproduced with permission from [2]. Copyright 2016 Elsevier.

Two majority voters (MVs) generating $\bar{A}B$ and $A\bar{B}$ are ORed together to obtain the output of Ex-OR gate. The three-input majority function implementation of an Ex-OR gate is

$$\begin{aligned} Y &= M(F_1(\overline{A}, O, B), 1, F_2(A, 0, \overline{B})) \\ &= M(\overline{A}B, 1, A\overline{B}) \\ &= \overline{A}B + A\overline{B}. \end{aligned} \tag{24.1}$$

24.1.1 The QCA 2-to-4 decoder

The Ex-OR gate is extended to design a 2-to-4 decoder. A 2-to-4 decoder consists of those two previous components as $\bar{A}B$ and $A\bar{B}$ along with two others, $\bar{A}B$ and $A\bar{B}$. The procedure for designing decoders results in a reduced cell count. The following equations implement the QCA decoder circuit.

A. Calculation for $\overline{A}\ \overline{B}$

$$\begin{aligned} P &= \overline{(\overline{A}B)}.\ \overline{A} \\ &= (\overline{\overline{A}} + \overline{B}).\ \overline{A} \\ &= (A + \overline{B}).\ \overline{A} \\ &= A.\ \overline{A} + \overline{A}.\ \overline{B} \\ &= \overline{A}.\ \overline{B}. \end{aligned} \tag{24.2}$$

B. Calculation for AB

$$\begin{aligned} P &= \overline{(\overline{B}A)}.\ A \\ &= (\overline{\overline{B}} + \overline{A}).\ A \\ &= (B + \overline{A}).\ A \\ &= A.\ \overline{A} + A.\ B \\ &= A.\ B. \end{aligned} \tag{24.3}$$

Table 24.2 and figure 24.2(a) show the truth table and the schematic diagram of the 2-to-4 decoder. An entire four clock zones or one clock cycle are required to synchronize the full circuit. In the plan the outputs of any majority gate acts as the input to the next gate at the same clock sector. Thus the interval of the whole circuit has been condensed to a large degree.

The majority functions relevant to this implementation are given in equations (24.4) and (24.5). The majority gate representation and the QCA layout of the circuit design are given in figures 24.1(b) and (c), respectively.

Table 24.2. The truth table for the 2-to-4 decoder.

Inputs		Output			
A	B	$\bar{A}.\,\bar{B}$	$\bar{A}.\,B$	$A.\,\bar{B}$	AB
0	0	1	0	0	0
0	1	0	1	0	0
1	0	0	0	1	0
1	1	0	0	0	1

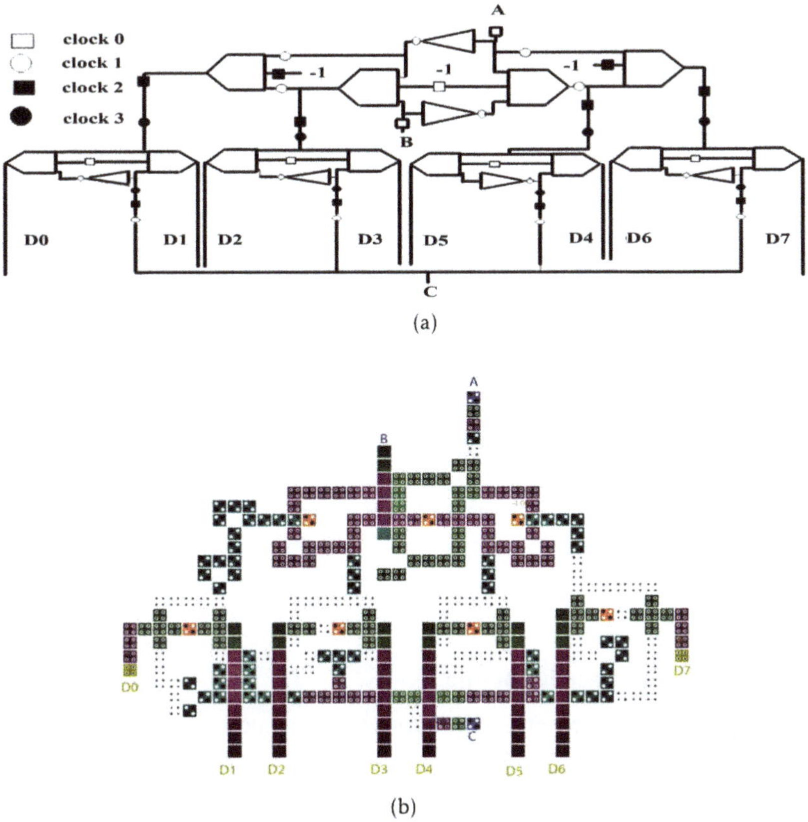

Figure 24.2. The 3-to-8 QCA decoder. (a) The three-input majority gate implementation of the 3-to-8 QCA decoder. (b) QCA layout. Reproduced with permission from [2]. Copyright 2016 Elsevier.

$$
\begin{aligned}
R &= F(\overline{A},\, 0,\, \overline{(\overline{A}B)}) \\
&= \overline{A}.\,(A + B) \\
&= \overline{A}\,\overline{B}
\end{aligned}
\tag{24.4}
$$

$$\begin{aligned} S &= F(A, 0, \overline{(\overline{B}A)}) \\ &= A.\,(A + \overline{B}) \\ &= AB. \end{aligned} \tag{24.5}$$

24.1.2 The QCA 3-to-8 decoder

The 2-to-4 decoder module is reached out to actualize the 3-to-8 decoder. Furthermore, the activity is performed with the acquired parts from 2-to-4 decoder modules which are used to accomplish the yields of the 3-to-8 decoder. The third contribution as C has been brought into the non-rearranged and modified structures through QCA wires to perform AND tasks with each of the recently arrived segments to determine the comparing yields. Here the wire intersections are required to expand the data sources and yields of the circuit outside the circuit. Now, it adds up to eight clock zones (two clock cycles) which is maintained according to the synchronization of the tasks. Comparative tasks are helpful in planning the 4-to-16 partner as well. For this situation and onwards crosses it cannot be used. The related majority gate execution and the QCA design of the 3-to-8 decoder circuit are given in figures 24.2(a) and (b), respectively. Furthermore, the equivalent of the 4-to-16 decoder is given in figure 24.3, and the programmable logic array (PLA)-based decoder circuit is illustrated in figure 24.4.

24.2 The QCA encoder

The encoder generates a high-performance convolutional code, i.e. turbo code, which closely approaches the channel capacity. The turbo code is a kind of error correction code (ECC) with a good performance of checking and correcting errors. It is widely used in mobile communications and satellite communications to protect information against distortion. It can also be applied to fault-tolerant computing. That is, the use of ECC leads to a scheme in which computation takes place in the encoded space, whereby errors are corrected locally, and encoding and decoding are only necessary at the beginning and end, respectively, of the computation.

24.2.1 The QCA turbo encoder design

The turbo encoder implements the convolution computation, feedback, and interleaver, as shown in figure 24.5.

1. *Convolution computation implementation.* In traditional circuits, the calculation components and registers of the resistor–capacitor (RC) encoder are isolated. The synchronization of the circuit is controlled by registers. However, in QCA the calculation components and interconnection generally present considerable delays that likewise assume a vital role in synchronization. The inborn shift registers and ‘processing-in-wire’ nature of QCA ensures that the RC encoder should be included in the calculation design with the register inserted.
2. *Feedback implementation.* The synchronization of feedback presents difficulty in executing an RC encoder. To meet this ongoing requirement, every

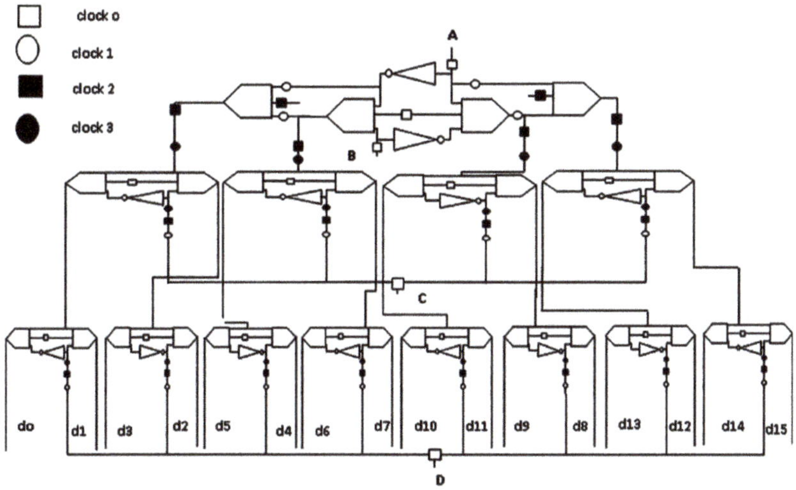

(a)

(b)

Figure 24.3. The QCA 4-to-16 decoder. (a) Three-input majority gate layout of the 4-to-16 QCA decoder. (b) QCA layout. Reproduced with permission from [2]. Copyright 2016 Elsevier.

novel information and comparing input bit should at the same time plot at the calculation unit. Consequently, the input must be implemented simultaneously. If this is the case, the numerous input circuits cannot offer sufficient delays to the current MUX based plan. Taking the case in figure 24.5(b), for example, the planning imperative of the feedback circle is one CLK delay ($1D$ = 1 CLK delay). The current MUX cannot be utilized to make information picking meaningful since it has expended something like one

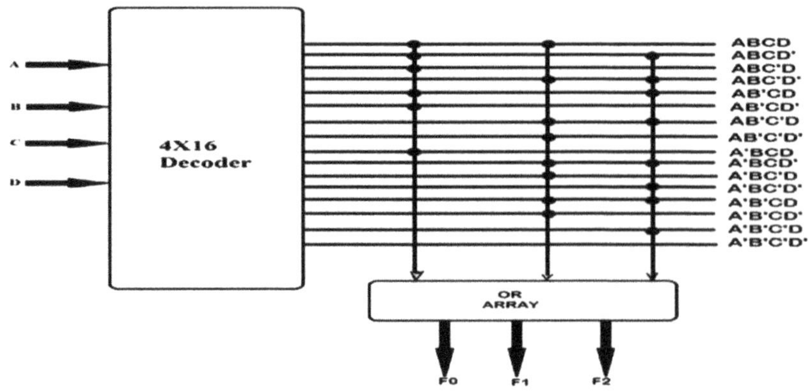

(a)

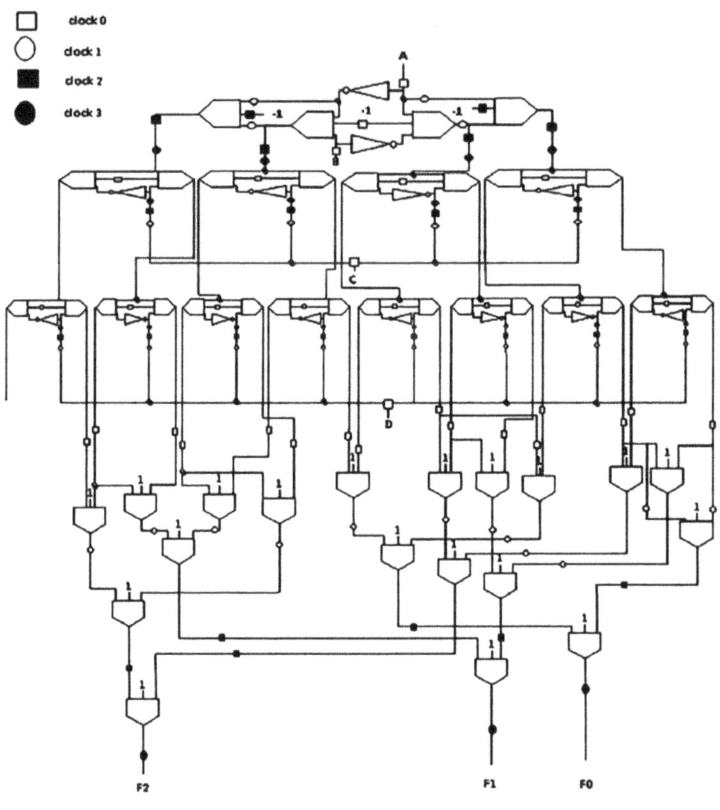

(b)

Figure 24.4. The QCA 4-to-16 PLA-based decoder. (a) The block diagram of a PLA-based adder. (b) Majority gate realization of a decoder-based PLA. (c) The QCA 4-to-16 PLA-based decoder. Reproduced with permission from [2]. Copyright 2016 Elsevier.

(c)

Figure 24.4. (Continued.)

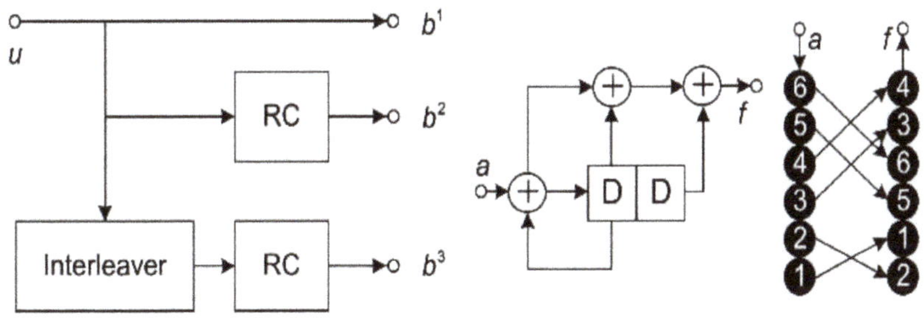

(a) (b) (c)

Figure 24.5. The turbo encoder. (a) Block diagram of a turbo encoder. (b) Block diagram of an RC encoder. (c) Model of an interleaver.

CLK delay. The delay expands the number of circles by utilizing a time-scaling strategy. Thus there will be an extended interval between two nearby active sources of information. Even though this approach does not change the planning of a protest segment, it causes a chain reaction in the general framework. In this way, a latency sparing substitution of MUX is the center of feedback execution.

3. *Interleaver implementation.* At the parallel-to-serial converting stage the right shift-register of the interleaver is about to accept the next group of transposed bits from the left shift-register. However, there are random states remaining in it due to the cyclic four-phase clocking mechanism. Therefore, the competition logjam between the random states and novel bits should be broken. A parallel-to-serial converter is also employed to construct a serial shift/copy/shift-register (SCSR) structure for the programmable array of

logic. It avoids competition by programming a suite of clock signals for the SCSR structure. However, this incurs high complexity of timing, in particular when many SCSR structures are used in a circuit. Hence, a simple parallel-to-serial converter is a major factor in achieving interleaver design.

24.2.2 The RC encoder with single-feedback

In this section, the design of the RC encoder in figure 24.6(a) is discussed. It is an extreme case of single-feedback as there is only one CLK delay for the loop. Therefore, the design can cope with this case, it would be fit for the other single-feedbacks.

Figure 24.6(a) demonstrates the changed block outline of the RC encoder in figure 24.5(b). It is linked by a convolution calculation element (set apart by a

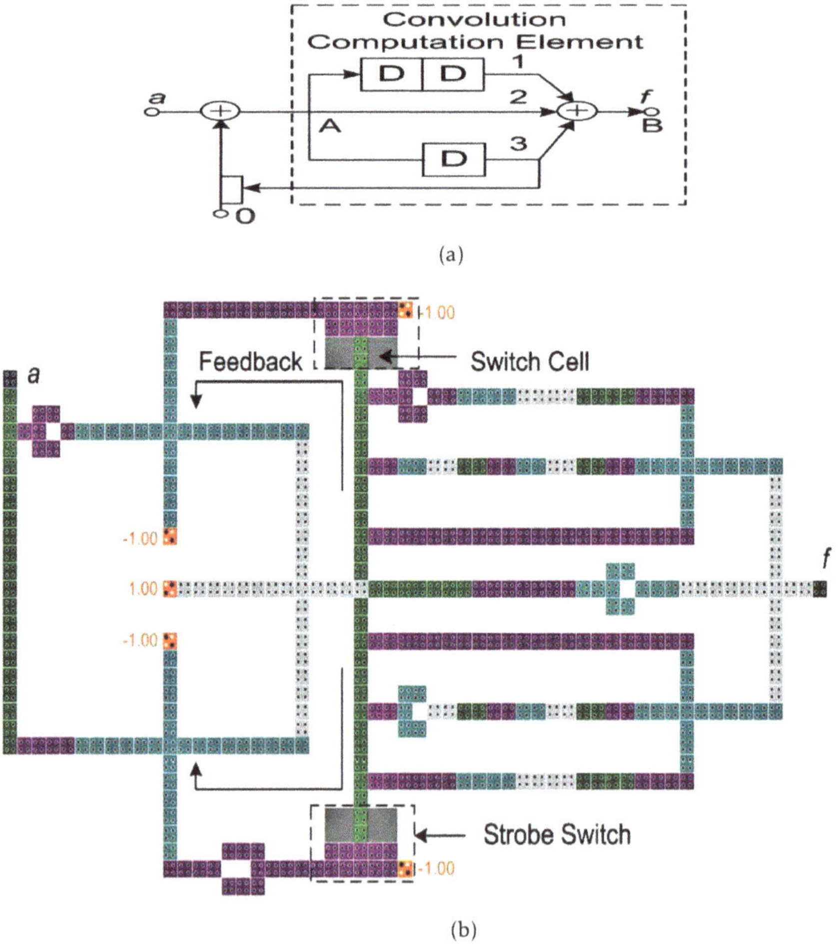

Figure 24.6. The QCA RC encoder with single-feedback. (a) Block diagram. (b) QCA layout. Reproduced with permission from [10]. Copyright 2015 IEEE.

dashed square shape) and a solitary input circle. To perform the convolution calculation, a fan-out is utilized to offer parallel paths, rather than the sequential to parallel structure in an ordinary circuit. Every one of the paths is relegated with different quantities of delays. Thus every bit goes through them at different speeds. In this way the bits obtained at different times can meet at the calculation unit. Code words are also obtained in progression.

Figure 24.6(a) demonstrates the QCA format of this RC encoder. The two arrows demonstrate the feedback paths that compare to the single-input in the square outline. The modulo-2 adder in the input circle works in the parallel frame rather than in the other plan since they expend more than one CLK, engendering delay. The strobe switches and switch cells are set apart by dashed square shapes and dark shadows, respectively.

In the feedback loop, the strobe switch is employed for data choosing. One end of the branch line is connected to a fixed 0 cell, while the main line is joined to the feedback path. The strobe switch alternates between 0 and the feedback signal. If there is no valid feedback signal, 0 is transmitted to the modulo-2 adder so as to protect the feed forward path from the interference caused by untimely feedback signal.

It is important to note that the delay in each path of the convolution computation element in figure 24.6(b) denotes the relative propagation delay from point A to point B. Given path 2 as the reference path, path 1 and path 3 consume two and one CLK delay more than path 2, respectively. When there are not sufficient delays to implement the three paths, the same number of delays can be supplemented to each path. This does not affect the final result except increasing the corresponding delays.

24.2.3 The RC encoder with multi-feedback

The contrasted and single-feedback use of multi-inputs in QCA raises more complex time issues, caused by the interconnection. In addition, there are many cases that cannot be executed directly for a multi-input plot. For example, figure 24.7(a) demonstrates a square outline of a run-of-the-mill RC encoder with multi-feedback. Each register is associated with a feedback path. It is difficult to meet the constant prerequisite if the multi-input is built directly using QCA devices. To structure the RC encoder for multi-feedback, two questions arise: (i) can any multi-input be changed into an equivalent single-feedback or linked single-feedback and (ii) how is this done? Considering the encoder in figure 24.7(a), for example, it is possible to obtain the exchange function $H(D) = \frac{1+D^2}{1+D+D^2+D^3+D^4}$. The numerator and denominator of the transfer function indicate the feed forward and feedback, respectively. Obviously, the transfer function of the equivalent single-feedback must be of the type of $H(D) = \frac{f(D)}{1+D^n}$, and $n > 4$. Here, the single-feedback transformation, it is computed as

$$\begin{aligned} H(D) &= \frac{1 + D^2}{1 + D + D^2 + D^3 + D^4} \cdot \frac{(1 + D)}{(1 + D)} \\ &= \frac{1 + D + D^2 + D^3}{1 + D^5}. \end{aligned} \tag{24.6}$$

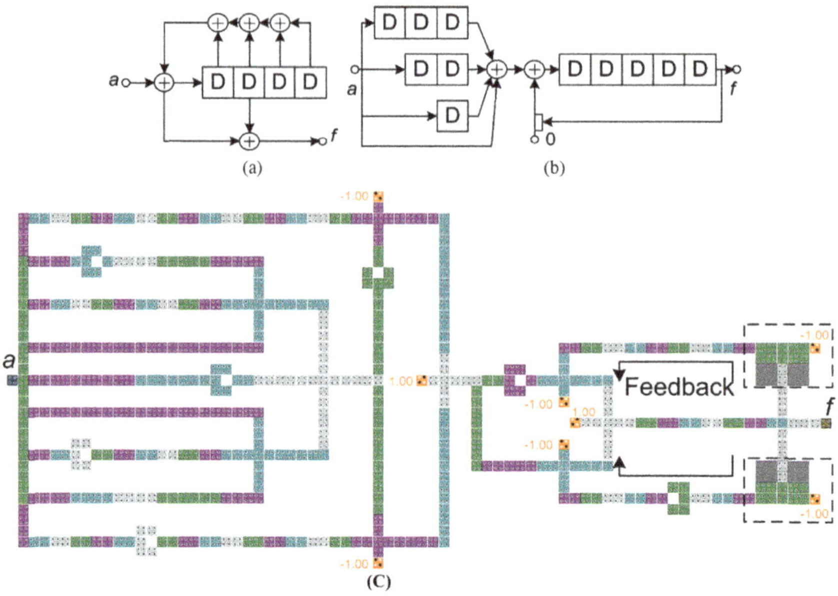

Figure 24.7. The QCA RC encoder with multi-feedback. (a) Classic block diagram. (b) Transformed block diagram. (c) QCA layout. Reproduced with permission from [10]. Copyright 2015 IEEE.

Thus the transformed block diagram and the corresponding layout are obtained, as shown in figures 24.7(b) and (c). In this way, a multi-feedback is transformed into the concatenation of a convolution computation element and a single-feedback loop. Similarly the transfer functions of RC encoder with multi-feedback are denoted as $H(D) = \frac{1}{1 + a_1D^1 + a_2D^2 + \cdots + a_nD^n}$, where $a_1, a_2, \cdots, a_n \in 0, 1$ and are not 0 at the same time.

Thus the first problem is equivalent to whether any transfer function $H(D) = \frac{1}{1 + a_1D^1 + a_2D^2 + \cdots + a_nD^n}$ in Galois field GF(16.6) can be transformed to the $H(D) = \frac{f(D)}{(1 + D^p)}$ or $H(D) = \frac{f(D)}{(1 + D^h)\cdots(1 + D^l)}$ type.

Property 24.1. For any Galois field GF(q), q = prime power, x^{q^n} = product of all monic and irreducible polynomials over GF(q), whose degree divides n.

Since the above polynomials are Boolean arithmetic, they are in field GF(2). In GF(2), x is a monic and irreducible polynomial of degree 1. Therefore, when $q = 2$, a similar property shows.

Property 24.2. For GF(2), $x^{2^{n-1}} - 1$ = the product of all monic and irreducible polynomials (except the polynomial x) over GF(2), whose degree divides n. In GF (2), modulo-2 '−' is equal to modulo-2 '+'. Thus $x^{2^{n-1}1} - 1$ is really equal to $x^{2^{n-1}} + 1$.

Except for the polynomial x, the other monic and irreducible polynomials are all in the form of $1 + a_1x^1 + a_2x^2 + \cdots + a_nx^n$. Moreover, any product of these polynomials is still in this form in GF(2). Therefore, this theorem can just settle the problem of multi-feedback discussed above.

According to the properties, a multi-feedback can be changed into a single-feedback when the denominator polynomial of its transfer function is irreducible. Otherwise, it can be transformed into concatenated single-feedbacks when the denominator polynomial is reducible. In the second case, the reducible polynomial should first be factorized into the product of irreducible polynomials. Then the multi-feedback can be further transformed into single-feedbacks according to the first case. The issues of the irreducibility test and the factorization of the polynomial in GF(2) are the common mathematics issues and can be settled by some calculators such as Magma. The relevant procedures are not discussed in detail here. To sum up, any multi-feedback of an RC encoder can be changed into the single-feedback type.

24.3 Summary

In this chapter, the decoder and encoder circuits are described using quantum dot cellular automata. 3-to-8 and 4-to-16 QCA decoder circuits are considered and presented explicitly with the functional descriptions. For the encoder circuit, a QCA turbo encoder and an RC encoder with single- and multi-feedback are described with block figures and QCA layouts.

Critical thinking questions

1. Describe the applications of QCA decoders and QCA encoders.
2. Is it possible to draw a QCA 5-to-32 decoder using a QCA 2-to-4 decoder and a QCA 3-to-8 decoder? If it is, then draw the circuit and explain the procedure.
3. What differentiates the QCA decoder and the QCA encoder from one another?
4. What is the process for designing a QCA turbo encoder?
5. Describe the differences between an RC encoder with single-feedback and an RC encoder with multi-feedback.

References

[1] WatElectronics 2019 Different types of encoder and decoder and its uses *WatElectronics.com* https://www.efxkits.us/different-types-encoder-decoder-applications/ (Accessed: 25 December 2018)

[2] De D, Purkayastha T and Chattopadhyay T 2016 Design of QCA based programmable logic array using decoder *Microelectron. J.* **55** 92–107

[3] Gladshtein M 2011 Quantum-dot cellular automata serial decimal adder *IEEE Trans. Nanotechnol.* **10** 1377–82

[4] Jagarlamudi H S, Saha M and Jagarlamudi P K 2011 Quantum dot cellular automata based effective design of combinational and sequential logical structures *World Acad. Sci. Eng. Technol. Int. J. Comput. Electr. Automat. Control Inf. Eng.* **5** 1529–33

[5] Kianpour M and Sabbaghi-Nadooshan R 2012 A conventional design for CLB implementation of a FPGA in quantum-dot cellular automata (QCA) *Proc. of the 2012 IEEE/ACM Int. Symp. on Nanoscale Architectures* pp 36–42

[6] Kianpour M and Sabbaghi-Nadooshan R 2014 A novel quantum-dot cellular automata CLB of FPGA *J. Comput. Electron* **13** 709–25

[7] Niemier M T and Kogge P M 1999 Logic in wire: using quantum dots to implement a microprocessor *Proc. 6th IEEE Int. Conf. on Electronics, Circuits and Systems* vol 3 pp 1211–5

[8] Thompson R M 2011 A look at the special number field sieve using ideals as implementable in the magma computational algebra software system *Doctoral Dissertation* San Diego State University

[9] Vetteth A, Walus K, Dimitrov V S and Jullien G A 2002 Quantum dot cellular automata carry-look-ahead adder and barrel shifter *IEEE Emerging Telecommunications Technologies Conf.* pp 2–4

[10] Zhang M, Cai L, Yang X, Cui H and Feng C 2015 Design and simulation of turbo encoder in quantum-dot cellular automata *IEEE Trans. Nanotechnol.* **14** 820–8

IOP Publishing

Quantum Computing (Second Edition)
A pathway to quantum logic design
Hafiz Md Hasan Babu

Chapter 25

The QCA multiplexer and demultiplexer

Learning objectives

- Describe the 2-to-1 QCA multiplexer with its circuit.
- Learn about the 1-to-2 QCA demultiplexer.
- Demonstrate the 4-to-1 QCA multiplexer's circuit clearly.
- Construct a 1-to-4 QCA demultiplexer.
- Define the QCA demultiplexer.
- Acquire knowledge about the techniques of QCA multiplexer and demultiplexer implementation.

The multiplexer (MUX) is the most common combinational circuit utilized in digital logic systems and it is an exceptionally valuable electronic circuit that is utilized in numerous applications, for example, signal directing, information interchange, and data transport control applications. An efficient quantum-dot cellular automaton (QCA) MUX/demultiplexer (DEMUX) design is important for current research networks. Multiplexers are utilized in different fields where various types of information need to be transmitted utilizing a solitary line. The multiplexer allows the transmission of different kinds of information utilizing a solitary transmission line. It can be utilized to activate an immense amount of memory in the PC simultaneously.

25.1 The QCA 2-to-1 multiplexer

In this section, a new approach is described to implement the 2-to-1 multiplexer. Any Boolean function can be implemented using a MUX. It has 2^n inputs, one output, and $n = 1$ select lines, which transfers one of input to the output based on the value of the select lines. In addition, the MUX has been constructed using the rotating majority gate (RMG) which presents less complexity and enables powerful techniques in terms of implementing logic.

A schematic representation of the 2-to-1 MUX is shown in figure 25.1(a). It has two data inputs I_0 and I_1, one select line S_0, and one output (MUX). The block

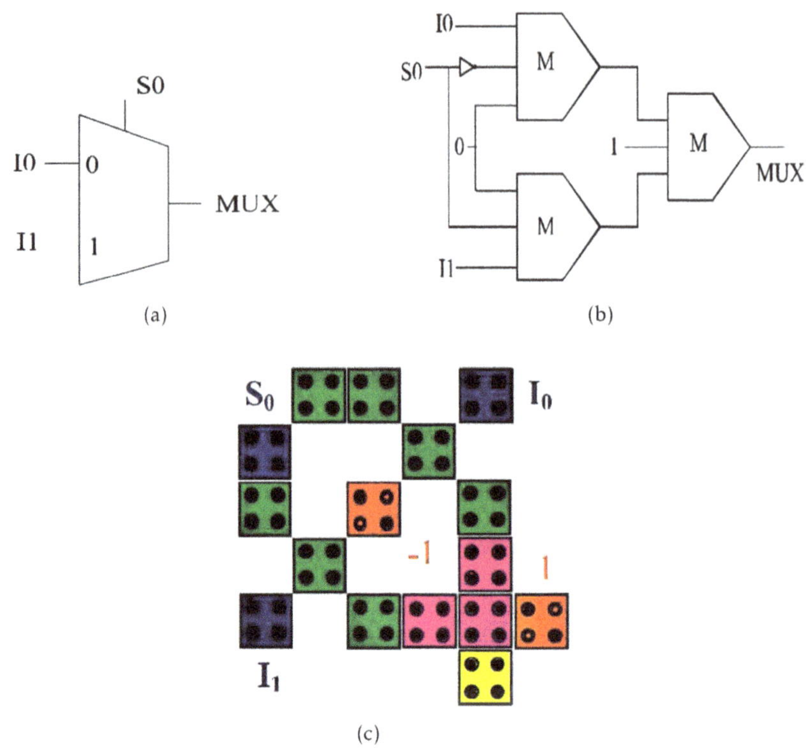

Figure 25.1. The 2-to-1 multiplexer. (a) Schematic symbol. (b) Majority logic diagram. (c) The QCA layout 2-to-1 MUX.

diagram is shown in figure 25.1(b). The output logic expression for the 2-to-1 MUX is

$$\text{Output} = I_0\overline{S_0} + I_1S_0. \tag{25.1}$$

From equation (25.1) it is clear that two AND gates, one OR gate, and an inverter are required to design the 2-to-1 MUX. The QCA equivalent block diagram is shown in figure 25.1(b). The layout of the 2-to-1 MUX design is presented in figure 25.1(c). The output (MUX) generates precise, highly polarized signals (shown inside the rectangles) leading to a high drivability for the circuit.

The implemented MUX has an area of 0.01 μm^2, a circuit complexity of 16 cells, and a latency of 0.5 clock delays. The QCA majority output logic expression of the 2-to-1 MUX is

$$\text{Output} = M_j\big(M_j\big(I_0, \overline{S}_0, 0\big), M_j\big(I_1, S_0, 0\big), 1\big). \tag{25.2}$$

25.2 The QCA 4-to-1 multiplexer

The diagram of the 4-to-1 MUX is shown in figure 25.2(a). It is created using the concatenation of three 2-to-1 MUXs. The first two MUXs have a shared control

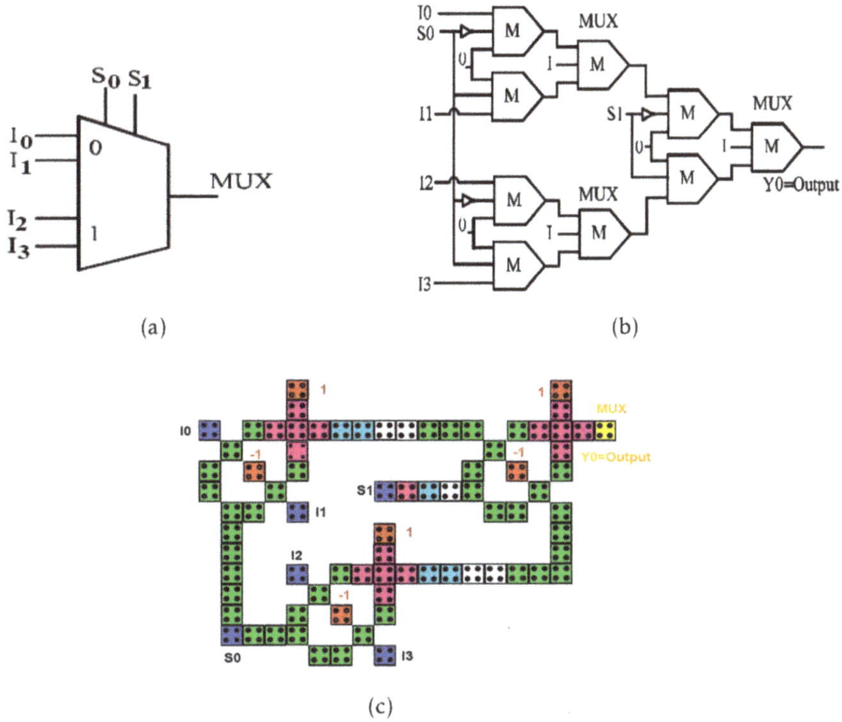

Figure 25.2. The 4-to-1 multiplexer. (a) Schematic symbol. (b) Majority logic diagram. (c) The QCA layout for the 4-to-1 MUX. (a) Schematic symbol. (b) Majority logic diagram. (c) The QCA layout 2-to-1 MUX.

signal link represented by S_0. The yields of the first two MUXs are fed to the third MUX, acting as an input where the regulator is provided by S_1. It has four data input lines I_0, I_1, I_2, and I_3 and a single output (Y_0). The block illustration of a 4-to-1 MUX is shown in figure 25.2(b). The output logic appearance for the 4-to-1 MUX is functions as

$$\begin{aligned}\text{Output} = M_j\big(M_j\big(M_j\big(M_j(I_0, \overline{S_0}, 0), M_j(I_1, S_0, 0), 1\big), \overline{S_1}\big),\\ M_j\big(M_j\big(M_j(I_2, \overline{S_0}, 0), M_j(I-3, S_0, 0), 1\big), S_1\big)\big).\end{aligned} \tag{25.3}$$

From equation (25.3) it is clear that six AND gates, four OR gates, and three inverters are required to design a 4-to-1 MUX. The QCA layout of the design is shown in figure 25.2(c). It has an area of 0.11 μm^2, circuit complexity of 79 cells, and a latency of 1.5 clock delays.

25.3 The QCA 1-to-2 demultiplexer

The DEMUX is a reverse process of the multiplexer. It produces parallel lines from a serial data input line. It has 2^n outputs and n select lines which transfer a single data input line corresponding to the particular output data line based on the value of select lines.

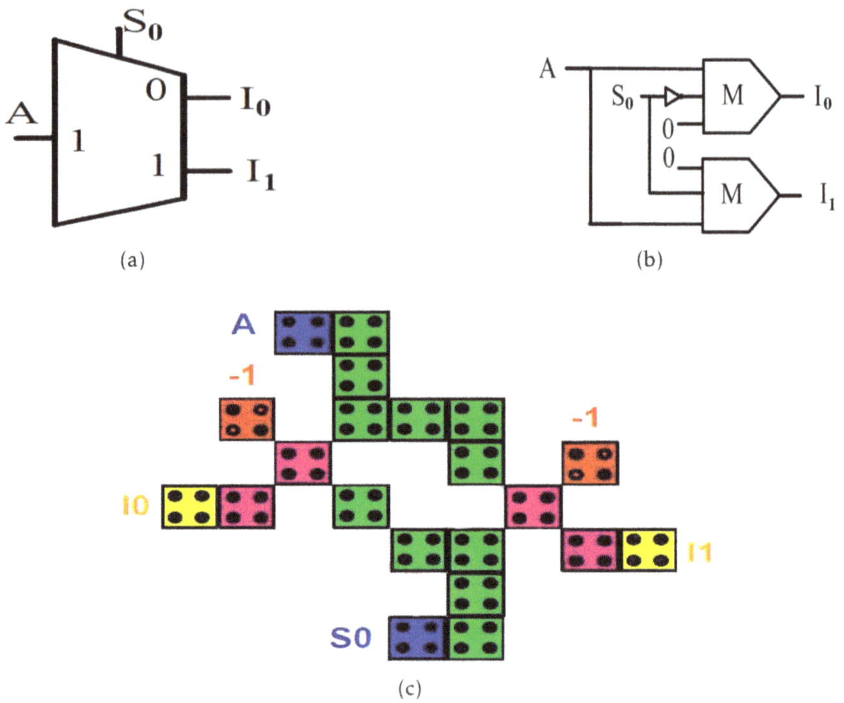

Figure 25.3. The 1-to-2 demultiplexer. (a) Schematic symbol. (b) Majority logic diagram. (c) The QCA layout. Reproduced with permission from [1]. Copyright 2018 Elsevier.

A schematic representation and the truth table of the 2-to-1 DEMUX is shown in figure 25.3(a). It has a single data input line A, two outputs (I_0, I_1), and a select line S_0. The QCA equivalent block diagram is shown in figure 25.3(b). The output logic expression for the 2-to-1 MUX is

$$I_0 = A\overline{S_0} \tag{25.4}$$

$$I_1 = AS_0. \tag{25.5}$$

From equations (25.4) and (25.5) it is understandable that two AND gates and an inverter are required to construct the 2-to-1 DEMUX. The QCA layout of the 2-to-1 DEMUX design is shown in figure 25.3(c). The DEMUX has an area of 0.03 μm^2, a circuit complexity of 21 cells, and a latency of 0.5 clock delays. It is clear that when input $A = 1$ and control $S_0 = 0$, then $I_0 = 1$. Again when input $A = 1$ and control $S_0 = 1$, then $I_1 = 1$. The QCA majority output logic expression of the 2-to-1 DEMUX are

$$I_0 = M_j\left(A,\ \overline{S_0},\ 0\right) \tag{25.6}$$

$$I_1 = M_j\left(A,\ S_0,\ 0\right). \tag{25.7}$$

25.4 The QCA 1-to-4 demultiplexer

A graphic illustration of the 1-to-4 DEMUX is shown in figure 25.4. It has a single data input line A, four outputs (I_0, I_1, I_2, I_3), and a select line S_0, S_1. The equivalent block diagram is shown in figure 25.4(b). The output logic expression for the 4-to-1 MUX is

$$I_0 = A\overline{S_0}.\overline{S_1} \tag{25.8}$$

$$I_1 = AS_0.\overline{S_1} \tag{25.9}$$

$$I_2 = A\overline{S_0}.\ S_1 \tag{25.10}$$

$$I_3 = AS_0.\ S_1. \tag{25.11}$$

From equations (25.8)–(25.11) it is clear that four three-input AND gates and two inverters are required to design the 1-to-4 DEMUX. The QCA equivalent block

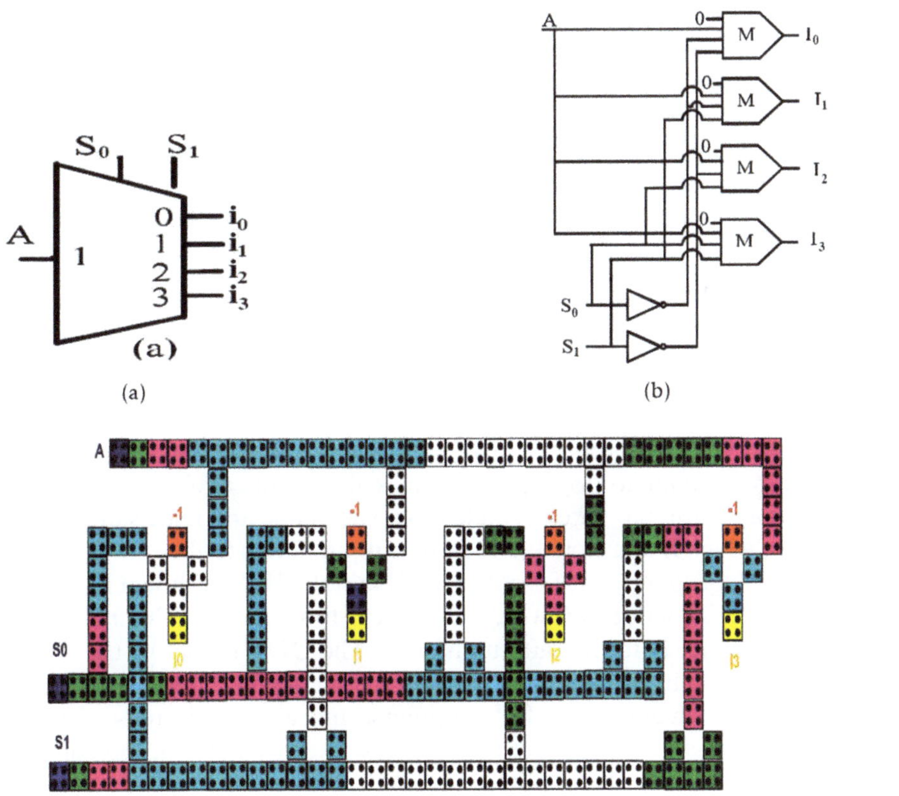

Figure 25.4. The 1-to-4 demultiplexer. (a) Schematic symbol. (b) Majority logic diagram. (c) The QCA layout. Reproduced with permission from [1]. Copyright 2018 Elsevier.

diagram is shown in figure 25.4(c). The DEMUX has an area of 0.18 μm^2, circuit complexity of 187 cells, and latency of 1.75 clock delays.

The QCA majority output logic expressions of the 1-to-4 DEMUX are

$$I_0 = M_j\left(A\overline{S_0},\ \overline{S_1},\ 0\right) \tag{25.12}$$

$$I_1 = M_j\left(AS_0,\ \overline{S_1},\ 0\right) \tag{25.13}$$

$$I_2 = M_j\left(A\overline{S_0},\ S_1,\ 0\right) \tag{25.14}$$

$$I_3 = M_j\left(AS_0,\ S_1,\ 0\right). \tag{25.15}$$

25.5 Multiplexing/demultiplexing using QCA

Multiplexing is the process in which data streams coming from different sources are combined and transmitted over a single data channel. In practice, a MUX is used to realize the concept of multiplexing. It allows one to select one of many possible sources. There are several data inputs and one of them is routed to the output (possibly the shared communication channel). It can be done using a selector line. The select inputs determine which data input gets through. The DEMUX performs the reverse process of multiplexing and routes the separated signals to their corresponding receiver. It allows one to select one of many possible destinations. A DEMUX has one data input and several data outputs. The data output can be selected using a selector line. Both the MUX and DEMUX are combined into a single device which has the capability to process outgoing and incoming signal lines, as shown in figure 25.5(a).

The idea of the proposed method is to design a novel ultra-high speed data multiplexing/demultiplexing circuit for computer network systems. The schematic logic symbol of the 2-to-1 MUX/DEMUX is shown in figure 25.5(a). It consists of a MUX, which has two inputs (I_0, I_1) and a single selector line S_0. The number of selector lines is equal to $[\log_2 n]$, where n is the number of inputs. Conversely, a DEMUX is a device that takes a single input signal line and selects one of many data output lines, which is connected to the single input line (possibly the shared channel). The majority logic diagram of the 2-to-1 MUX/DEMUX is shown in figure 25.5(b). In this case, a logic value of 0 would connect I_0 to the output channel; a logic value of 1 would connect I_1 to the output channel. The QCA layout of the 2-to-1 MUX/DEMUX is shown in figure 25.5(c). MUXs is used in computer memory to maintain a vast amount of memory in the computers, and also to decrease the number of copper lines necessary to connect the memory to other parts of the computer.

25.5.1 The effect of the selector line (S_0, S_1) on the 2-to-1 MUX/1-to-2 DEMUX

Here, S_0 and S_1 are the two selector lines at the both transmitter and receiver ends of the MUX and DEMUX.

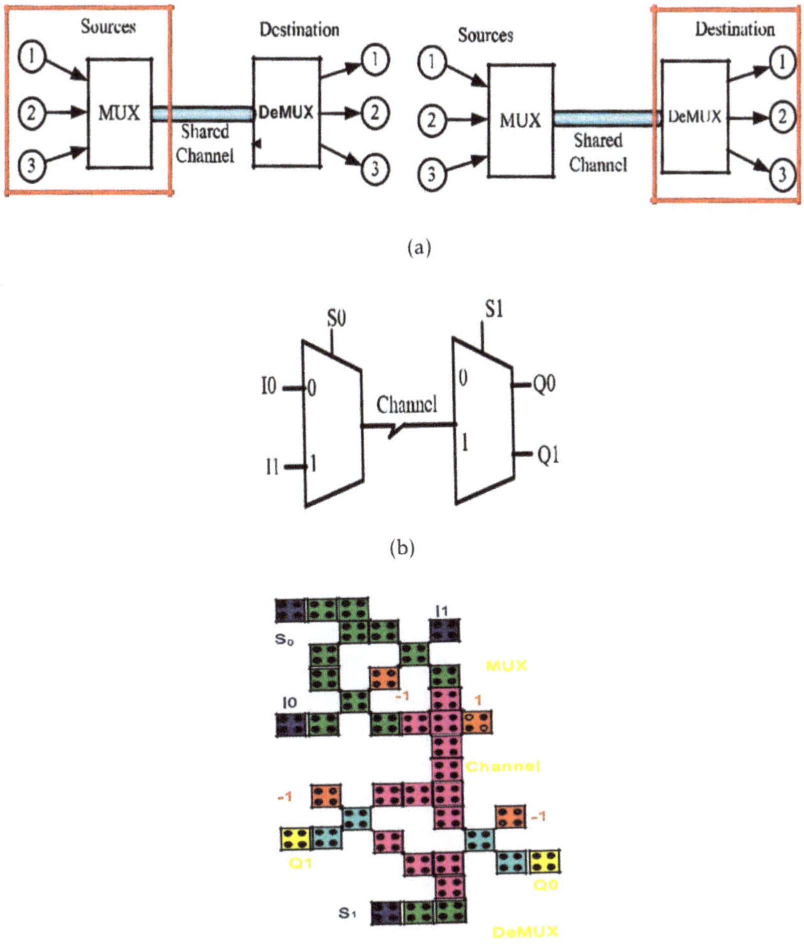

Figure 25.5. The 1-to-4 MUX/DEMUX using QCA. (a) Schematic symbol. (b) Majority logic diagram. (c) The QCA layout. Reproduced with permission from [1]. Copyright 2018 Elsevier.

- If S_0 and S_1 are low, the top AND gate of the MUX and DEMUX is open and I_0 is copied to Q_0.
- If S_0 and S_1 is high, the bottom AND gate of MUX and DEMUX is open and I_1 is copied to Q_1.
- Both the MUX and DEMUX are synchronized using the proper clocking signals.

25.6 Summary

This chapter presents an approach to deal with 2-to-1 MUX and 1-to-2 DEMUX circuits using a two-electron four-dot QCA. The 2-to-1 MUX has been used to execute the essential digital elements required for QCA. Any higher level MUX/

DEMUX can be designed utilizing the 2-to-1 MUX and 1-to-2 DEMUX, respectively. In this way, utilizing the 2-to-1 MUX and 1-to-2 DEMUX, a 4-to-1 MUX and a 1-to-4 DEMUX are defined.

Critical thinking questions

1. What are the QCA multiplexer and QCA demultiplexer? Explain them with diagrams?
2. What are the differences between a QCA multiplexer and a QCA demultiplexer?
3. How does the transport layer perform for QCA multiplexing and QCA demultiplexing?
4. Describe the characteristics of the QCA multiplexer and the QCA demultiplexer.
5. Using a 2-to-1 QCA multiplexer and a 4-to-1 QCA multiplexer, design a 8-to-1 QCA multiplexer.

References

[1] Ahmad F 2018 An optimal design of QCA based 2^n:1/1:2^n multiplexer/demultiplexer and its efficient digital logic realization *Microprocess. Microsyst.–Embedded Hardware Des.* **56** 64–75

[2] Amiri M A, Mahdavi M and Mirzakuchaki S 2008 QCA implementation of a MUX-based FPGA CLB *Int. Conf. on Nanoscience and Nanotechnology* (Piscataway, NJ: IEEE) pp 141–4

[3] Asfestani M N and Heikalabad S R 2017 A novel multiplexer-based structure for random access memory cell in quantum-dot cellular automata *Physica* B **521** 162–7

[4] Askari M and Taghizadeh M 2011 Logic circuit design in nano-scale using quantum-dot cellular automata *Eur. J. Sci. Res.* **48** 516–26

[5] Chandra D J and Debashis D 2016 Shannon's expansion theorem-based multiplexer synthesis using QCA *Nanomater. Energy* **5** 53–60

[6] Kim K, Wu K and Karri R 2007 The robust QCA adder designs using composable QCA building blocks *IEEE Trans. Comput.-Aided Des. Integr. Circuits Syst* **26** 176–83

[7] Mardiris V A and Karafyllidis I G 2010 Design and simulation of modular 2^n to 1 quantum-dot cellular automata (QCA) multiplexers *Int. J. Circuit Theory Appl.* **38** 771–85

[8] Momenzadeh M, Tahoori M B, Huang J and Lombard F 2004 Quantum cellular automata: new defects and faults for new devices *Proc. 18th Int. Parallel and Distributed Processing Symp.* (Piscataway, NJ: IEEE) pp 207–14

[9] Mukhopadhyay D and Dutta P 2012 Quantum cellular automata based novel unit 2:1 multiplexer *Int. J. Comput. Appl.* **43** 22–5

[10] Sabbaghi-Nadooshan R and Kianpour M 2014 A novel QCA implementation of mux-based universal shift register *J. Comput. Electron.* **13** 198–210

[11] Tahoori M B, Momenzadeh M, Huang J and Lombardi F 2004 Defects and faults in quantum cellular automata at nano scale *Proc. 22nd IEEE VLSI Test Symp.* (Piscataway, NJ: IEEE) pp 291–6

[12] Teodosio T and Sousa L 2007 QCA-LG: a tool for the automatic layout generation of QCA combinational circuits *Norchip 2007* (Piscataway, NJ: IEEE) pp 1–5

IOP Publishing

Quantum Computing (Second Edition)
A pathway to quantum logic design
Hafiz Md Hasan Babu

Chapter 26

QCA flip-flops

Learning objectives

- Explain the QCA D flip-flop in detail with the proper circuit.
- Discuss the QCA SR flip-flop with the circuit.
- Describe the working procedure of the QCA J–K flip-flop in detail.
- Acquire knowledge about the QCA T flip-flop.
- Design a QCA J–K flip-flop with its truth table.
- Construct a QCA T flip-flop using a QCA J–K flip-flop.

Computer scientists have been drawn to quantum-dot cellular automata (QCA) as a nanotechnology because of their remarkable qualities, including their small size and low power consumption. QCA flip-flops are sequential circuits whose output is dependent on both the input's history and its current state. Using this technology to design several QCA circuits and present logic gates in an ideal structure has been considered here. This chapter covers the QCA D flip-flops, J–K flip-flops, SR flip-flops, and T flip-flops, which are crucial components in the design of QCA nanotechnology circuits.

26.1 QCA D flip-flops

The QCA D flip-flop is also known as the delay (D) flip-flop because it adds a delay between the input and the output. It is one that recognizes the clock's edge and reflects it in the output:

$$Q_b = \overline{B.\ \text{CLK}.\ Q} = B.\ \text{CLK} + \overline{Q} \tag{26.1}$$

$$Q = \overline{A.\ \text{CLK}.\ Q_b} = A.\ \text{CLK} + \overline{Q_b} = A.\ \text{CLK} + \overline{B.\ \text{CLK}}.\ Q. \tag{26.2}$$

The equations from the previous section are used here to create the QCA D flip-flop. The circuits of all flip-flops in this chapter are designed using this equation. In order

doi:10.1088/978-0-7503-5683-1ch26

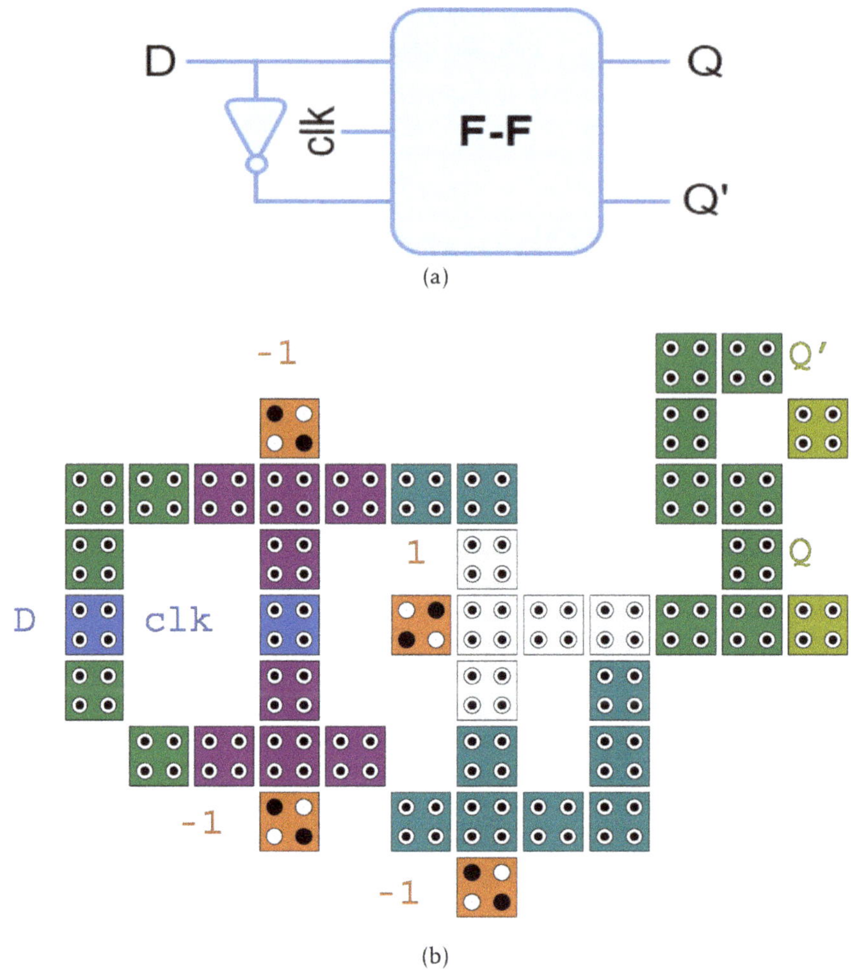

Figure 26.1. The QCA D flip-flop. (a) Schematic for the D flip-flop. (b) QCA layout.

to construct a QCA D flip-flop, A is changed to D and B to D', as illustrated in figure 26.1(b).

The clock pulse direction is shown by the symbols $\downarrow$ and $\uparrow$. These symbols are interpreted in D type flip-flops as edge-triggers.

The D flip-flop needs two inputs, a 'set' input and a 'reset' input, in order to operate. Due to the fact that the two input signals now complement one another, it is possible to set and reset the outputs using an inverter and just one input. The single input D of a D flip-flop is known as the 'data' input. Flip-flops are established when the data input is set to 1, and they change and become reset when the data input is set to 0. This would be useless, however, as each pulse delivered to this data input would result in a change in the flip flop's output. Only the D input condition is replicated to the output Q when the clock input is set to true. This serves as the

Table 26.1. The truth table of the QCA D flip-flop.

Clock	D	Q	Q'	Description
$\downarrow \gg 0$	X	Q	Q'	Memory no change
$\uparrow \gg 1$	0	0	1	Reset $Q \gg 0$
$\uparrow \gg 1$	1	1	0	Set $Q \gg 1$

Table 26.2. The truth table of a QCA J–K flip-flop.

J–K inputs		Outputs		SR inputs	
J	K	Q	Q'	S	R
0	0	0	0	0	X
0	0	1	1	X	0
0	1	0	0	0	X
0	1	1	0	0	1
1	0	0	1	1	0
1	0	1	1	X	0
1	1	0	1	1	0
1	1	1	0	0	1

foundation for the D flip-flop sequential device. The flip-flop's set and reset inputs are both set to 1 when the clock input is at 1. As a result, it will keep the data on its output that were there prior to the clock shift and not modify its state. The output is, to put it simply, 'latched' at 0 or 1 as shown in table 26.1.

26.2 QCA J–K flip-flops

The QCA J–K flip-flop overcomes the drawbacks of the QCA SR flip-flop. The QCA J–K flip-flop operates in a similar manner as the QCA SR flip-flop. The flip-flop is set and reset by the inputs J and K acting as inputs S and R, respectively. In contrast to QCA SR flip-flops, which produce invalid states as outputs when both inputs are set to 1, the QCA J–K flip-flops do not produce invalid states even when both the 'J' and 'K' flip-flops are set to 1, as shown in table 26.2. This is the only difference between the QCA J–K flip-flops and the QCA SR flip-flops (figure 26.2).

Q is the current state, and Q' is the state that results from applying the J and K inputs. There will be eight different combinations for the two inputs J and K. The appropriate Q' state is discovered for every combination of J, K, and Q. Q' merely denotes the values that the J–K flip-flop will eventually produce following the value of Q. Writing the values of S and R necessary to subtract each Q' from its associated

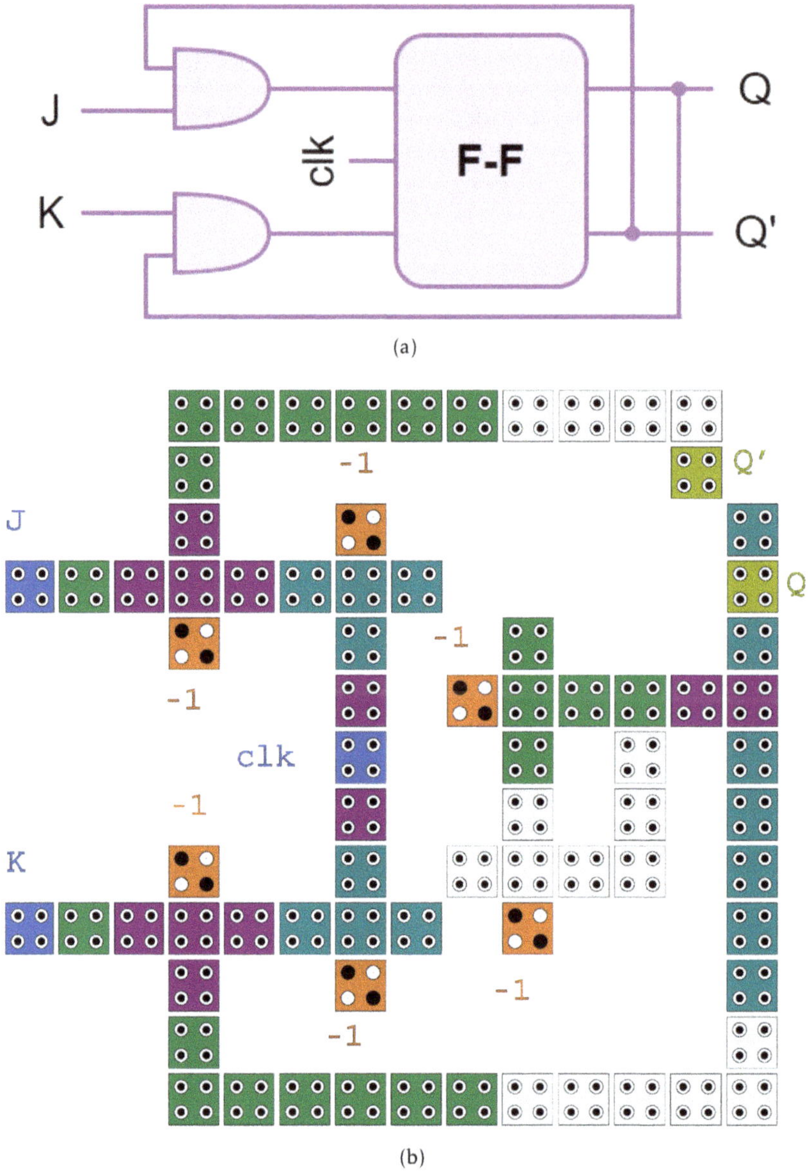

Figure 26.2. The QCA J–K flip-flop. (a) Schematic for the J–K flip-flop. (b) QCA layout for the J–K flip-flop.

Q completes the table. In other words, the values of S and R needed to switch the flip-flop's state from Q to Q' are recorded.

26.3 QCA SR flip-flops

The QCA SR flip-flop, often referred to as a QCA SR latch, is one of the most fundamental sequential logic circuits that is conceivable (figure 26.3).

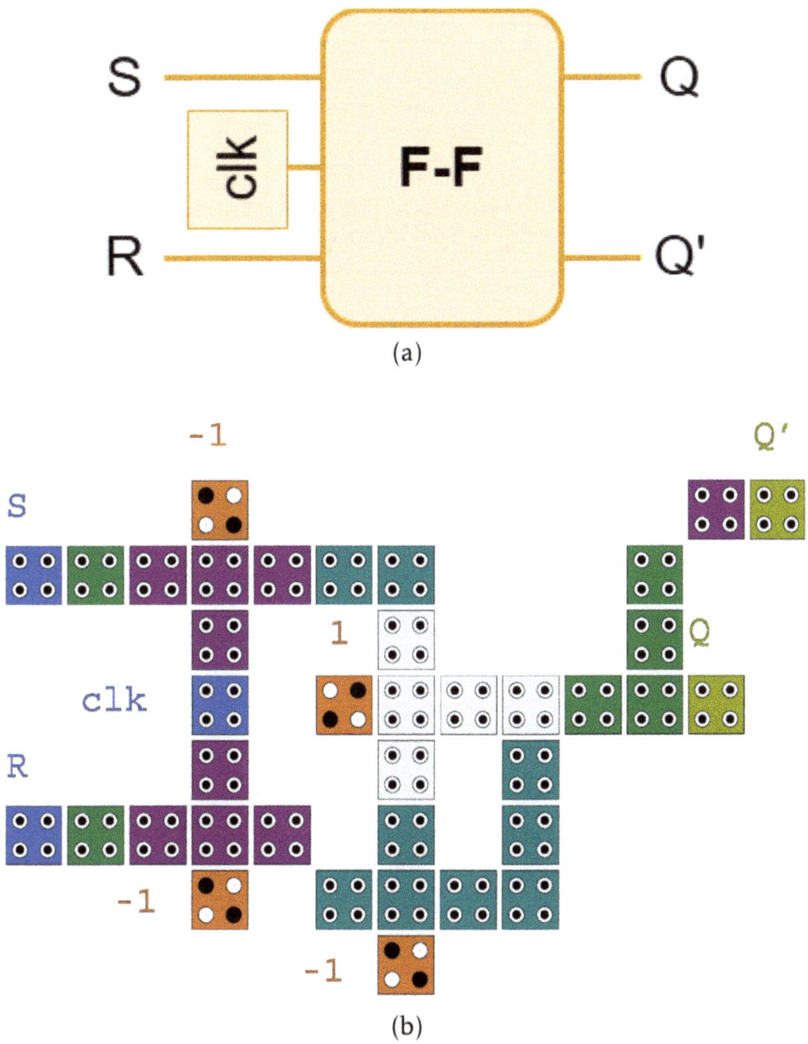

Figure 26.3. The QCA SR flip-flop. (a) Schematic for the SR flip-flop. (b) QCA layout for the SR flip-flop.

The QCA SR flip-flop has two complementing outputs, Q and Q', and two inputs, S and R, designated as set and reset, respectively, as shown in table 26.3.

The QCA SR flip-flop has two inputs, set and reset, and is a bistable device with one bit of memory. Inputs S and R set and reset devices, producing the outputs 1 and 0, respectively. A fundamental flip-flop, the NAND gate QCA SR flip-flop feeds information from both of its outputs back to its opposite input. The single data bit is stored in the memory circuit using this circuit. The QCA SR flip-flop hence has a total of three inputs, namely S and R, as well as current output Q. The present history or state is relevant to this output Q. The phrase 'flip-flop' refers to how the device really functions, since it may be 'flipped' into a logic set state or 'flopped' back into the opposing logic reset state.

Table 26.3. The truth table of a QCA SR flip-flop.

State	S	R	Q	Q'	Description
Set	1	0	0	1	Set $Q' \gg 1$
	1	1	0	1	No change
Reset	0	1	1	0	Reset $Q' \gg 0$
	1	1	1	0	No change
Invalid	0	0	1	1	Invalid condition

Table 26.4. The truth table of a QCA T flip-flop.

T	Q	Q'	
0	0	0	Unchanged/hold
0	1	1	Unchanged/hold
1	0	1	Toggle
1	1	0	Toggle

26.4 QCA T flip-flops

In the QCA T flip-flop there is only one data input (T), one clock input, two outputs (Q and Q'), as shown in table 26.4, and one basic flip-flop known as the T (trigger or toggle) QCA flip-flop. It is a modified version of the QCA J–K flip-flop. The J and K inputs are joined to form a single input called T, which is then used to build the QCA T flip-flops. This is why a QCA T flip-flop is often referred to as a single input QCA J–K flip-flop (figure 26.4).

The inputs J and K are related to obtaining the QCA T flip-flop. Both AND gates are disabled when $T = 0$. The output remains unchanged as a result. The output toggles when T equals 1. When the T input is set to false or 0, the next state of the T flip-flop is equivalent to the present state. In that situation, the next state will be 0 if the toggle input is set to 0 and the current state is also 0. The following state will be 1 if the toggle input is set to 0 and the current state is 1. When the toggle input is set to 1, the flip-flop's next state is the opposite of the one it is in right now. When the present state is 0 and the toggle input is set to 1, the following state will be 1, in that scenario. When the present state is set to 1 and the toggle input is set to 1, the next state will be 0 and vice versa, as shown in figure 26.4.

26.5 Applications

As complementary metal-oxide semiconductor (CMOS) transistor size continues to decrease, it will soon reach its limit. In order to continuously advance the development of electrical devices, a substitute device must be found. A potential tool for creating digital circuits is the QCA. QCA have the potential to be one of the most

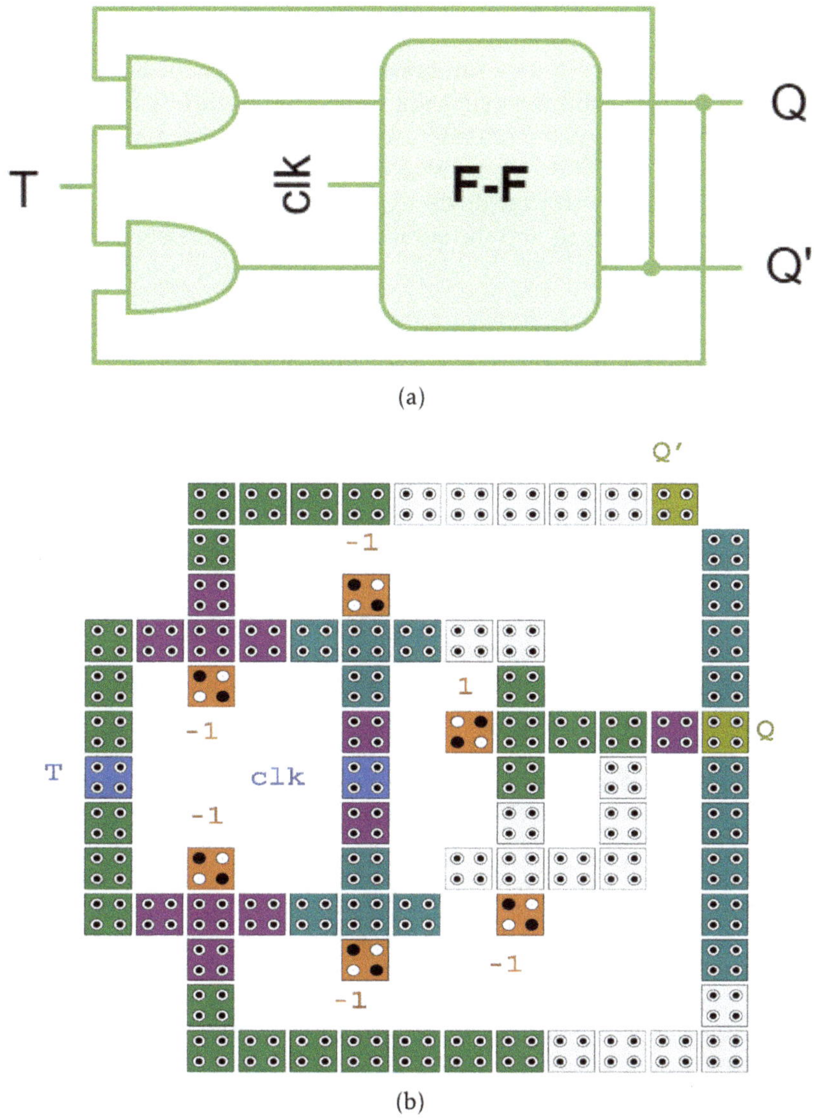

Figure 26.4. The QCA T flip-flop. (a) Schematic for the T flip-flop. (b) QCA layout for the T flip-flop.

promising nanotechnologies due to their advantages over transistor-based technology, including their higher speed, smaller size, and lower power consumption. Compared to traditional digital electronics devices, all QCA electrical devices have additional advantages. The benefits of QCA flip-flops outweigh those of conventional flip-flops significantly. They are mainly used in dividers for frequencies, counters, storage registers, shifts registers, data storage, latches, exchange of data, and memory.

26.6 Summary

The QCA SR flip-flop is used as a fundamental component of flip-flops, which are further used to build QCA flip-flops such as the D, T, and J–K QCA flip-flops. This is a novel method for designing nanoscale QCA flip-flops that requires less complicated hardware. The QCA flip-flops created with the method outlined here can be used to create any type of memory storage device. Using the QCA designer simulation tool, the layout has been developed and the simulation results have been confirmed.

Critical thinking questions

1. Describe the applications of QCA D flip-flops and QCA T flip-flops.
2. What are the benefits and drawbacks of QCA flip-flops?
3. Discuss the differences between the QCA SR flip-flop and the QCA J–K flip-flop.
4. What problem of QCA SR flip-flops is resolved by the QCA J–K flip-flop?
5. What are J and K in the QCA J–K flip-flop? Explain in detail.

References

[1] Abdullah-Al-Shafi M and Bahar A N 2017 Ultra-efficient design of robust RS flip-flop in nanoscale with energy dissipation study *Cogent Eng.* **4** 1391060

[2] Chakrabarty R, Mahato D K, Banerjee A, Choudhuri S, Dey M and Mandal N K 2018 A novel design of flip-flop circuits using quantum dot cellular automata (QCA) *IEEE 8th Annual Computing and Communication Workshop and Conf.* (Piscataway, NJ: IEEE) pp 408–14

[3] Gholamnia Roshan M and Gholami M 2018 Novel D latches and D flip-flops with set and reset ability in QCA nanotechnology using minimum cells and area *Int. J. Theor. Phys.* **57** 3223–41

[4] Hashemi S and Navi K 2012 New robust QCA D flip flop and memory structures *Microelectron. J.* **43** 929–40

[5] Jeon J-C 2020 Low-complexity QCA universal shift register design using multiplexer and D flip-flop based on electronic correlations *J. Supercomput.* **76** 6438–52

[6] Lim L A, Ghazali A, Yan S C T and Fat C C 2012 *Sequential Circuit Design Using Quantum-dot Cellular Automata (QCA)* (Piscataway, NJ: IEEE) pp 162–7

[7] Majeed A H, Alkaldy E, bin Zainal M S and Nor D B M D 2019 Synchronous counter design using novel level sensitive T-FF in QCA technology *J. Low Power Electron. Appl.* **9** 27

[8] Rezaei A 2017 Design of optimized quantum-dot cellular automata RS flip flops *Int. J. Nanosci. Nanotechnol.* **13** 53–8

[9] Sasamal T N, Singh A K and Ghanekar U 2019 Design of QCA-based D flip flop and memory cell using rotated majority gate *Smart Innovations in Communication and Computational Sciences* (Berlin: Springer) pp 233–47

[10] Shamsabadi A S, Ghahfarokhi B S, Zamanifar K and Movahedinia N 2009 Applying inherent capabilities of quantum-dot cellular automata to design: D flip-flop case study *J. Syst. Archit.* **55** 180–7

Chapter 27

QCA programmable logic devices

Learning objectives

- Provided a thorough demonstration of QCA programmable array logic (PAL).
- Learn about the QCA field-programmable gate array (FPGA).
- Describe the QCA programmable logic array (PLA) and explain how to build one.
- Build a QCA FPGA with an appropriate circuit.
- Explain the significance and uses of QCA programmable logic devices (PLDs).

Researchers in the VLSI domain are experimenting with quantum-dot cellular automata (QCA) to reduce the number of complementary metal-oxide semiconductor (CMOS) transistors. It is possible to research and create programmable logic devices (PLDs) using QCA circuits. This chapter describes simple QCA-based programmable logic devices such as programmable logic arrays (PLAs), programmable array logic (PAL), and field-programmable gate arrays (FPGAs). With the help of the QCA designer tool, the provided cell structures are copied, tested, and simulated. With regard to the area covered and cell complexity, these designs are good.

27.1 The QCA programmable array logic

A programmable AND array and fixed OR array are the main features of QCA PAL. It is employed to carry out logic functions. The block diagram of a QCA PAL is shown in figure 27.1.

It has m outputs and n inputs. A buffer gate and an inverter gate are present on each input. The normal input is supplemented by a buffer gate, while the complement or inverted input is supplemented with a NOT gate. The internal structure of the QCA PAL is depicted in figure 27.2.

The product terms can be programmed through the fuse link. This implies that the user can choose how the inputs and QCA AND gates are connected. The fuse link must be positioned at the interconnection if a specific input line is to be linked to

doi:10.1088/978-0-7503-5683-1ch27

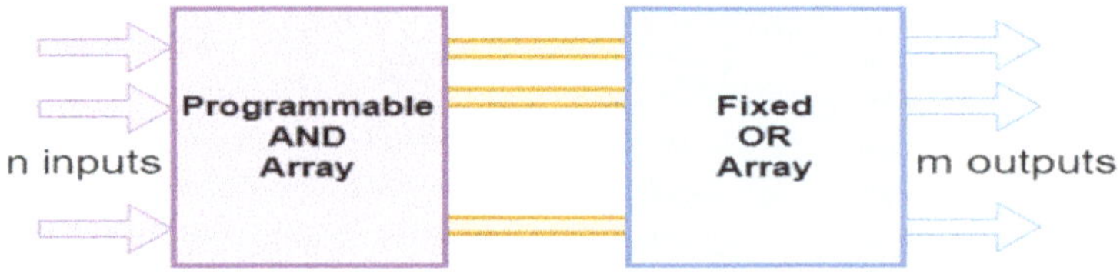

Figure 27.1. The block diagram for a QCA PAL.

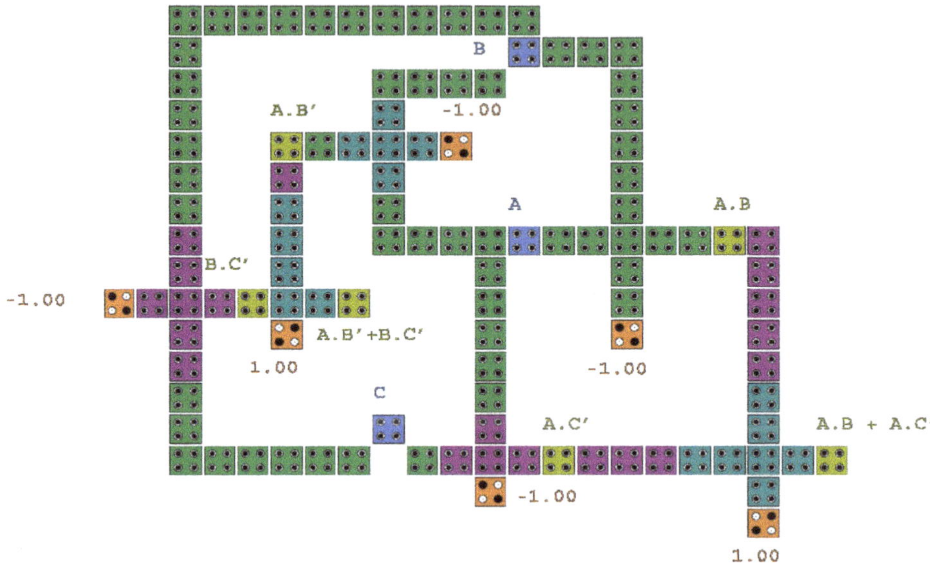

Figure 27.2. The QCA PAL layout.

the QCA AND gate. The fixed QCA OR gate is then supplied with the outputs of the QCA AND gate. The output line of the QCA AND gate is connected to the equivalent input of the QCA OR gate depending on the function that is desired. The result of this process is the realization of the logic function in the sum-of-products (SOPs) form.

27.2 The QCA programmable logic array

The simplest, smallest, and least expensive kind of programmable logic devices are called QCA PLAs. The AND, OR, and NOT gates can be replaced by SPLDs on circuit boards. This structure enables the construction of logic functions in the sum-of-products form. The block diagram of a QCA programmable logic array is shown in figure 27.3 and the QCA PLA layout is shown in figure 27.4.

There are n inputs, inverters, an input buffer, m outputs, and an output buffer in this device. The inputs of the QCA AND and QCA OR arrays both include fuses, which make them programmable.

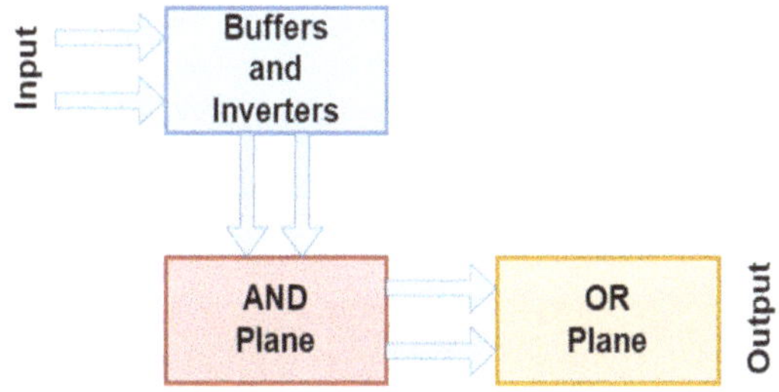

Figure 27.3. The block diagram of a QCA PLA.

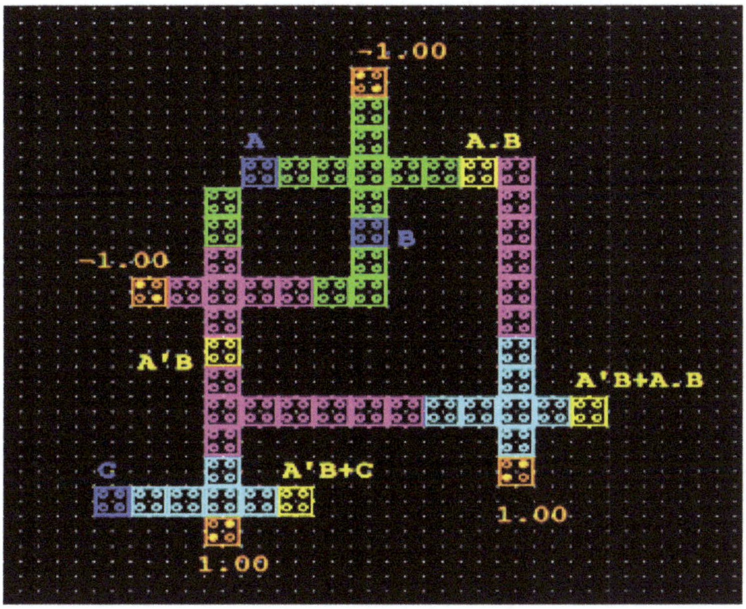

Figure 27.4. The QCA PLA layout.

The inputs of the QCA AND gates in this case are configurable. This means that each QCA AND gate contains inputs for variables that are both normal and complemented. It is possible to program any of those inputs depending on the requirement. Therefore, by employing these QCA AND gates, the necessary product words can be produced. In this case, QCA OR gates' inputs can be programmed as well. As a result, it would be easy to design any number of necessary product terms because each QCA OR gate accepts the outputs of all QCA AND gates as inputs. As a result, QCA PAL will produce results in the form of a sum of products.

27.3 The QCA field-programmable gate array

A matrix of reconfigurable QCA logic blocks coupled with programmable interconnects is the basis of QCA FPGAs, which are integrated circuits. The essential

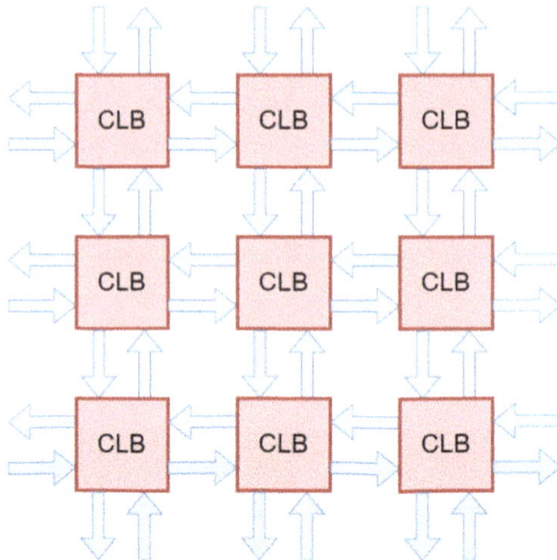

Figure 27.5. Architecture for a QCA FPGA.

building blocks of a QCA FPGA architecture are thousands of QCA customizable logic blocks (CLBs), which are connected to one another by a network of programmable interconnects known as a fabric. The block diagram of a QCA FPGA is shown in figure 27.5.

A QCA CLB only has one input and one output on each side. One input of a nearby QCA CLB is directly connected to a QCA CLB's four outputs. In a similar manner, a QCA CLB's four inputs are directly connected to its single output from adjacent QCA CLBs. Four QCA look-up tables (LUTs) are used in the proposed QCA CLB, and each one is situated on the appropriate side of the QCA CLB, as shown in figure 27.6. Each QCA LUT computes a separate QCA logic function with four inputs, as seen on the right side of figure 27.6. The outputs of nearby QCA CLBs on the 'north', 'south', 'east', and 'west' sides are connected to those four inputs (n, s, e, and w), in that order. All four QCA LUTs' corresponding inputs are connected to one another at the top level of the QCA CLB using a bus line. The north, south, east, and west QCA LUT outputs are represented by N, S, E, and W, respectively. As a result, the QCA CLB's outputs can each be any function of their neighbors' inputs.

Each QCA LUT has an output circuit, a sixteen-bit memory, and a 4-to-16 QCA decoder, as shown in figure 27.7(a). The decoder's functions include turning on one of the sixteen lines and choosing the memory cell located at the address given by the four inputs. A 4-to-16 QCA decoder is produced by combining one first-stage QCA decoder with four second-stage QCA decoders, as shown in figure 27.7(b).

A QCA LUT's primary components are memory cells. They are required to hold programmed values so that the QCA decoder can carry out QCA logic operations based on address lines (inputs n, s, e, and w).

Finally, three sub-components make up an LUT, such as a 4-to-16 QCA decoder, a sixteen-bit QCA memory, and an output QCA circuit. The QCA one-bit memory

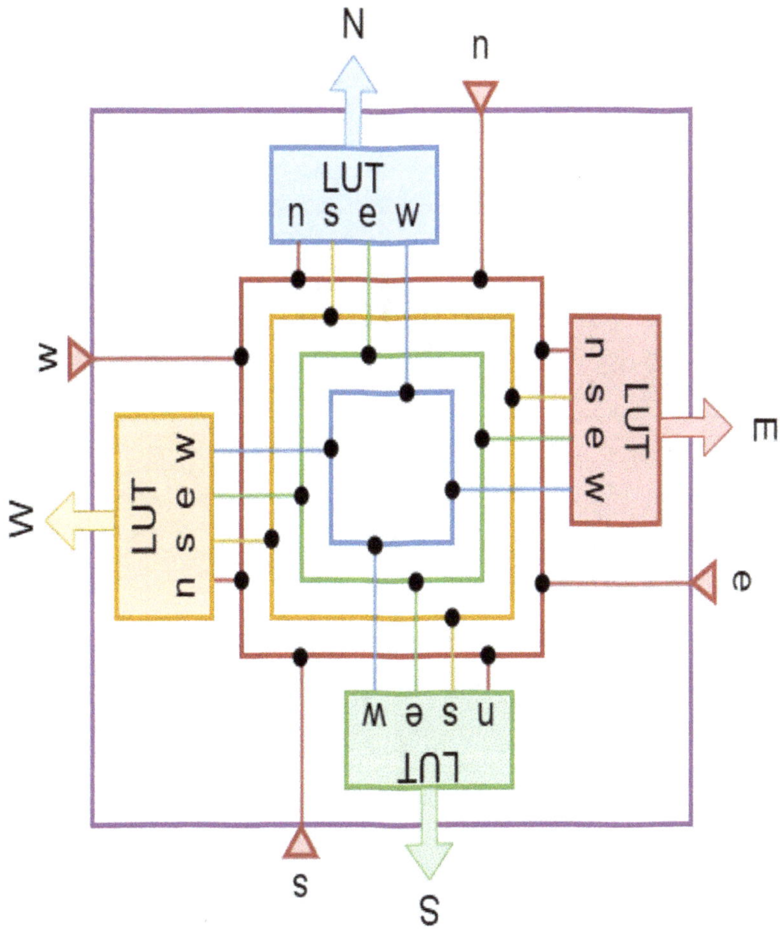

Figure 27.6. The block diagram for the QCA CLB.

cell is shown in figure 27.8 and a complete layout for the QCA LUT is shown in figure 27.9.

A QCA memory cell's enable line is linked to the decoder's corresponding enable line by a connecting wire. Likewise, a memory cell's output is coupled with the appropriate input of the output circuit. The time delay between the QCA decoder's inputs and its output is eleven clock cycles.

27.4 The importance and applications of QCA programmable logic devices

QCA PLDs are parts of electronic systems without a fixed purpose. QCA PLDs perform a variety of tasks, such as device-to-device interfacing, data

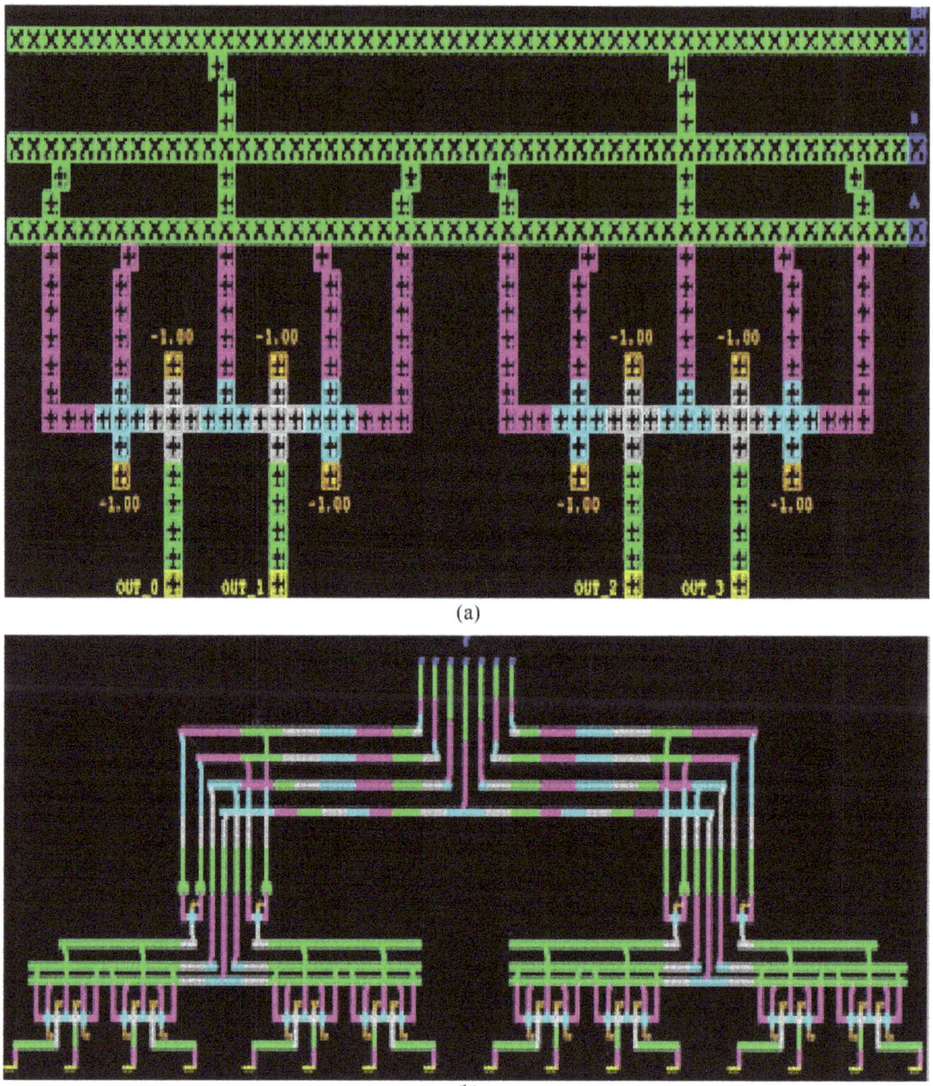

(a)

(b)

Figure 27.7. The QCA decoder. (a) Layout for the second-stage QCA decoder. (b) The layout for a 4-to-16 QCA decoder.

communication, signal processing, data display, timing and control operations, as well as nearly every other task that a system needs to do. QCA memory cells regulate and specify the operation of the logic as well as the connections between the different logic functions. QCA FPGAs are often utilized today because they offer a quicker

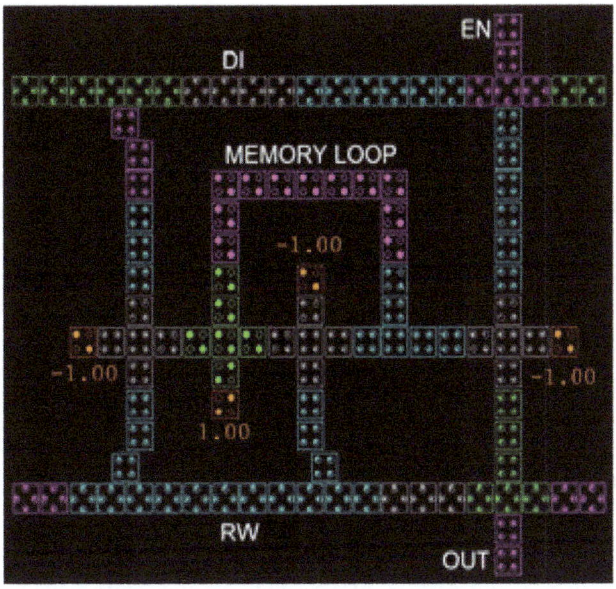

Figure 27.8. The QCA one-bit memory cell.

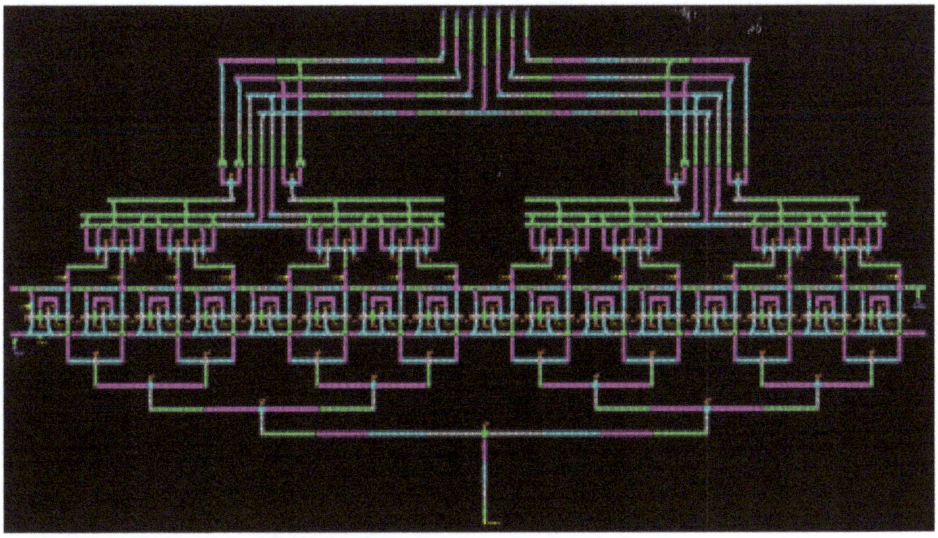

Figure 27.9. The layout for the QCA LUT.

design-to-market time than QCA application-specific integrated circuits (ASICs). They are suited for counters and quadrature decoders, parity checkers, checksums, error detection and correction, various counter and register types, LUTs, memory, and input/output (I/O) controllers for nanoprocessors.

27.5 Summary

This chapter provides logical circuit descriptions for QCA PAL, QCA PLAs, and QCA FPGAs. It also demonstrates how sophisticated QCA programming logic devices can be created and copied. These circuits' designs demonstrate extremely low power dissipation. QCA cells can be used to build QCA PLDs, paving the way for QCA nanocircuits.

Critical thinking questions

1. Describe the differences between QCA PLAs and QCA PAL.
2. Describe the uses for QCA PLAs and QCA PAL.
3. What are the benefits and drawbacks of QCA FPGAs?
4. Which programming module controls the QCA FPGA? Explain in detail.
5. Describe the core elements of a QCA FPGA.
6. What are the disadvantages of a QCA programmable logic array?

References

[1] Awasthi A, Saxena K K and Arun V 2021 Sustainable and smart metal forming manufacturing process *Mater. Today Proc.* **44** 2069–79

[2] De D, Purkayastha T and Chattopadhyay T 2016 Design of QCA based programmable logic array using decoder *Microelectron. J.* **55** 92–107

[3] Ganesh E N 2011 Implementation of programmable logic devices in quantum cellular automata technology *Comput. Sci. Telecommun.* **31** 51–65

[4] Lantz T and Peskin E 2006 A QCA implementation of a configurable logic block for an FPGA *IEEE Int. Conf. on Reconfigurable Computing and FPGAs* (Piscataway, NJ: IEEE) pp 1–10

[5] Thaddeus Niemier M, Rodrigues A F and Kogge P M 2002 A potentially implementable FPGA for quantum dot cellular automata *1st Workshop on Non-Silicon Computation* vol 69 pp 38–45

[6] Tougaw P D and Lent C S 1994 Logical devices implemented using quantum cellular automata *J. Appl. Phys.* **75** 1818–25

[7] Wang J, Sun L, Wang Y and Dai S 2021 A method based on LS-SVM to estimate time-domain Green function *J. Mar. Sci. Technol.* **26** 973–85

[8] Wang Y and Lieberman M 2004 Thermodynamic behavior of molecular-scale quantum-dot cellular automata (QCA) wires and logic devices *IEEE Trans. Nanotechnol.* **3** 368–76

IOP Publishing

Quantum Computing (Second Edition)
A pathway to quantum logic design
Hafiz Md Hasan Babu

Chapter 28

QCA RAM, ROM, and cache memory

Learning objectives

- Describe the significance of the QCA random access memory (RAM) cell.
- Construct a QCA read only memory (ROM) and explain it in detail.
- Explain the QCA RAM cell with the appropriate circuit.
- Explain the QCA cache memory.
- Become familiar with QCA ROM.
- Construct a QCA cache memory and explain how to do so in detail.

In this chapter, the QCA random access memory (RAM) cell, read only memory (ROM), and cache memory are discussed with graphical representation and QCA execution.

28.1 The RAM cell

Memory is one of the critical parts of any computerized framework. It is also of value to construct a fast and greatly upgraded QCA RAM.

Generally, there are two types of RAM cell designs in QCA which are categorized as loop-based and line-based according to the operational manner of the QCA circuits. In the loop-based RAM cell structures, the storing mechanism is achieved using a loop that contains all four clocking zones. In contrast, line-based RAM cell structures use a QCA line to save the previous value of the output. There is an early effort to present a loop-based QCA RAM cell and its schematic diagram is illustrated in figure 28.1. The presented architecture is based on the D latch and a loop which is used to save the memory content. In the QCA implementation, it also takes two clock cycles to transmit the input signal to the output.

However, a QCA RAM without set and reset ability might be implemented. As is shown in figure 28.1(a), this design has a loop-based mechanism and is constructed based on an SR latch. When select = 1 and write/read = 0 the content of the RAM cell is not changed and can be read, but when select = 1 and write/read = 1 the new

doi:10.1088/978-0-7503-5683-1ch28

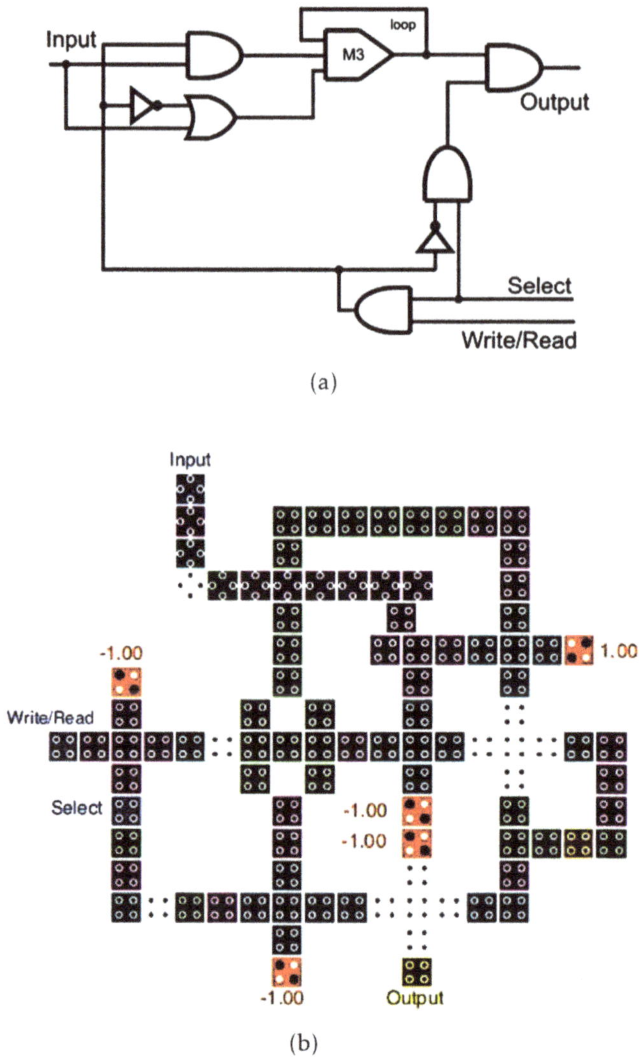

Figure 28.1. The RAM cell structure without set/reset. (a) Schematic symbol. (b) QCA implementation. Reproduced with permission from [1]. Copyright 2015 Elsevier.

input will be transmitted to the output. In this structure a coplanar wire crossing method has been employed. Furthermore this design has used only one cell in some of the clocking zones, which increases the noise sensitivity of the circuit, as shown in figure 28.1(b).

RAM cell engineering with set and reset capacity has been presented, and is shown in figure 28.1(b). This structure is made up of two 2-to-1 multiplexers. One of these multiplexers is used as the D latch by joining the yield flag to the main multiplexer's information input line. At the point when the select and compose/read signals are equal to 1, new information will be embedded in RAM cell, otherwise the

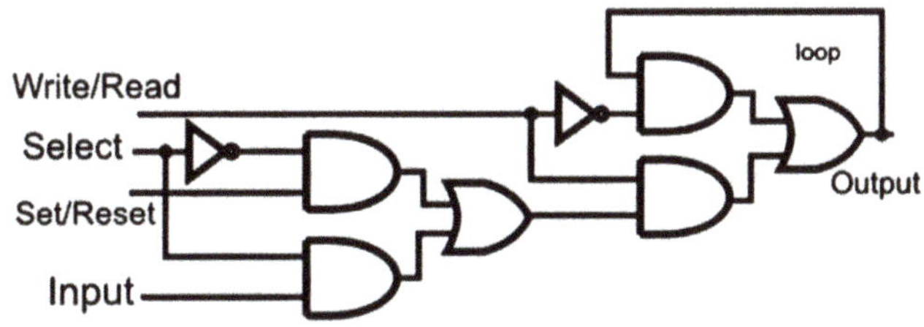

(a)

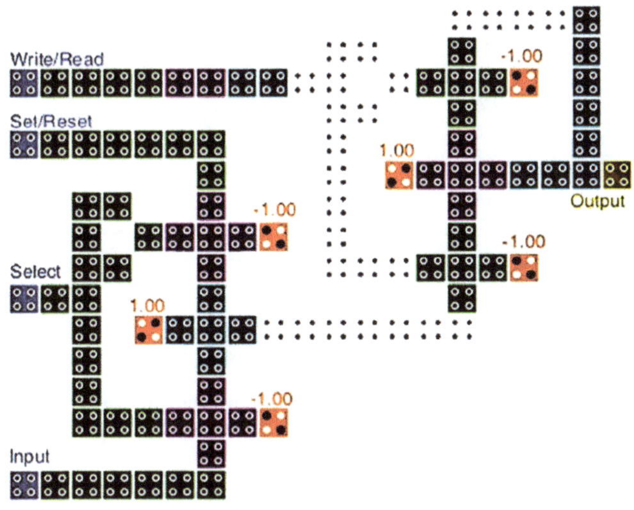

(b)

Figure 28.2. The RAM cell structure with set/reset. (a) Schematic symbol. (b) QCA implementation. Reproduced with permission from [1]. Copyright 2015 Elsevier.

reading task can be performed by setting the compose/read flag to 0. It utilizes a single multiplexed wire for set and reset tasks. This flag additionally decides the substance of the cell only when the select flag is deactivated by 0 and the compose/read flag is actuated by 1. The inertness of this RAM circuit is 1.75 clock cycles for transmission of the info flag to yield the cell, as shown in figure 28.2(b). As is shown in figure 28.2, another design of the RAM cell is given an indistinguishable task from the principal configuration in figure 28.1. This structure is based on a D latch with a non-robust three-input majority gate structure, and the set and reset capacity of the RAM is deficient in this design.

28.2 The QCA ROM

Nanotechnology is contributing to the introduction of new digital media with higher capacity and performance which allows digital data to be stored at the nano-scale. The QCA ROM is described a with a simple structure. The simplified ROM can store 2.5 Gbits of digital data per cm^2.

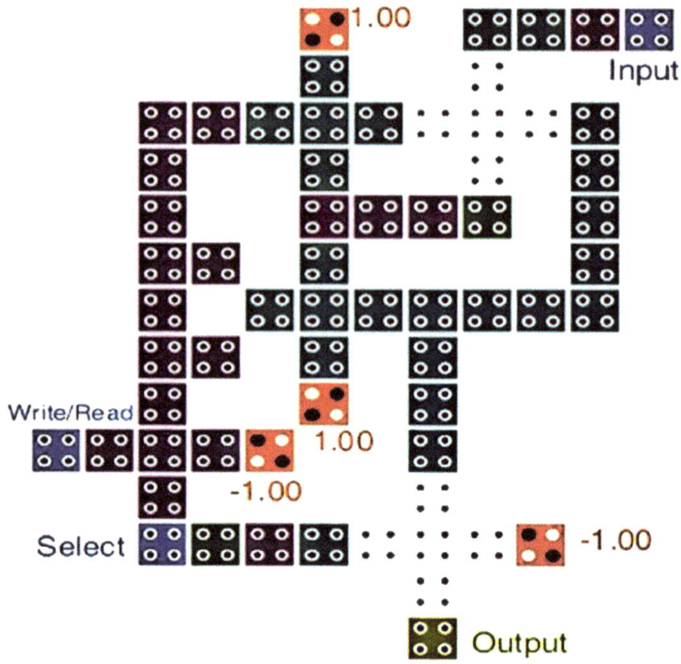

Figure 28.3. The QCA structure of a ROM. Reproduced with permission from [1]. Copyright 2015 Elsevier.

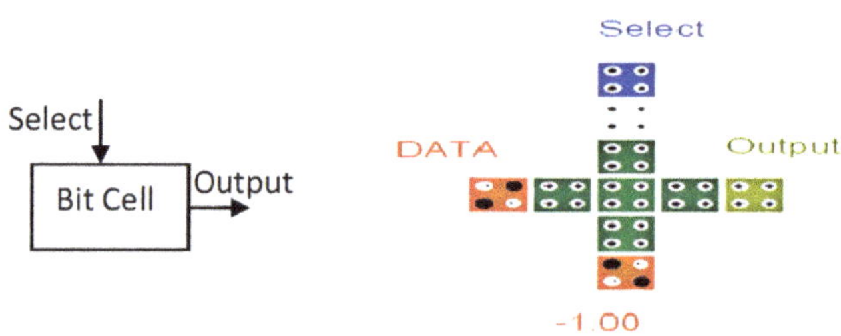

Figure 28.4. The block diagram of a ROM bit cell and its corresponding QCA structure. Reproduced with permission from [11]. Copyright 2008 IEEE.

Typical ROMs consist of data bit cells, decoders, and three-state buffers. Each bit cell has a select input and an output line. When the select input line is active, binary information appears at the output line. Figure 28.3 shows the QCA structure of a ROM. The select input has been added to the cell using an AND operation. An array of bit cells can be selected based on the address given by a decoder.

The block diagram of a simple ROM is shown in figure 28.4. Each ROM consists of some bit cells. The block diagram of each bit cell and its corresponding QCA structure are shown in figure 28.5. Current semiconductor memories achieve random access by connecting the memory cells to the bit lines in parallel. These kinds of ROM cell outputs are connected to the bit lines, which behave similarly to wires or gates. As a result, when the bit cell is not selected, the output of that cell is

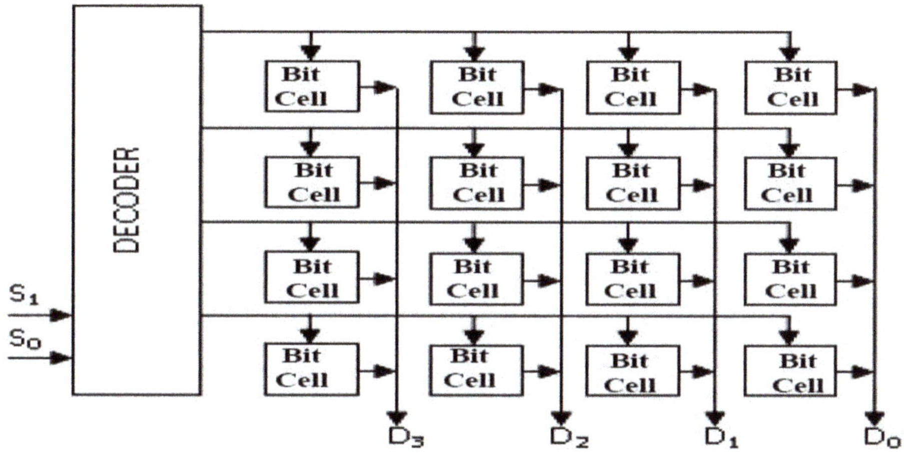

Figure 28.5. The block diagram of a typical 4 × 4 ROM. Reproduced with permission from [11]. Copyright 2008 IEEE.

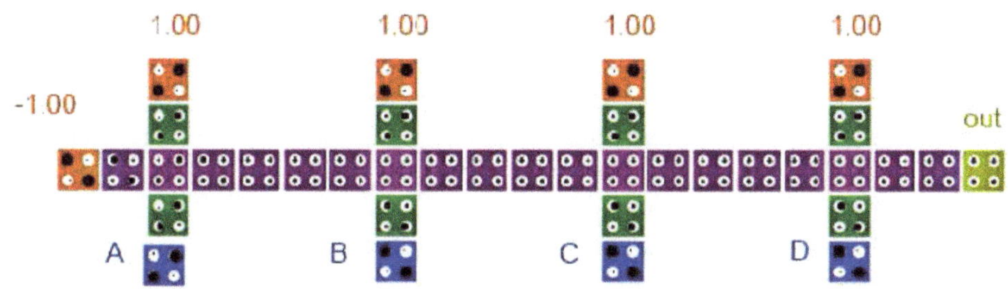

Figure 28.6. The QCA serial OR. Reproduced with permission from [11]. Copyright 2008 IEEE.

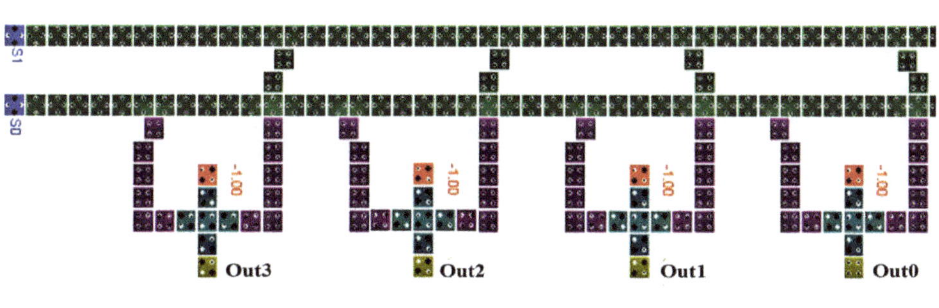

Figure 28.7. The QCA 2-to-4 decoder. Reproduced with permission from [11]. Copyright 2008 IEEE.

0 and when selected the output of the cell is 0 or 1 depending on the binary bit stored in the cell. Consequently, the bit value stored in the selected cell would come on to the output bit line of the ROM. This is a kind of serial OR that is very easy to implement using QCA cells. Figure 28.6 shows how the serial OR can be carried out on multiple inputs with QCA technology. Figure 28.7 shows a 2-to-4 QCA decoder layout.

28.3 The QCA cache memory

QCA cache memory is an additional memory system that temporarily stores information and commands to speed up processing. It is an extremely rapid memory type that keeps information and instructions that are accessed frequently so they are always available for use in subsequent processing. QCA static RAM (SRAM) is used as QCA cache memory due to its incredibly fast performance. Temporary data produced for the quantum computer can be stored in QCA cache memory because of this high speed. In this section, QCA cache memory is built using QCA SRAM which is based on the majority gate. A customized gate called the majority gate generates output that is equal to the majority of input logic levels. The majority of the gate's output will be 1 if the majority of its inputs are 1, otherwise, it will be 0. Figure 28.8 depicts the logic diagram for the majority gate-based memory cell that has been constructed as follows.

Basically, the write/read, select, input, and output signals are present in the QCA memory cell that has been created. When used in a large array, the select line can be utilized as a row select or a column select. For selecting the cell, the outputs of the row or column QCA decoders can be linked to the chosen line. No matter what other input lines are present, the output when select = 0 is 0. This indicates that the circuit will be in the hold state and that the specific cell will not be chosen for memory operation. When select is set to 1, the write/read input will be used to perform the write and read operations. The write operation will be carried out when write/read = 1. Because the majority of inputs to a majority gate are 1 and 0, respectively, the Q is 1 if the input is 1. Write/read = 0 and Q will be in hold mode during read operations, reading the previously stored data.

Using QCA technology, the majority gate-based one-bit cache memory arrangement in figure 28.9 is designed. To obtain the whole layout of the QCA memory cell, many fundamental gates, such as the QCA AND gate, QCA OR gate, QCA NOT gate, and majority gate, are developed and combined. An undistorted output is received because of how carefully all the cells' timing has been planned.

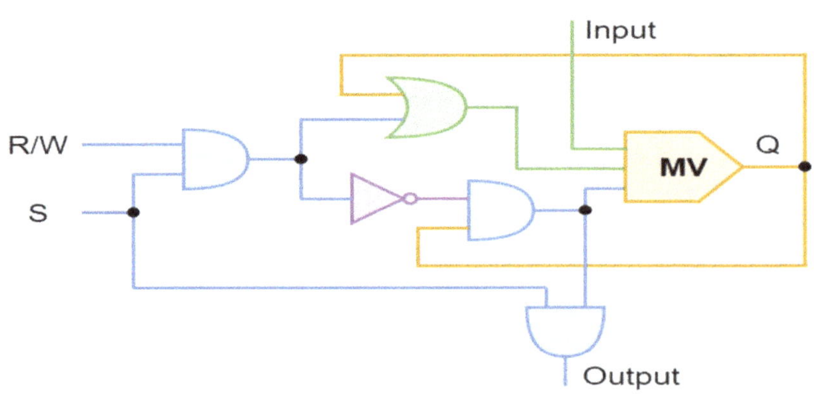

Figure 28.8. The architecture of a one-bit QCA cache memory using the majority gate.

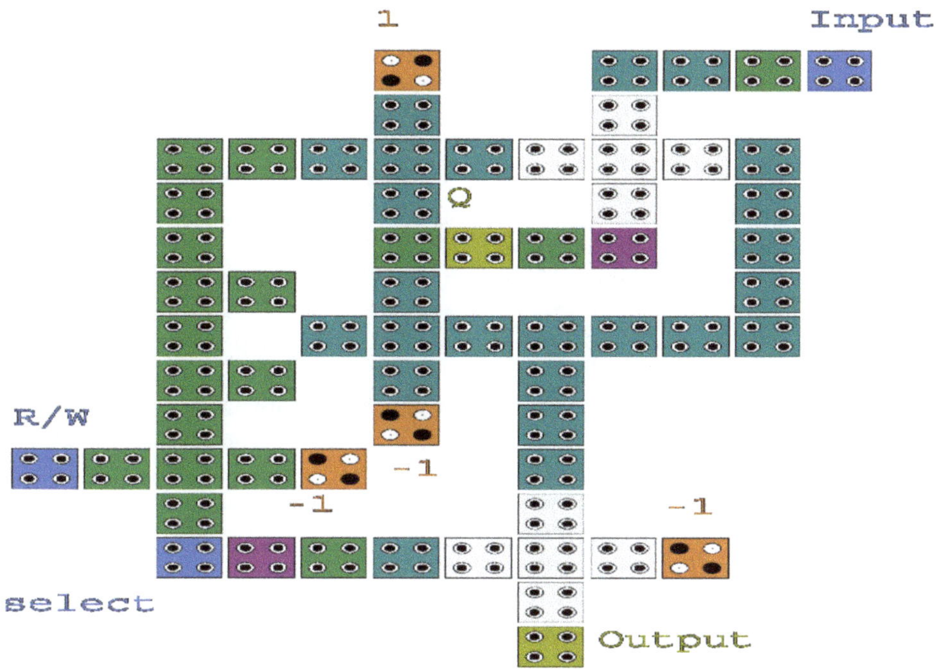

Figure 28.9. The QCA layout for a one-bit cache memory.

28.4 Summary

In this chapter, the structure of a QCA RAM is demonstrated. There are two kinds of RAM cell engineering, specifically circle/loop-based and line-based. In the QCA ROM structure it is shown how an improved ROM may be able to store 2.5 Gbits of computerized information per cm^2. The details of QCA cache memory is also discussed in this chapter.

Critical thinking questions

1. In QCA RAM, which control signals are chosen for the read and write operations? Explain in detail.
2. Describe the applications of QCA RAM and QCA ROM.
3. What are the benefits and drawbacks of using QCA cache memory?
4. For a QCA RAM chip with *n* input address lines, how many memory locations can it access?
5. What are the differences between QCA cache memory and QCA RAM?

References

[1] Angizi S, Sarmadi S, Sayedsalehi S and Navi K 2015 Design and evaluation of new majority gate-based RAM cell in quantum-dot cellular automata *Microelectron. J.* **46** 43–51

[2] Berzon D and Fountain T J 199 A memory design in QCAs using the squares formalism *Proc. 9th Great Lakes Symp. on VLSI* (Piscataway, NJ: IEEE) pp 166–9

[3] Dehkordi M A, Shamsabadi A S, Ghahfarokhi B S and Vafaei A 2011 Novel RAM cell designs based on inherent capabilities of quantum-dot cellular automata *Microelectron. J.* **42** 701–8

[4] Fazzion E, Fonseca O L H M, Nacif J A M, Neto P V, Fernandes A O and Silva D S 2014 A quantum-dot cellular automata processor design *Proc. 27th Symp. on Integrated Circuits and Systems Design* (New York: ACM) p 29

[5] Harish B G and Narashimaraja P 2020 Design of QCA based one-bit memory cell *RJET* **7** 6119–24

[6] Hashemi S and Navi K 2012 New robust QCA D flip flop and memory structures *Microelectron. J.* **43** 929–40

[7] Kim K, Wu K and Karri R 2005 Towards designing robust QCA architectures in the presence of sneak noise paths *Proc. Conf. on Design, Automation and Test in Europe-Volume 2* (Piscataway, NJ: IEEE) pp 1214–9

[8] Kubacki M and Sosnowski J 2019 Exploring operational profiles and anomalies in computer performance logs *Microprocess. Microsyst.* **69** 1–15

[9] Lin A 2010 *Carbon Nanotube Synthesis, Device Fabrication, and Circuit Design for Digital Logic Applications* (Redwood City, CA: Stanford University)

[10] Niamat M, Panuganti S and Raviraj T 2010 QCA design and implementation of SRAM based FPGA configurable logic block *53rd IEEE Int. Midwest Symp. on Circuits and Systems* (Piscataway, NJ: IEEE) pp 837–40

[11] Rahimi E and Nejad S M 2008 Quantum-dot cellular ROM: a nano-scale level approach to digital data storage *6th Int. Symp. on Communication Systems, Networks and Digital Signal Processing* (Piscataway, NJ: IEEE) pp 618–21

[12] Schulhof G, Walus K and Jullien G A 2007 Simulation of random cell displacements in QCA *ACM J. Emerging Technol. Comput. Syst.* **3** 2

[13] Shamsabadi A S, Ghahfarokhi B S, Zamanifar K and Movahhedinia N 2009 Applying inherent capabilities of quantum-dot cellular automata to design: D flip-flop case study *J. Syst. Archit.—Embedded Syst. Des.* **55** 180–7

[14] Shulaker M M, Hills G, Patil N, Wei H, Chen H-Y, Wong H-S P and Mitra S 2013 Carbon nanotube computer *Nature* **501** 526

[15] Taskin B and Hong B 2008 Improving line-based QCA memory cell design through dual phase clocking *IEEE Trans. Very Large Scale Integr. VLSI Syst.* **16** 1648–56

[16] Vankamamidi V, Ottavi M and Lombardi F 2005 A line-based parallel memory for QCA implementation *IEEE Trans. Nanotechnol.* **4** 690–8

[17] Vankamamidi V, Ottavi M and Lombardi F 2008 A serial memory by quantum-dot cellular automata (QCA) *IEEE Trans. Comput.* **57** 606–18

[18] Vetteth A, Walus K, Dimitrov V S and Jullien G A 2003 Quantum-dot cellular automata of flip-flops *ATIPS Lab.* **2500** 1–5

[19] Walus K, Vetteth A, Jullien G A and Dimitrov V S 2003 RAM design using quantum-dot cellular automata *NanoTechnology Conf.* vol 2 pp 160–3

[20] Walus K, Mazur M, Schulhof G and Jullien G A 2005 Simple 4-bit processor based on quantum-dot cellular automata (QCA) *16th IEEE Int. Conf. on Application-Specific Systems, Architecture Processors* (Piscataway, NJ: IEEE) pp 288–93

[21] Yang X, Cai L and Zhao X 2010 Low power dual-edge triggered flip-flop structure in quantum dot cellular automata *Electron. Lett.* **46** 825–6

[22] Yang X, Cai L, Zhao X and Zhang N 2010 Design and simulation of sequential circuits in quantum-dot cellular automata: falling edge-triggered flip-flop and counter study *Microelectron. J.* **41** 56–63

IOP Publishing

Quantum Computing (Second Edition)
A pathway to quantum logic design
Hafiz Md Hasan Babu

Chapter 29

The QCA processor circuit

Learning objectives

- Learn about QCA processors.
- Construct a QCA instruction register.
- Provide basic definitions of a QCA processor.
- Discuss design skills for a QCA accumulator and a QCA decoder.
- Construct the block diagram of a QCA processor.
- Explain a QCA program counter and a QCA multiplexer with the appropriate design architecture.
- Briefly discuss the fundamental parts of a QCA processor.
- Demonstrate a QCA arithmetic logic unit (ALU) with an appropriate circuit.
- Use a graphic to illustrate QCA random access memory (RAM).

Computer scientists are interested in quantum-dot cellular automata (QCA) as a new emerging nanotechnology to replace the current complementary metal-oxide semiconductor (CMOS) technology since they have special properties such as high frequency, extremely small feature size, and low power consumption. Any Boolean function may be represented using the majority gate and inverter, which are the fundamental components of a QCA. A processor based on QCA is fully implemented in this chapter. The input/output, elementary arithmetic, and logic operations are performed by a processor, which is a key component of QCA computers, which are nano-sized, use incredibly little power, and may even operate at terahertz clock rates.

29.1 Introduction

According to Gordon Moore of the Intel Corporation, the number of transistors in semiconductors should increase roughly every two years. This prediction, known as Moore's law, has amazingly held true for more than 40 years. In CMOS technology, the number of transistors has followed Moore's law. Reducing the transistor size will

doi:10.1088/978-0-7503-5683-1ch29

result in a circuit that is faster and uses less energy. However, as this technology develops below the sub-micron level, other issues appear. Several physical constraints, such as the quantum effect and unpredictable behavior at low currents, as well as technological constraints, such as power consumption, design complexity, and lithography complexity, preclude the development of micro-electronic systems that obey Moore's law.

Because of this, scientists have developed several technologies, including the single-electron transistor, nanocarbon tubes, molecular switches, and others. However, in recent years, numerous studies have focused on employing QCA technology to construct nano-scale circuits, and scientists and designers of digital circuits expect that CMOS will be replaced by this revolutionary technology. The fundamental functionality and functioning of a QCA cell have been demonstrated and implemented physically since in the late 1990s. Following that, other designs based on QCA have been presented. These designs range in size from tiny devices such as adders or XORs to large ones such as processors. QCA technology is used to create a QCA processor in this chapter.

29.2 Basic definitions

The term 'central processor' or 'main processor' refers to a central processing unit (CPU). A QCA computer's input/output operations are carried out using a particular type of quantum electronic circuit. Logic and basic math operations carry out the instructions of the computer program. The CPU manages all types of data transfer and instruction flow. The CPU consists of five main parts:

1. QCA control unit (CU).
2. QCA register.
3. QCA arithmetic logic unit (ALU).
4. QCA random access memory (RAM).
5. QCA buses.

1. *QCA CU.* This type of QCA circuit sends instructions to a computer processor. The control unit is made up of numerous selection circuits, including multiplexers and decoders.
2. *QCA register.* This is a part of a QCA processor that can store the smallest set of qubit data used in nanoprocessors. A register can store any type of data, including instructions, a qubit sequence, and a storage address.
3. *QCA ALU.* The program's execution of arithmetic operations is under the control of the arithmetic unit. The ALU unit performs all common arithmetic operations including subtraction, division, and multiplication.
4. *QCA RAM.* RAM, often known as volatile memory, is a part of the CPU that aids in boosting the efficiency of the system. Data access and storage on a short-term basis are the major functions of QCA RAM.
5. *QCA bus.* Whether they are quantum, super, or classical, buses are necessary for data transfer between processors and other parts of all computers. In order to understand in a better way how buses work in computers, a QCA

processor is presented below. The architecture of the fundamental components provides a brief description of the three types of buses: the address bus, the control bus, and the data bus.

29.3 The block diagram of a QCA processor

The CPU, which manipulates data and carries out commands, is the brain of a computer. Numerous entire circuits, including the instruction register (IR), program counter (PC), multiplexer (MUX), ALU, RAM, etc, are included in it. As it is a QCA processor, all circuits are constructed using QCA logic gates. A comprehensive block diagram of a QCA processor is shown in figure 29.1.

Only the most fundamental CPU components are visible in this entire two-bit QCA processor. To carry out useful work using QCA CPUs, they require two inputs, such as instructions and data. The purpose of the IR is to instruct the CPU on how to proceed with the data. A bit is used to represent instructions. Inputs to the CPU are kept in the memory. As can be seen in figure 29.1, the CPU functions are carried out in a cycle of fetching instructions as QCA RAM data from memory enters into the instruction register. It is decoded after it has been fetched before being executed. The 'fetch–decode–execute' cycle is the name of this process. Data are transmitted from memory to the instruction register, which initiates the cycle. It

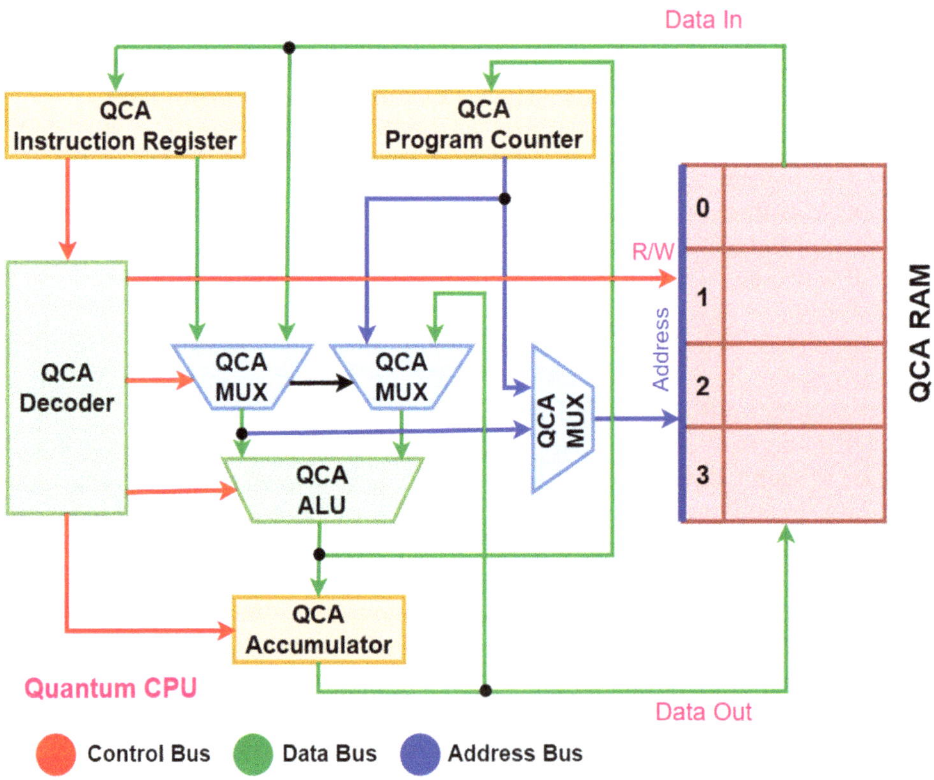

Figure 29.1. The QCA processor.

should be noted that data sent from memory will always use the data bus to convey data. Selecting the machine language from the IR allows the unique bit patterns to be removed before being delivered to the decoder. Decoding encoded data from one format to another is the decoder's main function. Because of the QCA decoder, the cycle's second stage can now function. The decoder indicates which bit pattern will be used and activates the circuit required to carry out the specific operation. The circuit will start working with the instruction if the operation was completed successfully.

29.4 The basic components of a QCA processor

A complete QCA processor has been designed using the components listed below, as shown in figure 29.1. The components of a QCA CPU include:

1. QCA RAM.
2. QCA PC.
3. QCA decoder.
4. QCA MUX.
5. QCA ALU.
6. QCA accumulator.

In the QCA CPU, data are also transferred from one component to another using buses. The data bus, address bus, and control bus are the three different types of buses. The data bus, which transports data back and forth between the CPU and RAM, is bidirectional. The address bus, which connects other components such as primary storage and input/output devices to the processor, is unidirectional and used to transfer memory addresses. The final bus is the control bus, which is used to connect processors to other parts that check to see if everything is moving from one place to another smoothly or not. These are additional crucial CPU elements that must function properly in order to perform useful work. Now, we will cover each element of a QCA CPU briefly.

29.4.1 The QCA RAM

For the purpose of emulating 4-to-2 QCA RAM, as shown in figure 29.2, two address lines are required, each of which must be in NOT form. This combination of address lines will be used as the input for two to four decoders, each of which consists of four QCA AND gates and has a single enable input. This decoder provides us with four select lines, and each select line will traverse every RAM cell. Keep in mind that the RAM will calculate words as $2\wedge k$, where k is the address line, $2 \wedge k$ is the total number of n-bit words, and $k \times 2 \wedge k$ is the decoder combination. This two-bit RAM comprises four distinct RAM cells, each of which has three inputs: In0 or In1, read/write inputs, and a line selector. A QCA OR gate, as shown in figure 29.3, which creates the final output, will take the output from four QCA RAM cells as its input. This is how a 4-to-2 bit QCA RAM is designed as a whole.

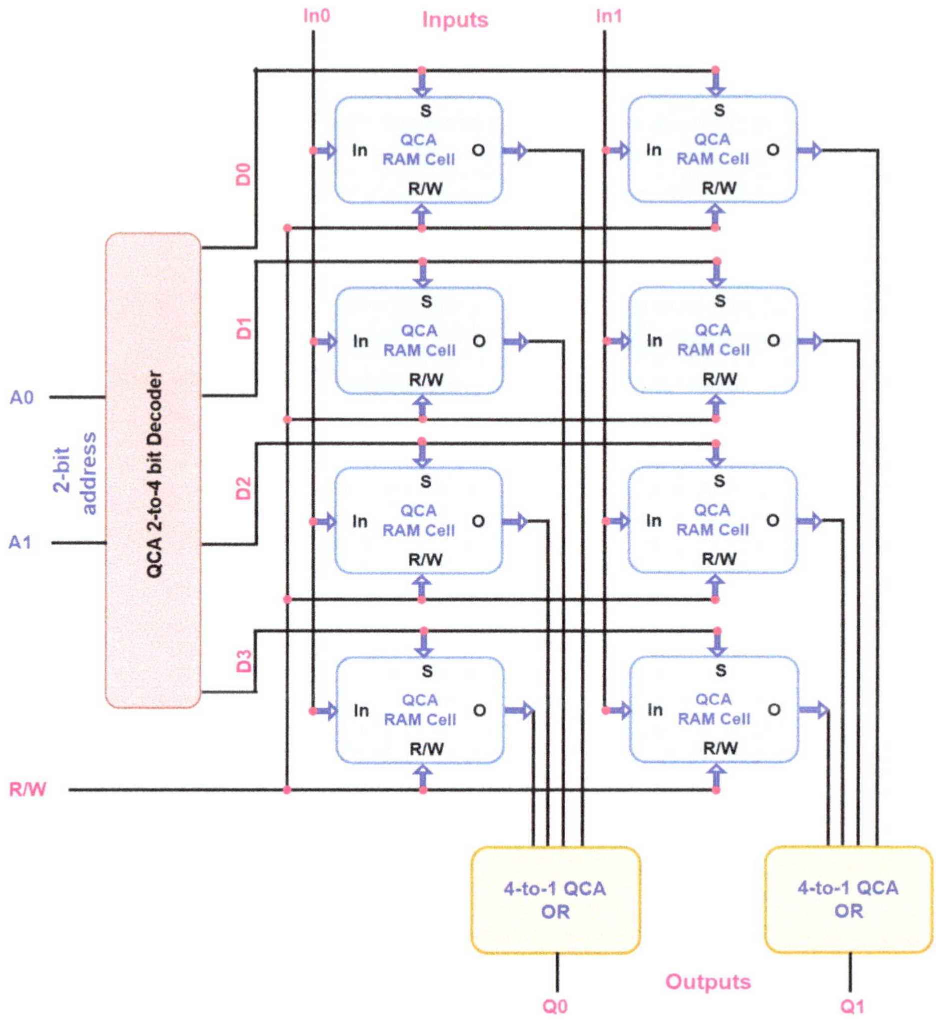

Figure 29.2. The 4-to-2 bit QCA RAM.

This figure illustrates a 4-to-2 bit QCA RAM implementation, which is made up of four distinct 'words' of memory, each of which is two bits wide. The QCA RAM cell is equipped with three inputs and one output. Figure 29.4 explains accurately the entire circuit of a QCA RAM cell. Two QCA RAM cells combined into one word is allowed for simultaneous access to both bits. Two address lines are necessary for memory with four words. The two-bit address lines A_0 and A_1 are the inputs that run through a 2-to-4 decoder and choose one of the four words. The input with memory support turns on the decoder. All of the decoder's output will be 0 if the memory enable is set to 0, which means that none of the memory addresses will be chosen in that situation. However, one of the four words is chosen when the memory enable values are 1. The value in the two address lines chooses the term. Once a word has

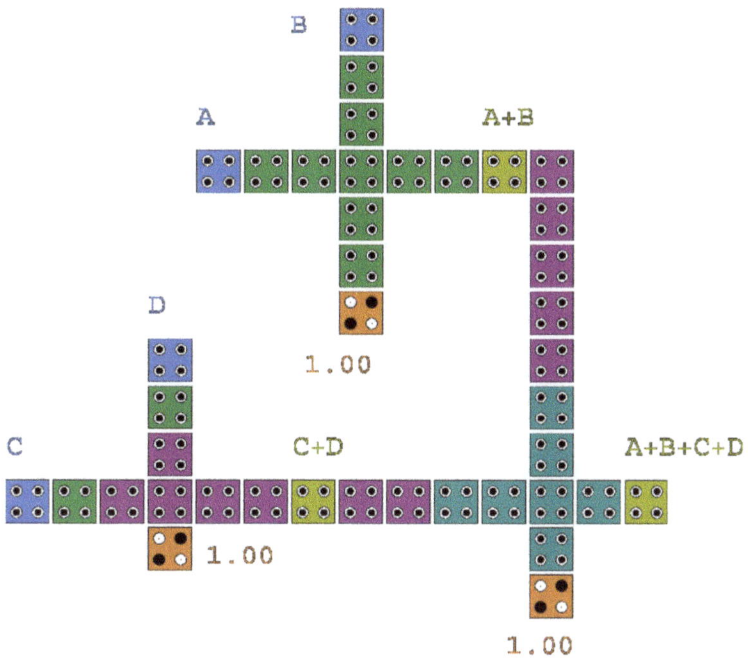

Figure 29.3. The 4-to-1 QCA OR.

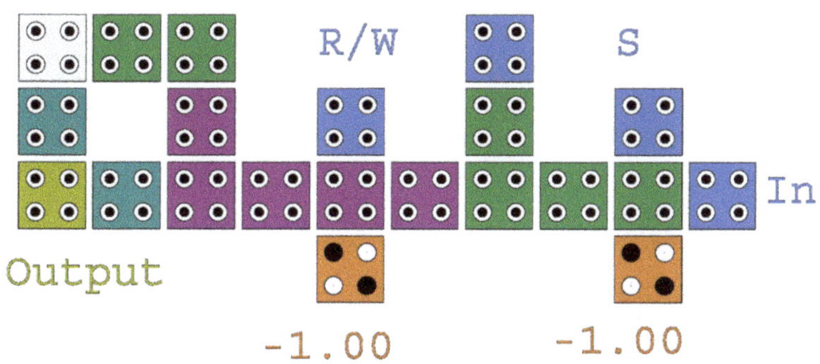

Figure 29.4. The QCA RAM cell.

been chosen, the operation is determined by the read/write input. The four bits of the chosen word are passed to the output Z_0 and Z_1 terminals of the QCA OR gates during the read operation.

The data, however, are transferred into the four QCA cells of the chosen word during the write operation from the input lines. Unselected QCA RAM cells become disabled and retain their previous bit forever. Although none of the words are chosen when the memory enable input that enters the decoder is equal to 0, all QCA cells stay intact regardless of the read/write input's value. The way RAM operates is described above. The QCA RAM cell is described below.

The QCA RAM cell is created utilizing an RS flip-flop. There will be $m \times n$ total QCA cells per word, where m stands for words with n bits. The QCA cell contains one output line with the label 'output' and three inputs with the labels 'select', 'read/write', and 'input'. For access to either reading or writing, use the 'choose' input. A memory action is carried out in the cell when the select line is high or 1. The cell is not interested in performing a read from or write to when the QCA cell's choose line is low or 0, though. The next input is labeled 'read/write', and a system clock will handle this input. The read/write line's clock will indicate 'read' if its value is 0 and perform the 'write' phase if it is 1. Take into consideration the chosen cell. The output value will only depend on the Q value of the flip-flop in this scenario if the clock value is 0, in which case the contents of the cell must be read. However, the output of the cell will be 0 if Q is low and 1 if Q is high.

29.4.2 The QCA instruction register

Figure 29.5 depicts the sixteen QCA AND operations that make up the QCA instruction register. This instruction is the part of a two-bit CPU. Hence the instruction value will be $2\wedge 2 = 4$.

An instruction register is a unique kind of register that is primarily used to hold the instructions that the quantum computer is currently executing. Every input is in QCA NOT form, which is used as a QCA 4-to-1 AND input in this case, as shown in figure 29.6. That instruction is now being performed and is stored in an instruction register. The instruction word is stored in the QCA IR. Any instruction that the QCA CPU retrieves from QCA memory is then stored momentarily in the QCA instruction register.

A bit word or code that specifies a particular operation to be carried out can be the instruction. The instruction is subsequently decoded and executed by the CPU.

29.4.3 The QCA accumulator

The two QCA D flip-flops in the two-bit QCA accumulator are each constructed from four QCA NAND gates, and there is one QCA AND gate in the QCA accumulator. The inputs of the QCA AND gate are LOAD and CLK, and the output that is acquired from the QCA AND gate serves as the input for each QCA D flip-flop. Following logical execution, the first and second QCA D flip-flops provide the first and second outputs, respectively. The circuit diagram is shown in figure 29.7.

This type of QCA register, known as a QCA accumulator, serves as a momentary storage area and maintains a value that is used as a bridge in logical and mathematical processes. For instance, in the operation '2 + 3 + 4', the QCA accumulator will initially store the value 2, then the value 5, then the value 9. When the output of the QCA AND gate is 1, the data will be deposited into the accumulator. No information will be saved in the accumulator if the output of the QCA AND gate is 0.

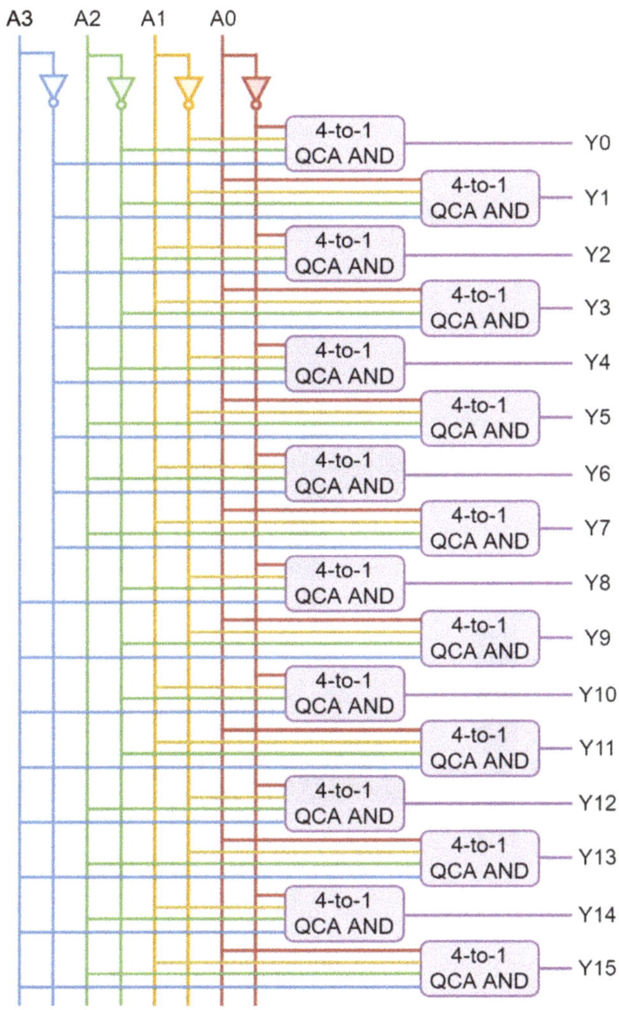

Figure 29.5. The four-bit QCA instruction register.

29.4.4 The QCA decoder

The 2-to-4 QCA decoders, which uses four QCA AND operations with a single enabling input, are utilized in two-bit QCA CPUs. D_0, D_1, D_2, and D_3 are outputs, and A_0 and A_1 are inputs with and without NOT form, respectively. The circuit of a 2-to-4 QCA decoder is shown in figure 29.8.

The QCA decoder is a combinational QCA circuit with n input lines and a maximum output capacity of $2 \wedge n$ lines. Each output has a single product, which is achieved through these QCA AND operations. The value for the two input variables A_0 and A_1 will be minimum when the enable input value is 1. But if the enable input is set to 0, then all of the decoder's outputs will be equal to 0, and if it is set to 1, then one of these four outputs will be active, namely 1.

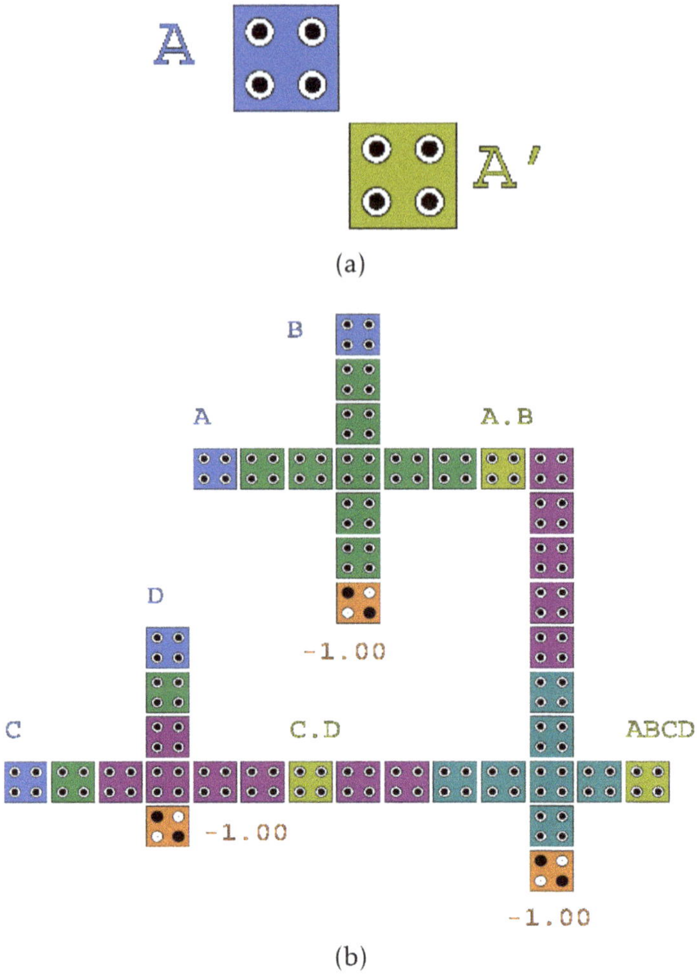

Figure 29.6. The QCA gate. (a) The QCA NOT gate. (b) The QCA 4-to-1 AND gate.

29.4.5 The QCA multiplexer

Two inputs, In1 and In2, one select input, *S*, and one output, *Y*, make up a 2-to-1 multiplexer (MUX). The output can be connected to either of the inputs, depending on the choose signal. One selection is required to perform these actions since there are only two feasible ways to connect the inputs to the outputs given the two input signals. The output will be switched to the In1 input if the selection line is low; if it is high, the output will be switched to the In2 input. The 2-to-1 multiplexer shown in figure 29.9 joins two one-bit inputs to a single destination.

29.4.6 The QCA program counter

Two QCA D flip-flops make up the QCA program counter, as shown in figure 29.10. The next instruction to be performed will be stored using this program counter.

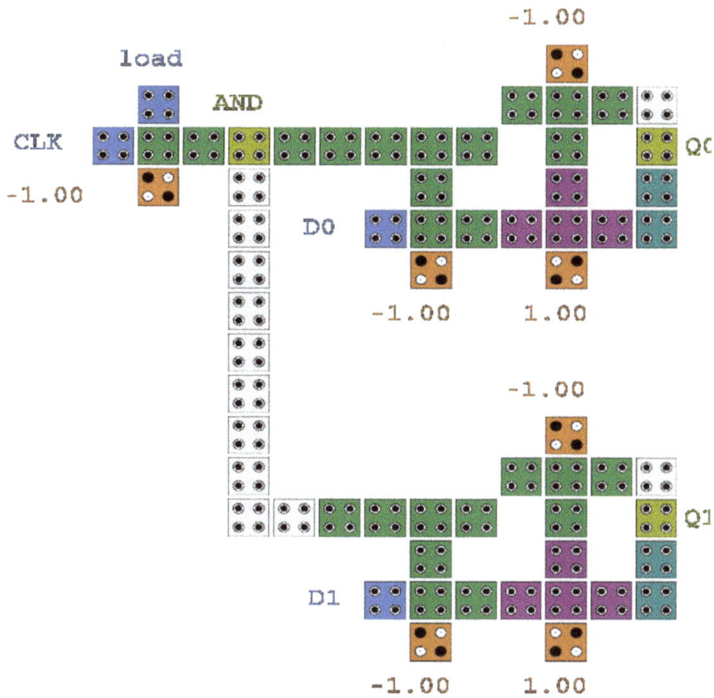

Figure 29.7. The two-bit QCA accumulator.

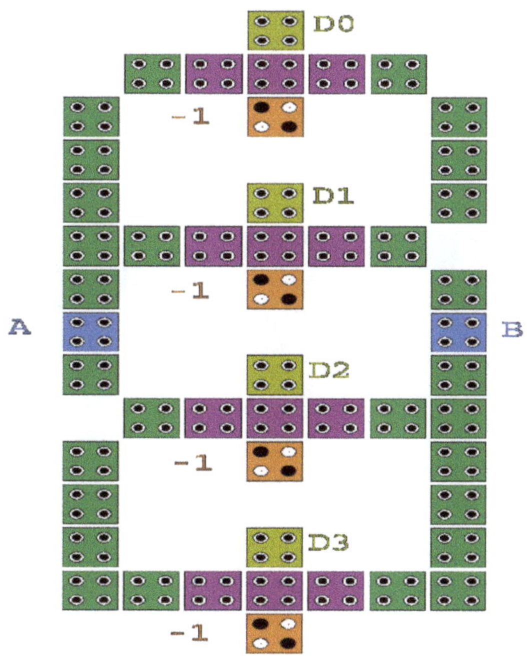

Figure 29.8. The 2-to-4 QCA decoder.

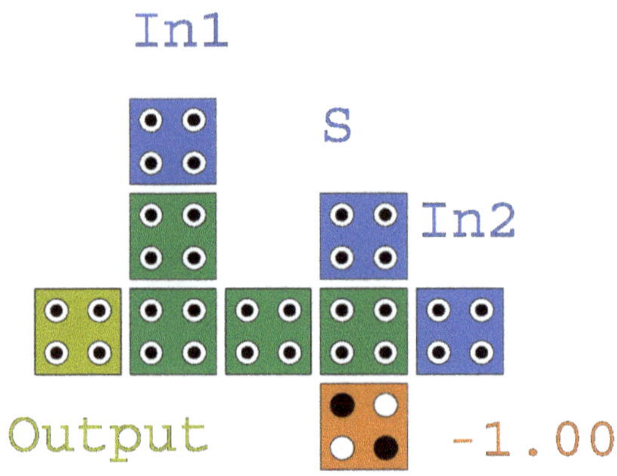

Figure 29.9. The 2-to-1 QCA MUX.

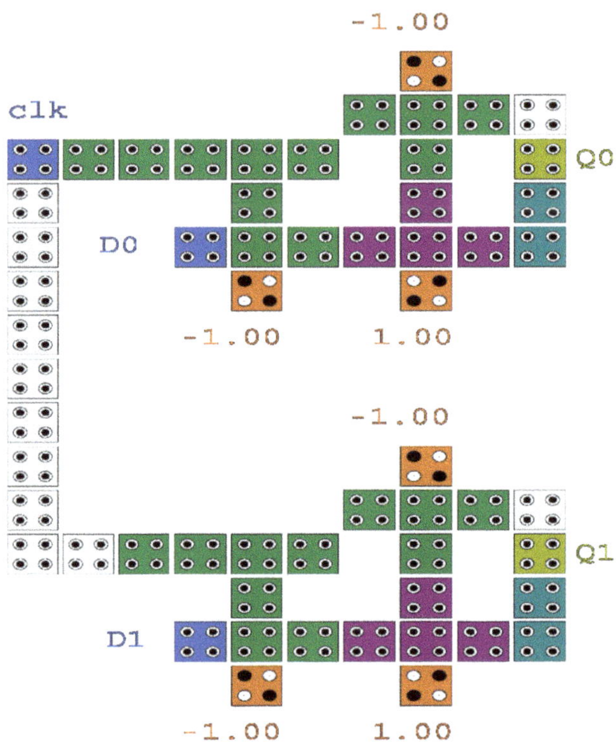

Figure 29.10. The two-bit QCA program counter register.

Upon completion of the current instruction, the program counter is raised by one. Each instruction and piece of data in memory has a unique address.

The QCA PC will first be loaded with address 2 if a program starts with an instruction located at memory location 2, for instance. The QCA PC is increased by one when this instruction is carried out, moving it to the address 3 next. The sequential memory location for storing the instructions in a program is followed at all times.

29.4.7 The QCA ALU

Any computer device's fundamental structure is the ALU. Addition, subtraction, increment, decrement, AND, OR, XOR, XNOR, and other arithmetic and logical operations are included in ALU. One of the multiple and combinational operation structures is the ALU, which has selection lines for carrying out various operations. The QCA ALU circuit is shown in figure 29.11.

The proposed QCA ALU has operations for QCA AND, QCA OR, QCA XOR, and QCA ADD. The QCA MUX 2:1, QCA XOR, and full adder enhanced structures are applied in this design. Instead of using any of the Boolean equations, the QCA MUX 2:1 and QCA XOR make use of a quintessential property of QCA cells. The suggested QCA ALU uses a complete QCA adder with little complexity, minimal latency, and extremely high speed. It also contains a three-input majority gate. QCA ALUs, which load data from QCA input registers, handle the majority of a QCA CPU's activities. A QCA register is a tiny bit of storage that comes with a QCA CPU. A QCA output register is where the QCA ALU saves the outcome of the operation the control unit instructs it to execute on the data. Between these QCA registers, the QCA ALU, and QCA memory, the QCA control unit transfers the data. Thus, the QCA ALU typically has storage locations for input operands, adding operands, the accumulated result (stored in an accumulator), and shifting results. Gated circuits regulate both the bit flow and the operations carried out on them in the QCA ALU's subunits.

29.5 Summary

One of the CMOS technology alternatives is the newly developed QCA nanotechnology. The circuits connected to the QCA flip-flops are another crucial component in the design of each processor's arithmetic and logic unit. The core QCA processor components are the QCA decoder, QCA accumulator, QCA ALU, and QCA RAM, which are presented in this chapter in a new configuration. The QCA processor discussed in this chapter has a quick response time and less complexity with low power consumption.

Critical thinking questions

1. How many locations in QCA memory can be accessed by a QCA RAM chip if it has n address input lines?
2. Describe the applications of a QCA RAM.
3. Describe the role of a QCA ALU in a QCA processor.
4. Describe the uses of the QCA instruction register.
5. Describe how the QCA control bus, QCA data bus, and QCA address bus are different from one another.
6. Using a QCA 2-to-1 MUX, construct a QCA 4-to-1 MUX.
7. What type of data are stored in a QCA instruction register?

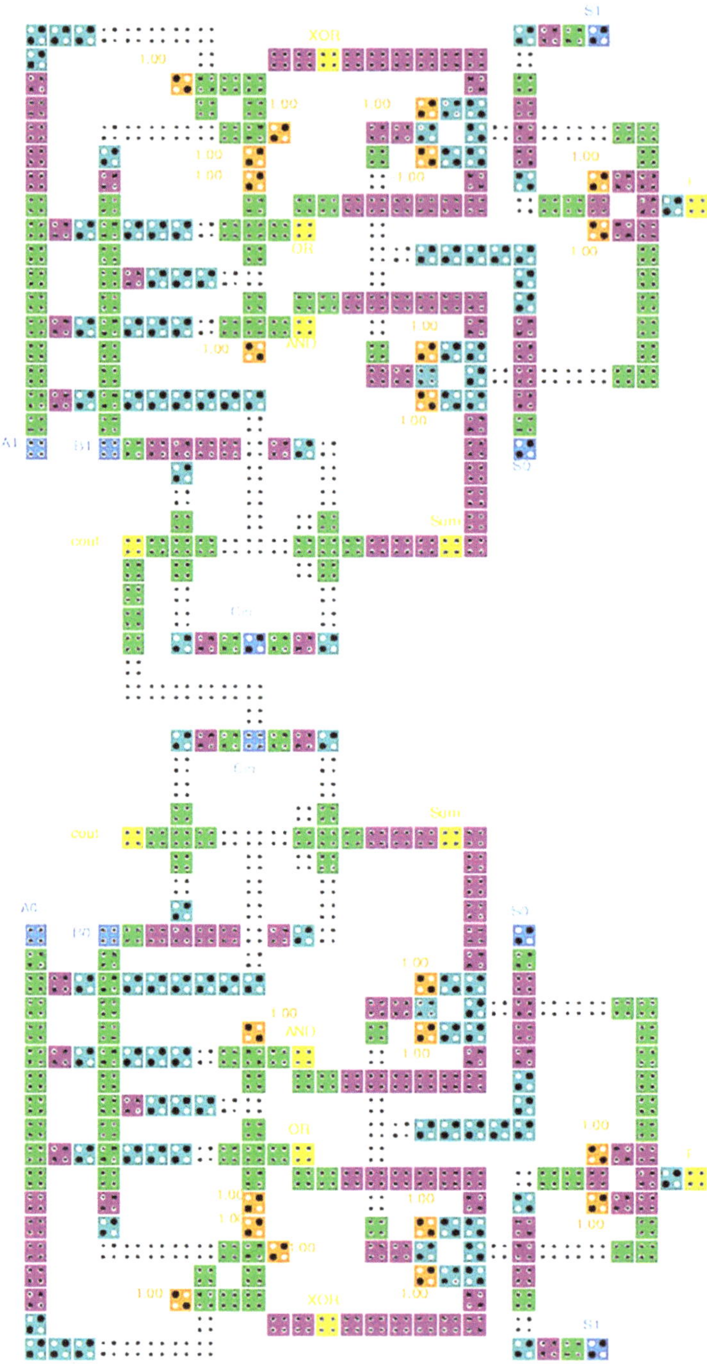

Figure 29.11. The two-bit QCA ALU.

8. How many enable lines are there in a QCA 2-to-4 decoder? Explain why a QCA decoder needs an enable line.
9. What are the typical values of a QCA RAM?
10. What does the QCA instruction register contain?
11. Using a circuit diagram, describe how a QCA decoder works.
12. How does QCA accumulator work for 7 + 2 + 5?
13. Write down the basic definitions of a QCA processor.
14. Illustrate the working principle of a QCA processor with the block diagram.
15. Give a succinct description of the QCA RAM with the relevant circuit.

References

[1] Angizi S, Sarmadi S, Sayedsalehi S and Navi K 2015 Design and evaluation of new majority gate-based RAM cell in quantum-dot cellular automata *Microelectron. J.* **46** 43–51

[2] Devadoss R, Paul K and Balakrishnan M 2011 P-QCA: a tiled programmable fabric architecture using molecular quantum-dot cellular automata *ACM J. Emerging Technol. Comput. Syst* **7** 1–20

[3] Fazzion E, Fonseca O L H M, Nacif J A M, Neto O P V, Fernandes A O and Silva D S 2014 A quantum-dot cellular automata processor design *27th Symp. on Integrated Circuits and Systems Design* (Piscataway, NJ: IEEE), pp 1–7

[4] Gholamnia Roshan M and Gholami M 2018 Novel D latches and D flip-flops with set and reset ability in QCA nanotechnology using minimum cells and area *Int. J. Theor. Phys.* **57** 3223–41

[5] Heikalabad S R and Gadim M R 2018 Design of improved arithmetic logic unit in quantum-dot cellular automata *Int. J. Theor. Phys.* **57** 1733–47

[6] Kumar M and Sasamal T N 2017 An optimal design of 2-to-4 decoder circuit in coplanar quantum-dot cellular automata *Energy Procedia* **117** 450–7

[7] Lent C S and Tougaw P D 1997 A device architecture for computing with quantum dots *Proc. IEEE* **85** 541–57

[8] Majeed A H, Alkaldy E, Zainal M S, Navi K and Nor D 2019 Optimal design of ram cell using novel 2:1 multiplexer in QCA technology *Circuit World* **46** 147–58

[9] Tougaw P D and Lent C S 1994 Logical devices implemented using quantum cellular automata *J. Appl. Phys.* **75** 1818–25

[10] Walus K, Vetteth A, Jullien G A and Dimitrov V S 2003 RAM design using quantum-dot cellular automata *NanoTechnology Conf.* 2 pp 160–3

[11] Walus K, Mazur M, Schulhof G and Jullien G A 2005 Simple 4-bit processor based on quantum-dot cellular automata (QCA) *IEEE Int. Conf. on Application-Specific Systems, Architecture Processors* (Piscataway, NJ: IEEE) pp 288–93

Chapter 30

Applications of QCA technology

Learning objectives

- Learning about the performance of QCA.
- Explain how QCA circuits consume less power.
- Discuss the use of QCA circuits for encryption.
- Explain the importance of any circuit with smaller size.
- Explain QCA for image processing.

Quantum-dot cellular automata (QCA) represent a developing approach that makes it possible to create nanoscale circuits. They use electrostatic and quantum phenomena as their foundation. There are many potential applications of QCA technology. A real-life QCA processor is shown in figure 30.1. Other QCA applications are discussed in this chapter.

30.1 High performance

In recent years, its high power consumption, slow speed, and density beyond 10 nm have restricted the adoption of complementary metal-oxide semiconductor (CMOS) technology. The QCA is employed for high-speed applications. It is an alternative of classical CMOS technology. Essentially, QCA is based on a cell. With the appropriate charge set-up, each cell can represent one bit. It comprises two charged electrons and four quantum dots. The two electrons can only be positioned in two quantum positions that are diametrically opposite because of the force of Coulombic repulsion. QCA has a higher capacity to deliver excellent performance in terms of clock frequency, and the employment of QCA on the nanoscale has a bright future.

30.2 Small size

The nanotechnology field of QCA is growing and they have the potential to be smaller than CMOS-based technologies. In addition, QCA have a lower cost and

doi:10.1088/978-0-7503-5683-1ch30

Figure 30.1. A real-life QCA processor.

they simplify the circuits. In QCA it is also possible to have huge circuits in a small space.

30.3 Low power consumption

QCA can cut down the amount of power used by digital circuitry. The cost will also be lower because less power is used. For instance, one of the crucial functions in a digital signal processing (DSP) unit is the low power multiply accumulator (MAC) in QCA technology, which can enhance voice processing, video coding, and digital filtering, among other functions. Voice processing using QCA technology is shown in figure 30.2.

30.4 Encryption and authentication

In this technology, crucial components include quantum cells and quantum-dot cellular automata at the nanoscale. The nanotechnology of quantum-dot cellular automata has enabled linkages in the security realm, such as in cryptographic circuits, where the security is improved. The data security using QCA technology is shown in figure 30.3.

30.5 Higher data speed

Higher data speeds between transceivers are made possible via optical networking, which is essential for decreasing data losses. One of the most promising

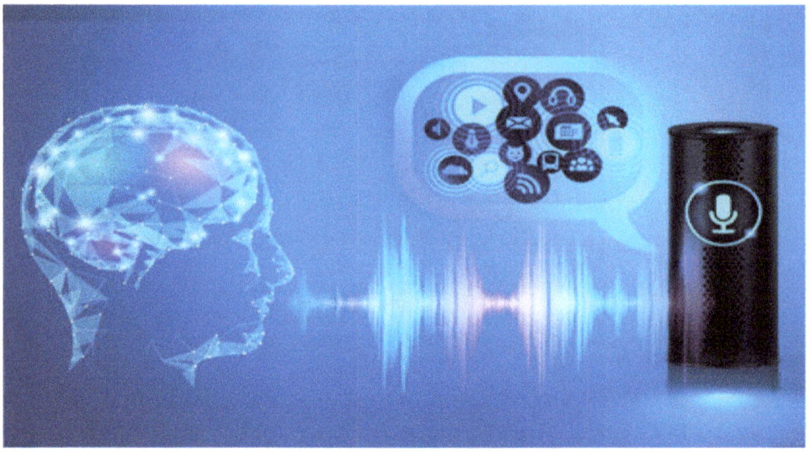

Figure 30.2. Voice processing using QCA technology.

Figure 30.3. Data security using QCA technology.

nanotechnologies is QCA, which requires smaller areas (60% less design area than CMOS technology) and creates circuits with high speed by requiring fewer cycles than other CMOS designs to minimize scaling difficulties. Thus a high data rate and high-speed network are the advantages of QCA technology. A high-speed QCA network is shown in figure 30.4.

30.6 Image processing

QCA represent a promising nanoelectronic technology that could be used in a variety of image processing applications. By utilizing parallel processing, greater silicon-area usage and clock speed optimization, QCA architecture offers better performance. Image processing using QCA is shown in figure 30.5.

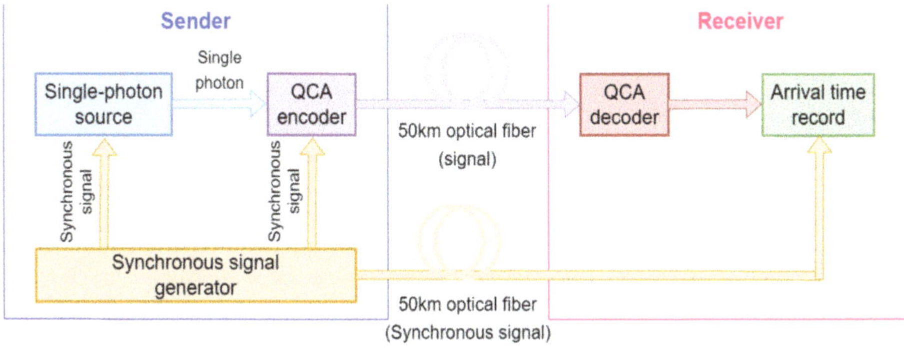

Figure 30.4. A higher data speed network.

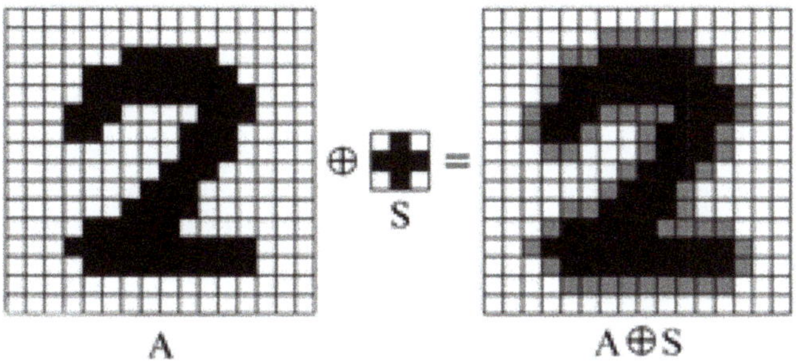

Figure 30.5. Image processing using QCA.

30.7 Summary

The topic of this chapter is the applications of QCA. QCA provide a few key benefits, such as faster speed, larger circuit density, and reduced power dissipation. Additionally, they are utilized in the construction of frequently used electronic modules. QCA have promise for greater use in all sectors because of its significant advantages.

Critical thinking questions

1. What factors contribute to the high performance of QCA technology?
2. Describe some of the useful features of QCA technology.
3. How does the QCA technology reduce the size of a device? Describe the steps.
4. What secure method does QCA technology use to transfer data?

References

[1] Amiri M A, Mirzakuchaki A and Mahdavi M 2011 A5/1 implementation in quantum cellular automata *Nanosci. Nanotechnol.* **1** 58–63

[2] Amiri M A, Mirzakuchaki S and Mahdavi M 2011 Cryptography in quantum cellular automata *Cellular Automata: Innovative Modelling for Science and Engineering* (Rijeka: InTech) pp 285–96

[3] Christie J A 2015 *Molecular Quantum Cellular Automata: Synthesis and Characterization* (Notre Dame, IN: University of Notre Dame)

[4] Hariprasad A and Rao Ijjada S 2019 Quantum-dot cellular automata technology for high-speed high-data-rate networks *Circuits Syst. Signal Process* **38** 5236–52

[5] Haris K, Efstratiadis S N, Maglaveras N and Katsaggelos A K 1998 Hybrid image segmentation using watersheds and fast region merging *IEEE Trans. Image Process.* **7** 1684–99

[6] Qadir F, Ahmad P Z, Wani S J and Peer M A 2013 Quantum-dot cellular automata: theory and application *Int. Conf. on Machine Intelligence and Research Advancement* (Piscataway, NJ: IEEE) pp 540–4

[7] Silva D S, Sardinha L H B, Vieira M A M, Vieira L F M and Neto O P V 2015 Robust serial nanocommunication with QCA *IEEE Trans. Nanotechnol.* **14** 464–72

Part IV

QCA fault tolerance

An overview of QCA fault-tolerant circuits

A system's fault tolerance is simply its capacity to function normally even when one or more of its components fail; it does not matter if it is a network, computer system, cloud cluster, or something else. Alternatively, fault tolerance describes how an operating system (OS) handles and accommodates hardware or software problems. When the fundamental components of the computer are faulty, fault-tolerant computing enables reliable quantum-dot cellular automaton (QCA) computation. Fault tolerance is the ability to identify and fix errors. Recently, QCA have emerged as a notable alternative to complementary metal-oxide semiconductor (CMOS) technology. They have a high integration density, potential as a computational fabric for nano-computing systems, ultra-high velocity, efficient energy, and low area for designing circuits. Fault-tolerant circuits offer reliability through computation redundancy cells. QCA-based circuits are unreliable and error-prone because they frequently experience numerous manufacturing flaws and variations. Therefore, designing fault-tolerant circuits is crucial for their reliable implementation. The main advantage of fault tolerance is to reduce or eliminate the possibility of systems failing due to a component error. Thus the following chapter on fault-tolerant QCA systems discusses a number of fault-tolerant designs, including fault-tolerant static random access memory (SRAM) cells, adders, multipliers, and subtractors.

IOP Publishing

Quantum Computing (Second Edition)
A pathway to quantum logic design
Hafiz Md Hasan Babu

Chapter 31

QCA fault-tolerant circuits

Learning objectives

- Understand the importance of QCA fault-tolerant circuits.
- Discuss the fault-tolerant QCA static random access memory (SRAM) cell in detail.
- Become familiar with the QCA fault-tolerant majority gate.
- Provide a figure to illustrate the fault-tolerant QCA subtractor.
- Construct a fault-tolerant QCA demultiplexer.
- Explain the fault-tolerant QCA multiplier.
- Describe a fault-tolerant QCA full-adder with a relevant diagram.

Quantum-dot cellular automaton (QCA) technology is a viable substitute for silicon transistors at the nanoscale due to its incredibly small size and incredibly low power consumption. Fault-tolerant QCA structures are a hot topic of study because the design of QCA circuits is constrained by the high defect rate during manufacture. That is why, in this chapter, fault-tolerant QCA gates, such as the 1-to-2 QCA demultiplexer, QCA full-adder, QCA static random access memory (SRAM) cell, and QCA subtractor will be presented.

31.1 The necessity of QCA fault-tolerant circuits

Researchers are looking into alternatives to complementary metal-oxide semiconductor (CMOS) transistors, such as graphene field-effect transistors (FET), silicon nanowire transistors, single electron transistors, carbon nanotube FETs, tunnel FETs, etc. QCA, a new binary computer paradigm, are one of the most popular nanotechnologies to replace CMOS technology.

The fact that QCA are a novel technology makes them vulnerable to different kinds of fabrication-related errors and process variances. As a result, QCA-based circuits present serious reliability-related problems due to their error-prone nature. Thus, the development of fault-tolerant QCA-based circuits is becoming increasingly necessary to address the reliability concerns.

doi:10.1088/978-0-7503-5683-1ch31

It has been discovered, however, that flaws created during deposition may be more harmful and may result in catastrophic errors in QCA circuits. The errors of cell deposition include cell omission, misalignment, displacement, and excess cell deposition, among others.

The capacity to continue operating properly in the face of a failure is known as fault tolerance. When one or more of a fault-tolerant design's components fail, the intended operation may still be carried out, although at a reduced level. This is done at the expense of either performance degradation or significant hardware overhead.

31.2 The fault-tolerant QCA majority gate

Two different forms of fault-tolerant QCA majority gates will be suggested in this chapter. Fault-tolerant three-input majority gates and fault-tolerant five-input majority gates make up the first group. Figure 31.1 depicts the fault-tolerant three-input majority gate.

To account for single-cell omission faults in this architecture, the baseline three-input majority is increased by five cells. Notably, there is no need for a multi-layer design for cascading because the output of our cell is accessible. Additionally, since no rotating cells are needed, the complexity and expense of manufacturing are reduced. This structure is assessed in this section using physical verification based on a single-cell omission fault.

In figure 31.2 a five-input majority gate is presented which is used to create the random access memory (RAM) cell. This five-input majority gate is the only one utilized for low-power applications.

The symmetrical design of the majority gate that is being proposed will help designers create better physical control. Additionally, this design can quickly modify different levels of any circuit with an excellent output derivation.

31.3 The fault-tolerant QCA 1-to-2 demultiplexer

Demultiplexer circuits are frequently used in communication systems, QCA technology, and digital electronics. The demultiplexer circuit transfers data from a single

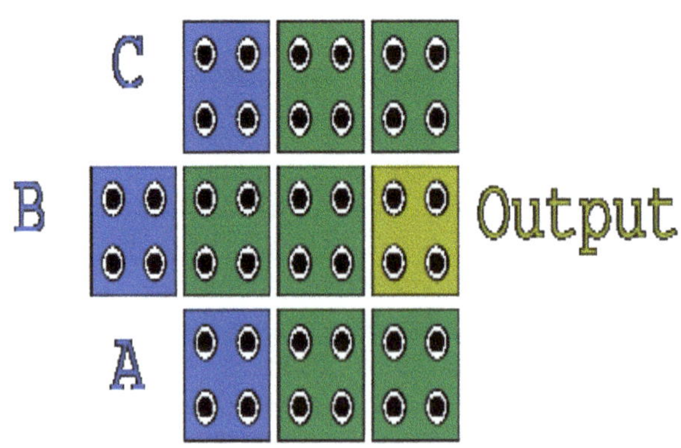

Figure 31.1. The fault-tolerant three-input majority gate.

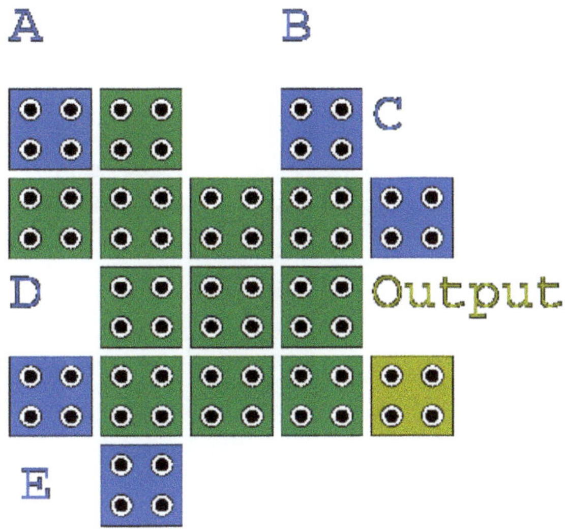

Figure 31.2. The fault-tolerant five-input majority gate.

input to several output lines. It is recognized as a distributor of data. Inverter and majority gates have been used to create the majority of QCA circuits, although they typically respond poorly when there are general defects. The demultiplexer transforms serial input data into several parallel output lines (figure 31.3).

The proposed demultiplexer has input, selector, and output lines (O_1, O_2) as well as one input line (IN). Typically, it moves the inputs to one of the outputs based on the value of the selector (S) line. One fault-tolerant inverter and two fault-tolerant majority gates are used to schematize the suggested circuit. Two clock phases are required to generate the outputs. The waves are created by the simulator software using logic 0 and 1. The values designated by the black rectangle signify the clock, the values designated by the red rectangle the output values, and the values designated by the blue rectangle the input values. The purpose of the introduced QCA construct for fault-tolerant 1-to-2 QCA demultiplexer construction is clearly evident once the data are analysed.

31.4 The fault-tolerant QCA full-adder

In building QCA arithmetic circuits, adders are essential, in particular the one-bit adder. Even though several complete adders have been built, their fault tolerance still needs more research. In this section, a full-adder is created using a three-input and a five-input majority gate, as shown in figure 31.4.

In this section, a new full-adder was created using a three-input and a five-input majority gate, as seen in figure 31.4. It is important to note that it works for a single missing cell defect.

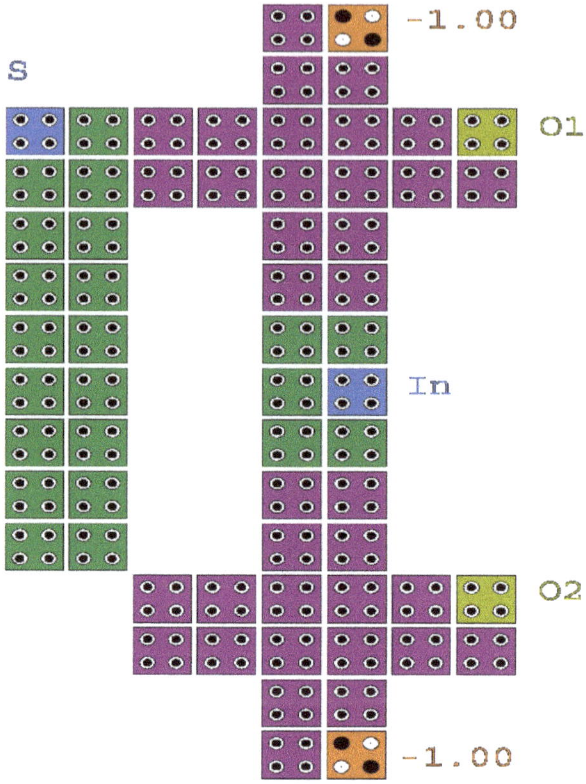

Figure 31.3. The fault-tolerant QCA 1-to-2 demultiplexer.

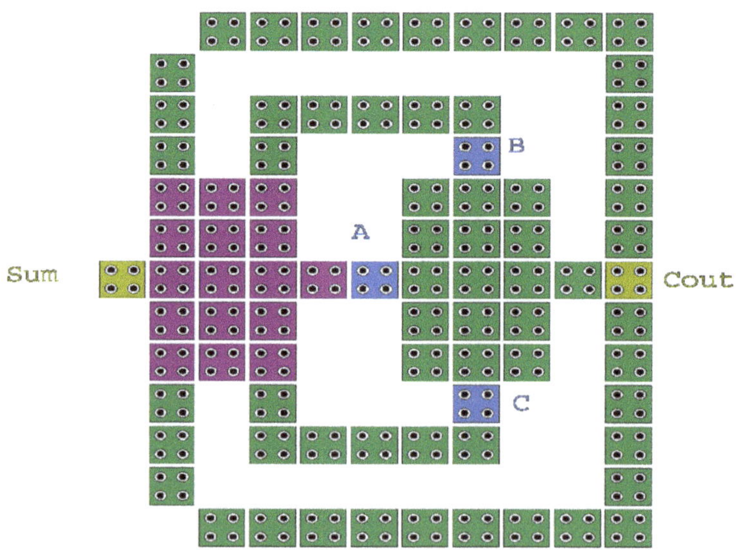

Figure 31.4. The fault-tolerant one-bit QCA adder.

31.5 The fault-tolerant QCA SRAM cell

One of the most important components of any electronic system is the RAM cell. The two forms of RAM cell architecture used in QCA are loop-based and line-based. The loop-based type of design is utilized entirely for SRAM memory cell design in the QCA field. The storing mechanism of this structure is operated by wrapping a small amount of data around a wire-loop made up of QCA cells. Popular logic operations, including the QCA D latch, QCA SR latch, QCA multiplexer, and majority gates, are frequently utilized in this scenario. A fault-tolerant QCA majority-based SRAM cell is shown in figure 31.5.

Two fault-tolerant line-based RAM cells with the capacity to set and reset values are suggested in this section. A three-input fault-tolerant majority gate is employed in this structure. Four majority gates with three inputs each and one with five inputs make up the design of the RAM. Additionally, there are four control lines in this design: set, reset, select, and write (read). In the initial state, it generates (set = 0 and rest = 0) when the select line is activated (1) and the write/(read) line is set to 1, then the input data will be sent to the output and, as a result, the write operation will be carried out. Additionally, the select and write/(read) signals are assigned to 1 and 0 for the read operation. The RAM cell's stored bit will be set to 1 in the set mode, where set = 1 and reset = 0. The RAM cell's bit will also be reset to 0 in the reset mode, where set = 0 and reset = 1.

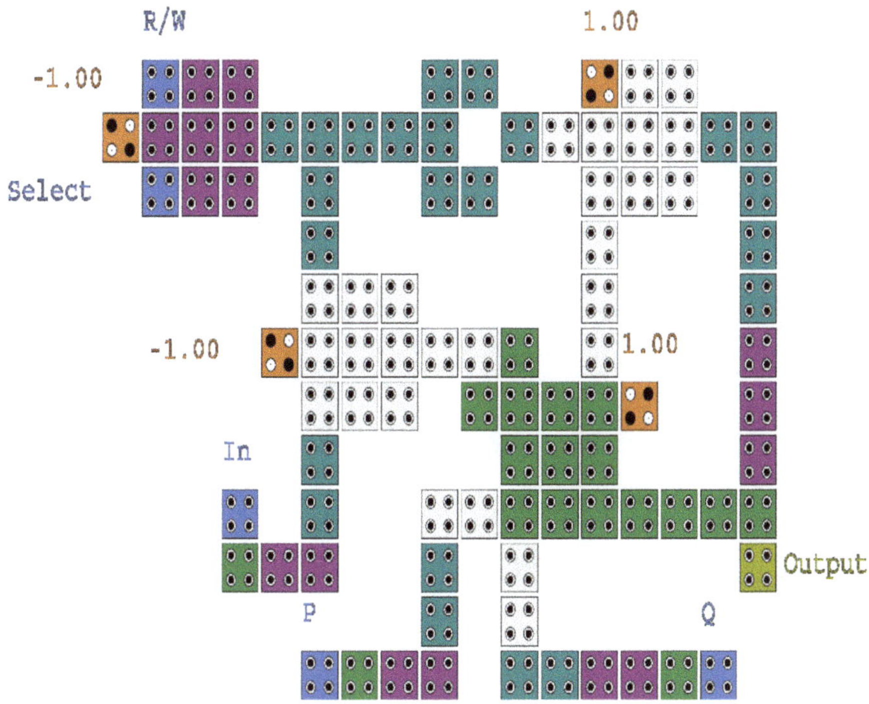

Figure 31.5. The fault-tolerant QCA majority-based SRAM cell.

31.6 The fault-tolerant QCA subtractor

A difference and borrow combination QCA circuit with two inputs and two outputs is called a QCA half-subtractor. The Boolean expression for a half-subtractor is as follows:

Difference = $A \oplus B$

Borrow = $\overline{AB}$.

Figure 31.6 depicts the block diagram of the half-subtractor.

It makes use of the parity-preserving Fredkin gate (FRG) and the reversible Feynman double gate (F2G). There are two inputs, A and B, as well as a control line, ctrl, that determine how it operates. Subtraction is carried out by the circuit when the control signal ctrl is at logic 1. The difference line is depicted as D, and its borrow signal as Br. While the garbage signals are g_1 to g_5, the remaining four inputs, which are referred to as constants, are compelled to logic 0. Figure 31.7 shows the F2G and figure 31.8 shows the FRG.

The output vector is $O(P, Q, \text{and } R)$, while the input vector is $I(A, B, \text{and } C)$. $P = A$, $Q = A \oplus B$, and $R = A \oplus C$ define the outputs. In figure 31.8 a QCA implementation of the FRG is presented. Only four majority gates are used in this FRG. This design reduces failure rates, ensures that there is no interference, and has a practical compact layout.

This FRG maximizes circuit density, uses the fewest possible gates, and focuses on designing the circuit with the fewest possible QCA cells. As a result, the design significantly decreases both the complexity and the area used. QCA cells can easily execute this design.

31.7 The fault-tolerant QCA multiplier

A multiplier is just a factor that increases the basic value of another thing. Partial product generation and the addition of partial products are the main causes of propagation delay in a multiplication operation (figure 31.9).

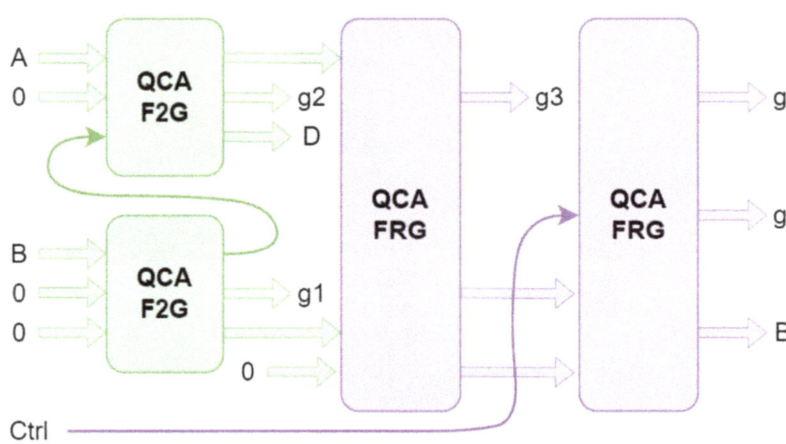

Figure 31.6. The block diagram of the QCA half-subtractor.

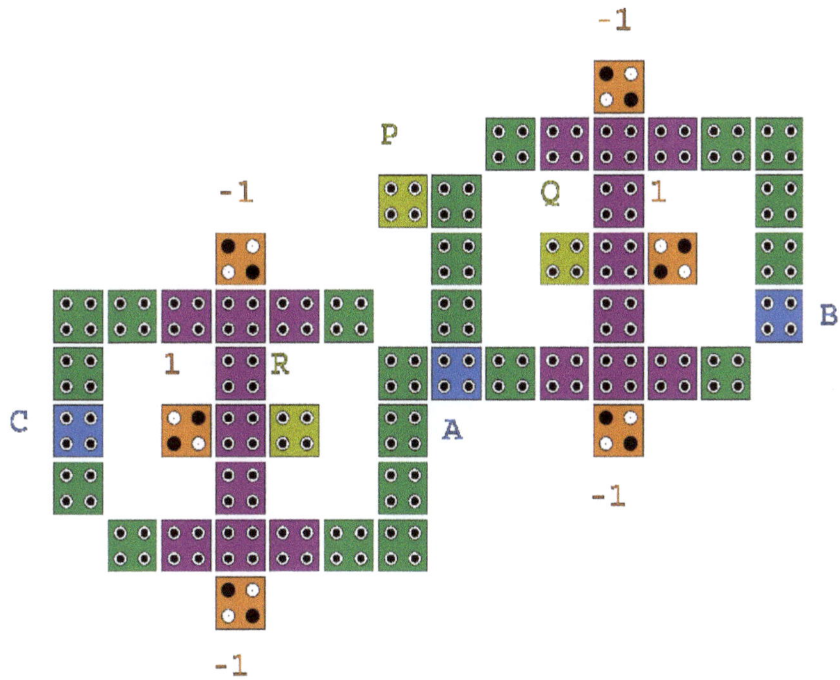

Figure 31.7. The QCA F2G.

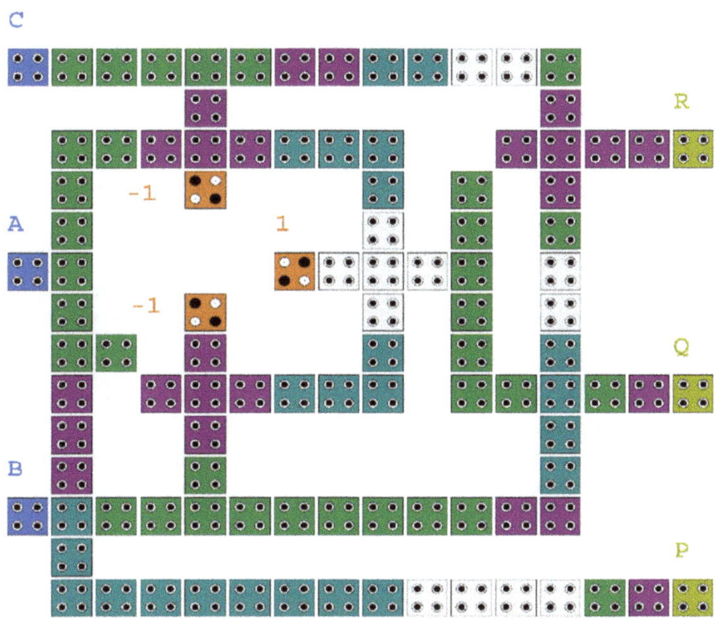

Figure 31.8. The QCA double FRG.

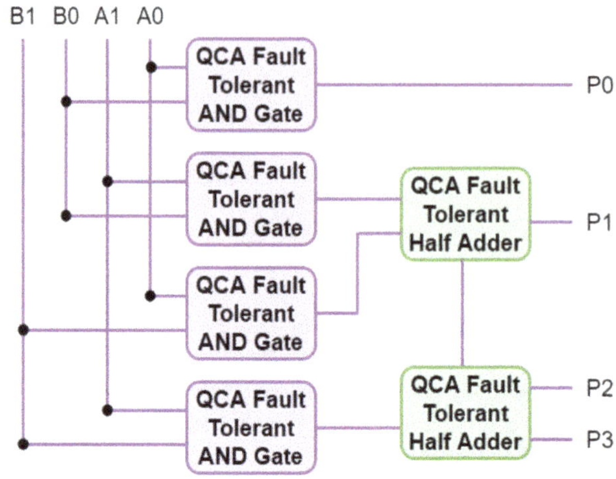

Figure 31.9. The block diagram of a QCA fault-tolerant multiplier.

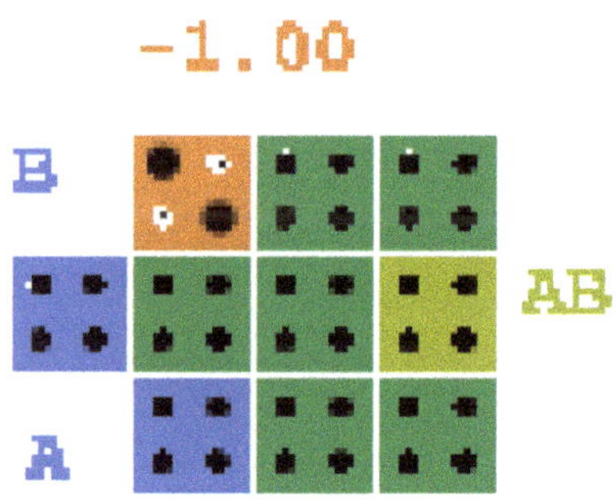

Figure 31.10. The QCA fault-tolerant AND gate.

Two bits are required for a two-bit fault-tolerant QCA multiplier. To construct the primary circuit for a two-bit multiplier, it primarily requires four fault-tolerant QCA AND gates and two fault-tolerant QCA half-adders (figure 31.10).

In this case, the multiplication will be done by the AND gates, and the partial product terms or carry will be added by half-adders (figure 31.11).

It is necessary to first carry out QCA AND operations. Consider the QCA numbers A_1, A_0 and B_1, B_0 which represent the multiplicand and the multiplier, respectively. In this circuit, the partial product terms are added by the half-adders, while the AND gate will execute the multiplication. These logical procedures result in a straightforward four-bit output.

31.8 Summary

In this chapter, two fault-tolerant majority systems are suggested: a three-input majority gate and a five-input majority gate that can tolerate a single omission defect. Comparing this design to other state-of-the-art implementations, it is simpler,

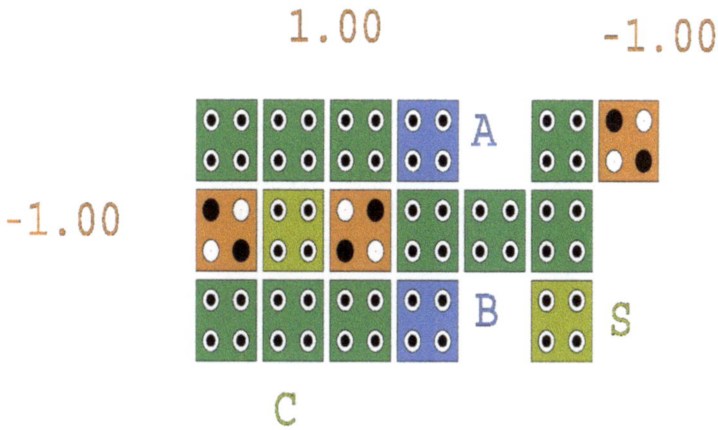

Figure 31.11. The QCA fault-tolerant half-adder.

takes up less space, and uses less energy. The fault-tolerant demultiplexer, complete adder, SRAM, subtractor, and multiplier presented in this chapter are revolutionary architectures.

Critical thinking questions

1. Why do QCA fault-tolerant circuits need to be built?
2. Is it possible to design a seven-input majority gate which is fault-tolerant? If so, draw the circuit.
3. What are the differences between regular QCA circuits and fault-tolerant QCA circuits?
4. Describe the benefits and drawbacks of a fault-tolerant QCA 1-to-2 demultiplexer over a conventional QCA 1-to-2 demultiplexer.

References

[1] Ahmadpour S-S, Mosleh M and Heikalabad S R 2019 Robust QCA full-adders using an efficient fault-tolerant five-input majority gate *Int. J. Circuit Theory Appl.* **47** 1037–56

[2] Ahmadpour S-S, Mosleh M and Heikalabad S R 2020 The design and implementation of a robust single-layer QCA ALU using a novel fault-tolerant three-input majority gate *J. Supercomput.* **76** 10155–85

[3] Bahar A N and Waheed S 2013 Double Feynman gate (F2G) in quantum-dot cellular automata (QCA) *Int. J. Comput. Sci. Eng.* **2** 351–5

[4] Du H, Lv H, Zhang Y, Peng F and Xie G 2016 Design and analysis of new fault-tolerant majority gate for quantum-dot cellular automata *J. Comput. Electron.* **15** 1484–97

[5] Hariprasad A and Ijjada S R 2019 Quantum-dot cellular automata technology for high-speed high-data-rate networks *Circuits Syst. Signal Process.* **38** 5236–52

[6] Kaur P and Dhaliwal B S 2012 Design of fault tolerant full adder/subtarctor using reversible gates *Int. Conf. on Computer Communication and Informatics* (Piscataway, NJ: IEEE) pp 1–5

[7] Khosroshahy M B, Moaiyeri M H and Abdoli A 2022 Design and energy analysis of a new fault-tolerant SRAM cell in quantum-dot cellular automata *Opt. Quant. Electron.* **54** 593

[8] Majeed A H, Alkaldy E, Zainal M S, Navi K and Nor D 2019 Optimal design of RAM cell using novel 2:1 multiplexer in QCA technology *Circuit World* **46** 147–58

[9] Mohammadi Z, Navi K and Sabbaghi-Nadooshan R 2020 Design of testable reversible latches by using a novel efficient implementation of Fredkin gate *Int. J. Electron.* **107** 859–78

[10] Raj M, Gopalakrishnan L and Ko S-B 2019 Fast quantum-dot cellular automata adder/subtractor using novel fault tolerant exclusive-OR gate and full adder *Int. J. Theor. Phys.* **58** 3049–64

[11] Seyedi S, Navimipour N J and Otsuki A 2021 Design and analysis of fault-tolerant 1:2 demultiplexer using quantum-dot cellular automata nano-technology *Electronics* **10** 2565

[12] Singh G, Raj B and Sarin R K 2018 Fault-tolerant design and analysis of QCA-based circuits *IET Circuits Devices Syst.* **12** 638–44

www.ingramcontent.com/pod-product-compliance
Ingram Content Group UK Ltd.
Pitfield, Milton Keynes, MK11 3LW, UK
UKHW051905100926
8556IPUK00005B/771

9 780750 356817